AF443298

Lucio Damascelli and Filomena Pacella
Morse Index of Solutions of Nonlinear Elliptic Equations

De Gruyter Series in Nonlinear Analysis and Applications

Editor-in Chief

Jürgen Appell, Würzburg, Germany

Editors

Catherine Bandle, Basel, Switzerland

Alain Bensoussan, Richardson, Texas, USA

Avner Friedman, Columbus, Ohio, USA

Mikio Kato, Tokyo, Japan

Wojciech Kryszewski, Torun, Poland

Umberto Mosco, Worcester, Massachusetts, USA

Louis Nirenberg, New York, USA

Simeon Reich, Haifa, Israel

Alfonso Vignoli, Rome, Italy

Vicenţiu D. Rădulescu, Krakow, Poland

Volume 30

Lucio Damascelli and Filomena Pacella

Morse Index of Solutions of Nonlinear Elliptic Equations

DE GRUYTER

Mathematics Subject Classification 2010
Primary: 3502, 35J61; Secondary: 35B99, 35B06

Authors
Prof. Dr. Lucio Damascelli
Università di Roma Tor Vergata
Dipartimento di Matematica
Via della Ricerca Scientifica
00133 Roma
Italy
damascel@mat.uniroma2.it

Prof. Dr. Filomena Pacella
Università di Roma Sapienza
Dipartimento di Matematica
Piazzale Aldo Moro 2
00185 Roma
Italy
pacella@mat.uniroma1.it

ISBN 978-3-11-053732-1
e-ISBN (PDF) 978-3-11-053824-3
e-ISBN (EPUB) 978-3-11-053743-7
ISSN 0941-813X

Library of Congress Control Number: 2019937572

Bibliographic information published by the Deutsche Nationalbibliothek
The Deutsche Nationalbibliothek lists this publication in the Deutsche Nationalbibliografie;
detailed bibliographic data are available on the Internet at http://dnb.dnb.de.

© 2019 Walter de Gruyter GmbH, Berlin/Boston
Typesetting: VTeX UAB, Lithuania
Printing and binding: CPI books GmbH, Leck

www.degruyter.com

Preface

Morse theory owes its name to the mathematician M. Morse who analyzed the relationship between the topology of a manifold and the number and the type of critical points of a smooth function defined on it [180]. In some sense, it can be viewed as a beautiful and natural extension of the principle which asserts that every continuous function on a compact space has a minimum and a maximum point. The basic ideas of Morse theory can be summarized by the following two statements.

Let M be a smooth finite dimensional manifold and $g : M \to \mathbb{R}$ a smooth real valued function.

(I) If for some $a < b$, the set $g^{-1}([a,b])$ is compact and does not contain any critical point of g, then the sublevel $g_a = \{x \in M : g(x) \le a\}$ is a deformation retract of $g_b = \{x \in M : g(x) \le b\}$

(II) If g has only one critical point p in $g^{-1}([a,b])$ which is nondegenerate, and $g^{-1}([a,b])$ is compact, then g_b is homotopically equivalent to g_a with a $m(p)$-cell attached, where $m(p)$ is the number of directions where the Hessian matrix of g is negative definite.

The number $m(p)$ is defined as the Morse index of the critical point p, and the word nondegenerate means that the Hessian matrix of g at the point p is nonsingular. So, on one side the topology of the manifold, in particular its homology, influences the number and type of critical points of any smooth nondegenerate function defined on it, on the other side from the knowledge of the critical points of a smooth nondegenerate function it is possible to recover the topological properties of the manifold.

Since the beginning, Morse theory has been proved useful in estimating the number and the type of critical points of functions related to many problems in analysis and differential geometry. Through the years and with the extension to an infinite dimensional setting, Morse theory has been successfully applied to many P. D. E. problems.

In this book, we focus on semilinear elliptic equations of the type

$$- \Delta u = f(x,u) \tag{1}$$

with Dirichlet or mixed boundary conditions. The Morse index $m(u)$ of a solution u of (1) is the maximal dimension of suitable linear spaces X where a certain quadratic form $Q_u(\psi)$, related to the linearization of (1), is negative definite. By linearization, we mean the linear operator $L_u = -\Delta - f'(x,u)$ where f' denotes the derivative of the function f with respect to the second variable. The linear spaces X involved in the definition are prescribed by the boundary conditions assigned and the type of solutions considered. Typically, X will be subspaces of a suitable Sobolev space where the weak solutions of (1) belong. Since the weak solutions of (1) can be viewed as critical points

https://doi.org/10.1515/9783110538243-201

of a functional J, the quadratic form $Q_u(\psi)$, corresponds to the one associated with the second derivative of J, so we are back to the original definition of the Morse index.

However, let us observe that the definition of the Morse index by linearization can be given independently of the variational structure of (1). This means that the definition survives for problems for which there is not a "natural" variational formulation. This is, for example, the case when systems are considered, instead of scalar equations (see Chapter 7), or fully nonlinear problems are studied (cf. [35]).

In this book, we study the Morse index of solutions of (1) having two purposes in mind. On one side, we want to describe some interesting applications, and on the other side we would like to focus on the computation or estimates of the Morse index itself. This last question is a delicate issue and it is obviously a prerequisite to the first one. When the domain where the problem is posed is a bounded domain of $\mathbb{R}^N$, it amounts to calculate or estimate the number of negative eigenvalues of the linear Schrödinger-type operator $L_u = -\Delta - V$, where $V = V(x) = f'(x, u)$. The spectral theory of this kind of linear operators is fascinating and difficult. A good knowledge of the properties of the potential function V is necessary to deduce information on the spectrum. Since V depends on the solution itself, this in turn means having precise information on solutions of (1). We present both some classical results for least energy solutions and some recent developments for radial solutions of Lane–Emden problems.

Concerning applications, the ones included in this book reflect the taste of the authors and the work done by them through the years. In particular, we devote the last part of the book to the study of symmetry properties of solutions of (1) by Morse index.

Symmetry naturally arises in many problems and it is often observed that symmetric solutions tend to minimize some energy associated. The connection between the Morse index of a solution and its symmetry was first shown in [186] driven from the idea that solutions with low Morse index should have "some symmetry." In [186], this was proved for convex nonlinearities and Morse index one solutions, which are "almost" minima of the energy functional, since its Hessian matrix is negative definite only in one direction. Hence we are back to the principle of minimizing the energy by symmetry. It was subsequently shown that, for nonlinearities which are either convex or have convex derivative, solutions of (1) are symmetric as long as their Morse index is less than or equal to the dimension of the space $\mathbb{R}^N$. Later these results were extended to systems and unbounded domains.

Morse theory has also strong connections with bifurcation, in particular, through degree theory. Plenty of results are available in the literature. We have chosen to present here a result on the existence of branches of nonradial solutions bifurcating from radial ones, when the domain is an annulus. This is achieved by detecting the change of Morse index of radial solutions, varying the bifurcation parameters. This is again a connection between Morse index and symmetry and shows that "break of symmetry" arises when the Morse index increases.

The book is organized in eight chapters. We have made an effort to keep the exposition as simple and as self-contained as possible. Therefore, the first three chapters, in particular, Chapter 1 and Chapter 3, are somehow introductory, and could be used for a graduate course in P. D. E. Chapters 4–7 are more advanced and contain recent results. In the short Chapter 8, some developments for unbounded domains are outlined.

The aim of Chapter 1 is to review different forms of weak and strong maximum principles as well as the classical eigenvalue theory for boundary value problems involving elliptic operators. We will consider mainly weak solutions and mixed Dirichlet–Neumann boundary conditions (with nonlinear Neumann boundary conditions on part of the boundary), therefore we include a section on Sobolev spaces of functions vanishing only on some portion of the boundary of the domain where they are defined. In particular, we will deal with functions defined in cylindrically symmetric domains. We give great emphasis on the variational characterization of the eigenvalues, which is crucial to study the Morse index.

We also include a section on systems, describing the related maximum principles and spectral theory. This last one works well for symmetric systems to which the theory of compact self-adjoint operators can be applied, but not in the case of general systems. Nevertheless, we show how to exploit a symmetric system, naturally associated to a generic one, in order to get information and, in particular, to check the validity of the maximum principle.

The material of this chapter concerning scalar equations is largely classical, but the proofs of many well-known results are not easy to be found or are spread in several textbooks. We feel that it is useful to present them all together. The results on systems are taken from recent papers and to present them together with the scalar case allows the reader to have a unified vision.

In Chapter 2, we introduce the classical Morse theory for functions on finite dimensional manifolds as well as its extension to the infinite dimensional case. In particular, we will describe the steps to show that a mountain pass critical point of a functional in Banach spaces has Morse index less than or equal to one. This is a very important result for applications to semilinear elliptic equations, which has been proved in different ways and is reported and used many times but whose clear proof is not easy to be found.

In Chapter 3, we introduce the Morse index of solutions of semilinear elliptic equations. We distinguish between the cases of positive or sign changing solutions. In the first case, we show the existence of a positive solution with Morse index one and present an application to an uniqueness result. In the second case, after proving the existence of a sign changing solution with Morse index two, we give some estimates for the Morse index of nodal radial solutions from which symmetry breaking results derive. The material of this chapter is not fully included in any book in nonlinear analysis but is mostly taken from several research articles. We believe that it could serve as a beginner reference for the study of solutions of semilinear elliptic equations and their Morse index.

In Chapter 4, we review some recent results on sharp computations of the Morse index for radial solutions. To the purpose not to get distracted by too many technical details, and thus lose the main thread of the arguments, some proofs which are long and involve heavy computations have been omitted, giving precise references to the interested reader. We hope to have made clear the strategy for the evaluation of the negative eigenvalues of the linear operators involved.

In Chapter 5, we describe one of the many applications of Morse theory to bifurcation. We describe some results about bifurcation from radial positive solutions of Lane–Emden Dirichlet problems in an annulus. As compared to the original paper [129], we simplify and detail some of the proofs.

Finally, in Chapter 6 and Chapter 7, we present the results obtained by the authors about symmetry of solutions via Morse index bounds.

Chapter 6 deals with the case of scalar equations. It is mostly taken from [186] and [190] but we present the results in an unified context simplifying some proofs and adding recent developments for nonlinear mixed boundary value problems from [78].

Chapter 7 deals with the case of systems. We present some results by the authors contained in [72] and [77] and we simplify some proofs of both papers.

The bibliography that we have included is fairly extensive though, as is to be expected, far from exhaustive. Further references to a rich literature can be found in the works cited here.

L. Damascelli and F. Pacella

Contents

Notation

- $\mathbb{K}$ will denote either the field $\mathbb{R}$ or the field $\mathbb{C}$.
- $\mathbb{N} = \{0, 1, 2, \dots\}$ is the set of natural numbers, $\mathbb{N}^+ = \mathbb{N} \setminus \{0\} = \{1, 2, \dots\}$.
- $\mathbb{R}^N = \{x = (x_1, \dots, x_N) : x_i \in \mathbb{R}\}$ is the N-dimensional Euclidean space, $N \geq 1$.
- $\mathbb{R}^N_+ = \{x = (x_1, \dots, x_N) = (x', x_N) \in \mathbb{R}^N = \mathbb{R}^{N-1} \times \mathbb{R} : x_N > 0\}$.
- $\mathbb{R}^N_0 = \{(x', 0) : x' \in \mathbb{R}^{N-1}\} = \partial \mathbb{R}^N_+$. It will be often identified with $\mathbb{R}^{N-1}$.
- $Q(x, r) = Q^N(x, r) = \{y \in \mathbb{R}^N : |y_i - x_i| < r, i = 1, \dots, N\}$ for $r > 0$, $x \in \mathbb{R}^N$.
- $B(x, r) = B^N(x, r) = B_r(x) = \{y \in \mathbb{R}^N : |y - x| < r\}$ for $r > 0$, $x \in \mathbb{R}^N$.
- $B_r = B_r(0)$.
- $S(x, r) = S^{N-1}(x, r) = S_r(x) = \{y \in \mathbb{R}^N : |y - x| = r\}$ for $r > 0$, $x \in \mathbb{R}^N$.
- $S^{N-1} = S^{N-1}(0, 1) = \{y \in \mathbb{R}^N : |y| = 1\}$ is the unit sphere or equivalently the set of directions in $\mathbb{R}^N$.
- Ω denotes an open set in $\mathbb{R}^N$, $N \geq 2$ and, unless otherwise stated, it will be assumed that Ω is at least a *Lipschitz domain*, i. e., it is connected and has a locally Lipschitz boundary $\partial \Omega$. By this, we mean the following:

  for any $x = (x', x_N) \in \partial \Omega$, there exists $r > 0$ and a Lipschitz function $\beta : Q' = Q^{N-1}(x', r) \to \mathbb{R}$ such that, up to rotations and relabelings of the coordinates, it holds

$$\Omega \cap Q(x, r) = \{y \in Q(x, r) : y_N > \beta(y_1, \dots, y_{N-1})\},$$

$$\partial\Omega \cap Q(x, r) = \{y \in Q(x, r) : y_N = \beta(y_1, \dots, y_{N-1})\}.$$

  Ω is a $C^{k,\alpha}$ domain if the functions β belongs to the class $C^{k,\alpha}$ (see below).
- If Ω_1 and Ω_2 are open sets, $\Omega_1 \subset\subset \Omega_2$ means that $\overline{\Omega_1}$ is a compact subset of Ω_2.
- $\mathrm{meas}_N(A)$, or simply $\mathrm{meas}(A)$ or even $|A|$ denotes the N-dimensional Lebesgue measure of a measurable set $A \subset \mathbb{R}^N$.
- If Ω is a bounded domain with Lipschitz boundary, we will consider the $(N-1)$ Hausdorff measure on the boundary $\partial\Omega$, and denote by: $\mathrm{meas}_{N-1}(D)$, the $(N-1)$-dimensional measure of a measurable set $D \subset \partial\Omega$.
- A property holds *almost everywhere* (we will write shortly *a. e.*) in an open set $\Omega \subset \mathbb{R}^N$, respectively on $\partial\Omega$ if it holds in $\Omega \setminus S$ with $\mathrm{meas}_N(S) = 0$, respectively in $\partial\Omega \setminus T$ with $\mathrm{meas}_{N-1}(T) = 0$.
- $C^k(\Omega)$ is the space of functions with continuous derivatives up to the order k in Ω.
- If Ω is bounded, the Banach space $C^k(\overline{\Omega})$ consists of the functions $u \in C^k(\Omega)$ such that for any multiindex $\beta = (\beta_1, \dots, \beta_N) \in \mathbb{N}^N$ with $|\beta| = \beta_1 + \cdots \beta_N \leq k$ the derivatives $D^\beta u = \frac{\partial^{|\beta|} u}{\partial x_1^{\beta_1} \dots \partial x_N^{\beta_N}}$ are (restrictions of functions) continuous in $\overline{\Omega}$, with the norm

$$\|u\|_{C^k(\overline{\Omega})} = \sum_{|\beta| \leq k} \sup_{x \in \Omega} |D^\beta u(x)|.$$

https://doi.org/10.1515/9783110538243-202

- $C_c^k(\Omega)$ is the subspace of $C^k(\Omega)$ of the functions with continuous derivatives up to the order k in Ω whose support is a compact subset of Ω.
- $C^\infty(\Omega)$ is the space of functions with continuous derivatives of any order in Ω.
- For $0 < \alpha \le 1$, $C^{k,\alpha}(\overline{\Omega})$ is the subspace of $C^k(\overline{\Omega})$ of the functions u whose derivatives of order k are α-Hölder continuous in $\overline{\Omega}$, and if Ω is bounded it is a Banach space with the norm

$$\|u\|_{C^{k,\alpha}(\overline{\Omega})} = \|u\|_{C^k(\Omega)} + \sum_{|\beta|=k} \sup_{x,y\in\Omega, x\neq y} \frac{|D^\beta u(x) - D^\beta u(y)|}{|x-y|^\alpha}.$$

- $C_c^\infty(\Omega)$ is the space of functions having continuous derivatives of any order in Ω whose support is a compact subset of Ω.
- $L^p(\Omega)$, $1 \le p < \infty$, denote the Lebesgue space of measurable functions u in Ω such that $\|u\|_{p,\Omega}^p = \int_\Omega |u(x)|^p \, dx < \infty$.
- $L^\infty(\Omega)$ is the spaces of (essentially) bounded measurable functions u in Ω such that $\|u\|_{\infty,\Omega} = \operatorname{ess\,sup}_\Omega |u(x)| < \infty$.
- If Ω is a bounded domain and $u \in L^1(\Omega)$, the *integral mean* or *average* of u in Ω is defined by the number $\frac{1}{\operatorname{meas}(\Omega)} \int_\Omega u(x) \, dx$, and it is denoted by one of the symbols

$$(u)_\Omega = \fint_\Omega u(x) \, dx = \frac{1}{\operatorname{meas}(\Omega)} \int_\Omega u(x) \, dx.$$

- If Ω is a bounded domain with Lipschitz boundary and Γ is a measurable subset of $\partial\Omega$, we denote by $L^p(\Gamma)$, $1 \le p \le \infty$ the Lebesgue spaces with respect to the $(N-1)$-dimensional Hausdorff measure on Γ.
- $W^{k,p}(\Omega)$, $1 \le k \le \infty$, $1 \le p \le \infty$ is the Sobolev space of functions $u \in L^p(\Omega)$ whose distributional derivatives $D^\beta u$, $|\beta| \le k$, belong to $L^p(\Omega)$ with the norm

$$\|u\|_{k,p,\Omega} = \left(\sum_{j=0}^{k} \sum_{|\beta|=j} \int_\Omega |D^\beta u|^p \, dx \right)^{\frac{1}{p}} \quad \text{if } 1 \le p < \infty,$$

$$\|u\|_{k,\infty,\Omega} = \max_{0\le|\beta|\le k} \|D^\beta u\|_{L^\infty(\Omega)} \quad \text{if } p = \infty.$$

- If $u \in W^{1,p}(\Omega)$ the gradient of f is denoted by one of the symbols

$$Du = \nabla u = (u_{x_1}, \ldots, u_{x_N}).$$

- $H^k(\Omega) = W^{k,2}(\Omega)$. It is a Hilbert space with the scalar product

$$(u,v) = \sum_{j=0}^{k} \sum_{|\beta|=j} \int_\Omega D^\beta u \, D^\beta v \, dx.$$

- $W_0^{k,p}(\Omega)$, $1 \le p < \infty$ is the closure of $C_c^\infty(\Omega)$ (or $C_c^k(\Omega)$) in the Sobolev space $W^{k,p}(\Omega)$.

- $H_0^k(\Omega) = W_0^{k,2}(\Omega)$.
- If Γ is a relatively open subset of $\partial\Omega$, $W_0^{k,p}(\Omega \cup \Gamma)$, $1 \le p < \infty$, is the closure of $C_c^\infty(\Omega \cup \Gamma)$ (or $C_c^k(\Omega \cup \Gamma)$) in the Sobolev space $W^{k,p}(\Omega)$. We often consider the case when Γ is a smooth embedded $(N-1)$-submanifold in $\mathbb{R}^N$.
- $H_0^k(\Omega \cup \Gamma) = W_0^{k,2}(\Omega \cup \Gamma)$.
- Trace $(u) \in L^p(\partial\Omega)$ is the trace on $\partial\Omega$ of a function $u \in W^{1,p}(\Omega)$. It belongs to the fractional space $W^{1-\frac{1}{p},p}(\partial\Omega)$ and to other Lebesgue spaces, as we will recall. We will write sometimes $u]_{\partial\Omega}$ or simply u instead of Trace (u).
- For a function $v : \Omega \to \mathbb{R}$, we write

$$v^+ = \max\{v, 0\} \quad \text{and} \quad v^- = -\min\{v, 0\} = \max\{-v, 0\}$$

for the positive and negative part of u.
- If V is a Banach space, we will use the same notation that we use in $\mathbb{R}^N$ for the ball centered at $x \in V$ with radius $r > 0$, i. e., we write

$$B(x, r) = B_r(x) = \{y \in V : \|y - x\| < r\} \quad \text{for } r > 0, \ x \in V \text{ and } B_r = B_r(0).$$

- V^* denotes the dual space of a Banach space V.
- We say that the Banach space W is continuously injected (or embedded) in the Banach space V and we write $W \hookrightarrow V$ if $W \subset V$ and there exists a constant $C > 0$ such that $\|w\|_V \le C\|w\|_W$ for any $w \in W$. More generally, we use the same notation if there exists an injective continuous linear operator $T : W \to V$.

1 Preliminaries

Here we will state and/or prove some results needed to understand the content of the book.

We will start by recalling the basic theory for Sobolev spaces of functions vanishing on some part of the boundary of a domain in Section 1.1. Next, we describe in Section 1.2 several forms of maximum principles. In Section 1.3 we recall the spectral theory of bounded self-adjoint operators in Hilbert spaces and apply it in Section 1.4 to some elliptic operators. We extend the classical theory for the Dirichlet boundary value problem to the case of mixed Dirichlet–Neumann problems, providing the construction and variational characterization of the eigenvalues and their main properties. A section on maximum principles and spectral theory for elliptic cooperative systems ends the chapter.

1.1 Sobolev spaces

Limiting ourselves to first-order spaces, we recall some well-known properties of Sobolev spaces. For most of the proofs, we refer to [2, 41, 115, 116, 149, 174, 218, 227] or any book on Sobolev spaces and PDEs. We will detail a few proofs, in the case the proof itself is needed in the sequel or it is not easily available in textbooks.

In the sequel, we refer to the notation for all the undefined symbols.

Let Ω be an open set in $\mathbb{R}^N$, $N \geq 2$. A useful characterization of the Sobolev spaces $W^{1,p}(\Omega)$ is the following.

Theorem 1.1. *A function $v \in L^p(\Omega)$ belongs to $W^{1,p}(\Omega)$ if and only if it has a representative $\bar{v}$ which, for any $i = 1, \ldots, N$, is absolutely continuous on almost all segments in Ω parallel to the x_i-axis that intersect Ω^1 and whose (classical) partial derivatives $\frac{\partial v}{\partial x_i}$ belong to $L^p(\Omega)$. As a consequence, we have that:*

- *if Ω is connected, $v \in W^{1,p}(\Omega)$ and $\nabla v = 0$ in Ω, then v is constant in Ω;*
- *if $v \in W^{1,p}(\Omega)$, then the positive and negative parts $v^+ = \max\{v, 0\}$ and $v^- = -\min\{v, 0\} = \max\{-v, 0\}$, belong to $W^{1,p}(\Omega)$, with*

$$\nabla v^+(x) = \begin{cases} \nabla v(x) & \text{if } v(x) > 0 \\ 0 & \text{if } v(x) \leq 0 \end{cases}$$

$$\nabla v^-(x) = \begin{cases} 0 & \text{if } v(x) \geq 0 \\ -\nabla v(x) & \text{if } v(x) < 0 \end{cases}$$

Moreover, $\nabla v = 0$ a. e. on $[v = 0] = \{x \in \Omega : v(x) = 0\}$.

---

1 By this we mean that if, e. g., $i = N$, the set of the points $x' \in \mathbb{R}^{N-1}$, such that the line $L_{x'} = \{(x', t) : t \in \mathbb{R}\}$ intersects Ω in at least one compact interval $\{x'\} \times [a, b]$ and $\bar{v}(x', .)$ is not absolutely continuous on $[a, b]$, is a set of $(N-1)$-measure zero.

https://doi.org/10.1515/9783110538243-001

We now recall some of the classical Sobolev inequalities.

Theorem 1.2 (Sobolev inequalities in $C_c^1(\mathbb{R}^N)$). *Let $v \in C_c^1(\mathbb{R}^N)$. Then:*

1. *if $1 \leq p < N$, and $p^* = \frac{Np}{N-p}$, then there exists a constant C, depending on p, N such that*

$$\|v\|_{L^{p^*}(\mathbb{R}^N)} \leq C\|\nabla v\|_{L^p(\mathbb{R}^N)} \tag{1.1}$$

2. *if $p = N$ and $q \in [N, \infty)$, then there exists a constant C, depending on p, N, q such that*

$$\|v\|_{L^q(\mathbb{R}^N)} \leq C\|v\|_{W^{1,N}(\mathbb{R}^N)} \tag{1.2}$$

3. *if $p > N$ and $y = 1 - \frac{N}{p}$, then there exists a constant C, depending on p, N such that*

$$\|v\|_{C^{0,y}(\mathbb{R}^N)} \leq C\|v\|_{W^{1,p}(\mathbb{R}^N)} \tag{1.3}$$

If Ω is an open set in $\mathbb{R}^N$, we denote by $W_0^{1,p}(\Omega)$ the closure in $W^{1,p}(\Omega)$ of the set $C_c^\infty(\Omega)$ (or $C_c^1(\Omega)$) of the smooth functions whose support is a compact subset of Ω. We also denote by $W_{\text{comp}}^{1,p}(\Omega)$ the set of the functions in $W^{1,p}(\Omega)$ whose support is a compact subset of Ω.

Theorem 1.3 (Sobolev inequalities in $W_0^{1,p}(\Omega)$).

i) *If $\Omega = \mathbb{R}^N$, then $W_0^{1,p}(\mathbb{R}^N) = W^{1,p}(\mathbb{R}^N)$, i. e., if $v \in W^{1,p}(\mathbb{R}^N)$ there exists a sequence $v_n \in C_c^1(\mathbb{R}^N)$ that converges to v in the $W^{1,p}$ norm.*

ii) *If $\Omega \subset \mathbb{R}^N$, then $W_{\text{comp}}^{1,p}(\Omega) \subset W_0^{1,p}(\Omega)$, i. e., any $v \in W^{1,p}(\Omega)$ such that $\text{supp}(v)$ is a compact subset of Ω is the limit in $W^{1,p}(\Omega)$ of a sequence $v_n \in C_c^1(\Omega)$ and, therefore, $W_{\text{comp}}^{1,p}(\Omega)$ is dense in $W^{1,p}(\Omega)$.*

iii) *If $\Omega \subset \mathbb{R}^N$ and $v \in W_0^{1,p}(\Omega)$, then the trivial extension of v to $\mathbb{R}^N$ belongs to $W_0^{1,p}(\mathbb{R}^N) = W^{1,p}(\mathbb{R}^N)$.*

As a consequence, we get Sobolev inequalities in $W_0^{1,p}(\Omega)$:
If $\Omega \subseteq \mathbb{R}^N$ and $v \in W_0^{1,p}(\Omega)$ (in particular, if $\Omega = \mathbb{R}^N$ and $v \in W^{1,p}(\mathbb{R}^N)$), then:

1. *if $1 \leq p < N$, then $v \in L^{p^*}(\Omega)$, where $p^* = \frac{Np}{N-p}$, and there exists a constant C, depending only on p, N such that*

$$\|v\|_{L^{p^*}(\Omega)} \leq C\|\nabla v\|_{L^p(\Omega)} \tag{1.4}$$

for any $v \in W_0^{1,p}(\Omega)$.
Moreover, if $q \in [p, p^*]$ then $v \in L^q(\Omega)$ and there exists a constant $C > 0$, depending on p, N, q, such that

$$\|v\|_{L^q(\Omega)} \leq C\|v\|_{W^{1,p}(\Omega)} \tag{1.5}$$

2. *if $p = N$, then $v \in L^q(\Omega)$ for any $q \in [N, \infty)$ and, for any such q, there exists a constant C, depending on p, N, q such that*

$$\|v\|_{L^q(\Omega)} \le C_1 \|v\|_{W^{1,N}(\Omega)} \tag{1.6}$$

for any $v \in W_0^{1,N}(\Omega)$

3. *if $p > N$, then $v \in L^\infty(\Omega) \cap C^{0,\gamma}(\Omega)$, where $\gamma = 1 - \frac{N}{p}$, and there exists a constant C, depending on p, N such that*

$$\|v\|_{C^{0,\gamma}(\Omega)} \le C \|v\|_{W^{1,p}(\Omega)} \tag{1.7}$$

for any $v \in W_0^{1,p}(\Omega)$.

Note that only (1.4) is a *"pure" Sobolev inequality*, i. e., on the right-hand side only the L^p norm of the gradient appears.

When Ω is bounded, we can substitute on the right-hand side the L^p norm of the gradient in all the previous inequalities, thanks to the fundamental Poincaré's inequality, that we prove below using Sobolev's inequalities.

Theorem 1.4 (Poincaré's inequality in $W_0^{1,p}(\Omega)$). *Let Ω be a bounded domain, $1 \le p < \infty$. Then there exists a constant $C > 0$, depending on p, N, such that for any $v \in W_0^{1,p}(\Omega)$,*

$$\|v\|_{L^p(\Omega)} \le C |\Omega_v|^{\frac{1}{N}} \||\nabla v|\|_{L^p(\Omega)} \tag{1.8}$$

where $\Omega_v = \{x \in \Omega : v(x) \ne 0\} \subset \Omega$ and $|\Omega_v| = meas_N(\Omega_v)$.

Proof. If $1 \le p < N$, denoting by the same symbol the trivial extensions of functions to the whole $\mathbb{R}^N$, we have, by Hölder's inequality with exponents $\frac{p^*}{p}$ and $\frac{p^*}{p^*-p}$, and by Sobolev inequalities,

$$\int_\Omega |v|^p \, dx = \int_{\Omega_v} |v|^p \, dx \le |\Omega_v|^{1-\frac{p}{p^*}} \left(\int_\Omega |v|^{p^*} \, dx \right)^{\frac{p}{p^*}} \le C |\Omega_v|^{\frac{p}{N}} \int_\Omega |\nabla v|^p \, dx$$

If instead $p \ge N \ge 2$ and $q = \frac{Np}{N+p}$, then $1 \le q < N$ and $p = q^*$, so that by Sobolev's inequality and Hölder's inequality with exponents $\frac{p}{q}, \frac{p}{p-q}$, we get (recalling that $\nabla v = 0$ a. e. on $[v = 0]$)

$$\int_\Omega |v|^p \, dx = \int_\Omega |v|^{q^*} \, dx \le C \left(\int_\Omega |\nabla v|^q \, dx \right)^{\frac{q^*}{q}} = C \left(\int_{\Omega_v} |\nabla v|^q \, dx \right)^{\frac{q^*}{q}}$$

$$\le C \left(\int_\Omega |\nabla v|^p \, dx \right)^{\frac{q^*}{p}} |\Omega_v|^{(1-\frac{q}{p})\frac{q^*}{q}} = C |\Omega_v|^{\frac{p}{N}} \int_\Omega |\nabla v|^p \, dx \qquad \square$$

If Ω is bounded (or more generally with finite measure), the Poincaré's inequality and a "pure" Sobolev inequality are not true in the space $W^{1,p}(\Omega)$ as it can be easily observed considering the constant functions. Instead there are unbounded domains Ω where a pure Sobolev's inequality holds true in $W^{1,p}(\Omega)$ (see [169] and the references therein for related questions). In particular, we consider the case of half-spaces. This in turn will give Poincaré's type inequalities for a class of functions defined in cylindrically symmetric domains that will be extensively used in the sequel.

Let us recall the notation: $\mathbb{R}^N_+ = \{x = (x', x_N) \in \mathbb{R}^N : x_N > 0\}$ and $\mathbb{R}^N_0 = \{(x', 0) : x' \in \mathbb{R}^{N-1}\} = \partial\mathbb{R}^N_+$. The last one will be often identified with $\mathbb{R}^{N-1}$.

Let us denote by $\tilde{v}$ the reflection through the hyperplane $x_N = 0$ of a function v defined in $\mathbb{R}^N_+$:

$$\tilde{v}(x', x_N) = \begin{cases} v(x', x_N) & \text{if } x_N \geq 0 \\ v(x', -x_N) & \text{if } x_N < 0 \end{cases}$$

It provides an *extension operator* in $W^{1,p}(\mathbb{R}^N_+)$, in the following sense. Let $1 \leq p < \infty$ and $v \in W^{1,p}(\mathbb{R}^N_+)$. Then $\tilde{v} \in W^{1,p}(\mathbb{R}^N)$ with

$$\frac{\partial\tilde{v}}{\partial x_i}(x', x_N) = \begin{cases} \frac{\partial v}{\partial x_i}(x', x_N) & \text{if } x_N > 0 \\ \frac{\partial v}{\partial x_i}(x', -x_N) & \text{if } x_N < 0 \end{cases}, \quad 1 \leq i \leq N-1,$$

$$\frac{\partial\tilde{v}}{\partial x_N}(x', x_N) = \begin{cases} \frac{\partial v}{\partial x_N}(x', x_N) & \text{if } x_N > 0 \\ -\frac{\partial v}{\partial x_N}(x', -x_N) & \text{if } x_N < 0 \end{cases},$$

so that there exists a constant C such that

$$\|\tilde{v}\|_{W^{1,p}(\mathbb{R}^N)} \leq C\|v\|_{W^{1,p}(\mathbb{R}^N_+)} \tag{1.9}$$

As a consequence, it provides *approximations* in $W^{1,p}(\mathbb{R}^N_+)$ in the sense that $C^\infty_c(\overline{\mathbb{R}^N_+})$ is dense in $W^{1,p}(\mathbb{R}^N_+)$.

Moreover, Sobolev inequalities hold in $W^{1,p}(\mathbb{R}^N_+)$ and, by density, it is easy to prove *trace Sobolev inequalities* as well.

Theorem 1.5 (Sobolev inequalities in $W^{1,p}(\mathbb{R}^N_+)$). *Let $1 \leq p < \infty$ and $v \in W^{1,p}(\mathbb{R}^N_+)$.*
1. *If $1 \leq p < N$ and $p^* = \frac{Np}{N-p}$, then $v \in L^{p^*}(\mathbb{R}^N_+)$ and there exists a constant $C > 0$, depending on p, N, such that*

$$\left(\int_{\mathbb{R}^N_+} |v|^{\frac{Np}{N-p}}\, dx\right)^{\frac{N-p}{Np}} \leq C\left(\int_{\mathbb{R}^N_+} |\nabla v|^p\, dx\right)^{\frac{1}{p}} \tag{1.10}$$

Moreover if $q \in [p, p^*]$ then $v \in L^q(\mathbb{R}^N_+)$ and there exists a constant $C > 0$, depending on p, N, q, such that

$$\left(\int_{\mathbb{R}^N_+} |v|^q\, dx\right)^{\frac{1}{q}} \leq C\|v\|_{W^{1,p}(\mathbb{R}^N_+)} \tag{1.11}$$

2. If $p = N$ and $q \in [N, \infty)$, then $v \in L^q(\mathbb{R}_+^N)$ and there exists a constant $C > 0$, depending on p, N, q, such that

$$\left(\int_{\mathbb{R}_+^N} |v|^q \, dx \right)^{\frac{1}{q}} \leq C \|v\|_{W^{1,p}(\mathbb{R}_+^N)} \tag{1.12}$$

3. If $p > N$, any function $v \in W^{1,p}(\mathbb{R}_+^N)$ is bounded and (has a representative) continuous in $\overline{\mathbb{R}_+^N}$. Moreover, if $\gamma = 1 - \frac{N}{p}$, there exists a constant $C > 0$, depending on p, N, such that

$$\|v\|_{C^{0,\gamma}(\overline{\mathbb{R}_+^N})} \leq C \|v\|_{W^{1,p}(\mathbb{R}_+^N)} \tag{1.13}$$

Theorem 1.6 (Trace inequalities in $C_c^1(\overline{\mathbb{R}_+^N})$.).
1. If $1 \leq p < N$ and $p^\sharp = \frac{Np-p}{N-p}$, there exists a constant $C > 0$, depending on p, N, such that

$$\left(\int_{\mathbb{R}_0^N = \partial \mathbb{R}_+^N} |v|^{p^\sharp} \, dx' \right)^{\frac{1}{p^\sharp}} \leq C \left(\int_{\mathbb{R}_+^N} |\nabla v|^p \, dx \right)^{\frac{1}{p}} \tag{1.14}$$

for any $v \in C_c^1(\overline{\mathbb{R}_+^N})$.
Moreover, if $q \in [p, p^\sharp]$, there exists a constant $C > 0$, depending on p, N, q, such that

$$\left(\int_{\mathbb{R}_0^N = \partial \mathbb{R}_+^N} |v|^q \, dx' \right)^{\frac{1}{q}} \leq C \|v\|_{W^{1,p}(\mathbb{R}_+^N)} \tag{1.15}$$

for any $v \in C_c^1(\overline{\mathbb{R}_+^N})$.
2. If $p = N$ and $q \in [N, \infty)$, there exists a constant $C > 0$, depending on p, N, q, such that

$$\left(\int_{\mathbb{R}_0^N = \partial \mathbb{R}_+^N} |v|^q \, dx' \right)^{\frac{1}{q}} \leq C \|v\|_{W^{1,N}(\mathbb{R}_+^N)} \tag{1.16}$$

for any $v \in C_c^1(\overline{\mathbb{R}_+^N})$.

Proof. Let $1 \leq p \leq N$. If $v \in C_c^1(\overline{\mathbb{R}_+^N})$ and $q \geq 1$, we have

$$|v(x', 0)|^q = - \int_0^{+\infty} \frac{\partial}{\partial x_N} \left(|v(x', x_N)|^q \right) dx_N$$

$$= -q \int_0^{+\infty} |v(x', 0)|^{q-2} v(x', 0) \frac{\partial v}{\partial x_N} (x', x_N) \, dx_N$$

so that integrating in the x'-variable and using Hölder's inequality we get

$$\int_{\mathbb{R}^{N-1}} |v(x',0)|^q \, dx' \le q \int_{\mathbb{R}^N} |v(x)|^{q-1} |\nabla v(x)| \, dx$$

$$\le q \left(\int_{\mathbb{R}^N} |v(x)|^{(q-1)p'} \, dx \right)^{\frac{1}{p'}} \left(\int_{\mathbb{R}^N} |\nabla v(x)|^p \, dx \right)^{\frac{1}{p}}$$

where p' is the conjugate exponent of p, i. e., $\frac{1}{p} + \frac{1}{p'} = 1$.

If $1 \le p < N$, then $(q-1)p' \in [p, p^*]$ if and only if $q \in [p, p^\sharp]$, while if $p = N$ then $(q-1)p' \in [p, \infty)$ if and only if $q \in [p, \infty)$, and, if this is the case, by the Sobolev inequalities we obtain

$$\int_{\mathbb{R}^{N-1}} |v(x',0)|^q \, dx' \le C \left(\int_{\mathbb{R}^N} |\nabla v(x)|^p \, dx \right)^{\frac{q-1}{p}} \left(\int_{\mathbb{R}^N} |\nabla v(x)|^p \, dx \right)^{\frac{1}{p}} = C \|\nabla v\|_{L^p}^q \qquad \square$$

If $p > N$, by considering the reflection through the hyperplane $x_N = 0$, it is immediate to see that any function $v \in W^{1,p}(\mathbb{R}^N_+)$ is bounded and (has a representative) Hölder continuous in $\overline{\mathbb{R}^N_+}$, so that we can consider the *trace* or restriction of a function v to the boundary hyperplane $x_N = 0$.

In the case when $1 \le p \le N$, we can also define the trace of a Sobolev function on the boundary hyperplane $x_N = 0$ using the previous theorem.

Indeed, thanks to the previous inequalities, the restriction operator T which maps the function $v = v(x', x_N) \in C^1_c(\overline{\mathbb{R}^N_+})$ (endowed with the $W^{1,p}$ norm) to the function $Tv = Tv(x')$ with $Tv(x') = v(x', 0) \in L^p(\mathbb{R}^{N-1})$, extends by density to a bounded linear operator

$$T : W^{1,p}(\mathbb{R}^N_+) \to L^p(\mathbb{R}^{N-1})$$

which coincides with the pointwise trace on the boundary for any $v \in W^{1,p}(\mathbb{R}^N_+) \cap C^0(\overline{\mathbb{R}^N_+})$.

We will refer to Tv as the **trace** of v on $\mathbb{R}^N_0$ (which we identify with $\mathbb{R}^{N-1}$) and denote it by

– Trace(v) or simply $v]_{\mathbb{R}^N_0}$ or even v if its meaning is clear from the context.

The previous inequalities extend immediately to traces in $W^{1,p}(\mathbb{R}^N_+)$ and we get the following theorem.

Theorem 1.7 (Trace inequalities in $W^{1,p}(\mathbb{R}^N_+)$). *Denoting by the same symbol v both a function and its trace on the boundary hyperplane $\mathbb{R}^N_0$, we have:*

1. *If $1 \le p < N$ and $p^\sharp = \frac{Np-p}{N-p}$ then there exists a constant $C > 0$, depending on p, N, such that*

$$\left(\int\limits_{\mathbb{R}^N_0 = \partial \mathbb{R}^N_+} |v|^{p^\sharp} \, dx' \right)^{\frac{1}{p^\sharp}} \le C \left(\int\limits_{\mathbb{R}^N_+} |\nabla v|^p \, dx \right)^{\frac{1}{p}} \tag{1.17}$$

for any $v \in W^{1,p}(\mathbb{R}^N_+)$.
Moreover, for any $q \in [N, p^\sharp]$, there exists a constant $C > 0$, depending on p, N, q, such that

$$\left(\int\limits_{\mathbb{R}^N_0 = \partial \mathbb{R}^N_+} |v|^q \, dx' \right)^{\frac{1}{q}} \le C \|v\|_{W^{1,p}(\mathbb{R}^N_+)} \tag{1.18}$$

for any $v \in W^{1,p}(\mathbb{R}^N_+)$.

2. *If $p = N$, then for any $q \in [N, \infty)$ there exists a constant $C > 0$, depending on p, N, q, such that*

$$\left(\int\limits_{\mathbb{R}^N_0 = \partial \mathbb{R}^N_+} |v|^q \, dx' \right)^{\frac{1}{q}} \le C \|v\|_{W^{1,N}(\mathbb{R}^N_+)} \tag{1.19}$$

for any $v \in W^{1,N}(\mathbb{R}^N_+)$.

3. *If $p > N$ and $v \in W^{1,p}(\mathbb{R}^N_+)$, then v has an Hölder continuous representative and is bounded in $\overline{\mathbb{R}^N_+}$. Moreover, if $\gamma = 1 - \frac{N}{p}$, there exists a constant $C > 0$, depending on p, N, such that*

$$\|v\|_{C^{0,\gamma}(\mathbb{R}^N_0)} \le C \|v\|_{W^{1,p}(\mathbb{R}^N_+)} \tag{1.20}$$

for any $v \in W^{1,p}(\mathbb{R}^N_+)$.

Using partitions of unity and local flattenings, many of the previous results can be used to deduce corresponding inequalities in the spaces $W^{1,p}(\Omega)$ when Ω is a bounded Lipschitz domain.

Theorem 1.8 (Extensions, approximations and traces). *Let Ω be a bounded Lipschitz domain, and $1 \le p < \infty$. Then:*

1. *$C^\infty(\overline{\Omega}) \cap W^{1,p}(\Omega)$ is dense in $W^{1,p}(\Omega)$.*
2. *There exists a bounded linear operator $E : W^{1,p}(\Omega) \to W^{1,p}(\mathbb{R}^N)$ such that $Ev(x) = v(x)$ a. e. in Ω for any $v \in W^{1,p}(\Omega)$ (and $\|Ev\|_{W^{1,p}(\mathbb{R}^N)} \le C\|v\|_{W^{1,p}(\Omega)}$ for some constant C). We will refer to it as an **extension operator**.*
3. *There exists a bounded linear operator $T : W^{1,p}(\Omega) \to L^p(\partial\Omega)$ such that $Tv(x) = v(x)$ for any $x \in \partial\Omega$ if $v \in W^{1,p}(\Omega) \cap C^0(\overline{\Omega})$ (and $\|Tv\|_{L^p(\partial\Omega)} \le C\|v\|_{W^{1,p}(\Omega)}$ for some constant C).*

As before, we will refer to the operator T as the **trace operator** and denote by Trace$(v) \in L^p(\partial\Omega)$ the trace Tv of a function $v \in W^{1,p}(\Omega)$. Note that the trace of $v \in W^{1,p}(\Omega)$ actually belongs to the fractional space $W^{1-\frac{1}{p},p}(\partial\Omega)$ and to other Lebesgue spaces, as we will recall.

In the sequel, we will write sometimes $v]_{\partial\Omega}$ or simply v instead of Trace(v).

Using the extension operator E, it is easy to see that the embeddings theorems for Sobolev spaces hold in every Lipschitz domain. Moreover, some embeddings are compact.

Theorem 1.9 (Sobolev embeddings and Rellich–Kondrachov theorem). *Let Ω be a bounded Lipschitz domain. Then*

- *If $1 \le p < N$ and $p^* = \frac{Np}{N-p}$, then $W^{1,p}(\Omega) \hookrightarrow L^{p^*}(\Omega)$, i.e., there exists a constant $C > 0$ depending on p, N, Ω, such that*

$$\|v\|_{L^{p^*}(\Omega)} \le C\|v\|_{W^{1,p}(\Omega)} \tag{1.21}$$

for any $v \in W^{1,p}(\Omega)$.

Moreover, if $1 \le q < p^*$ then $W^{1,p}(\Omega)$ is continuously and compactly embedded in $L^q(\Omega)$, i.e., $W^{1,p}(\Omega) \hookrightarrow L^q(\Omega)$, and any bounded sequence v_n in $W^{1,p}(\Omega)$ has a subsequence v_{k_n} that converges in $L^q(\Omega)$ to a function $v \in L^{p^*}(\Omega)$ (with $v \in W^{1,p}(\Omega)$ if $p > 1$).

- *If $p = N$, then $W^{1,N}(\Omega)$ is continuously and compactly embedded in $L^q(\Omega)$ for any $q \ge 1$.*
- *If $p > n$, then $W^{1,p}(\Omega)$ is continuously and compactly embedded in $C^0(\overline{\Omega})$.*

Analogously, the following embeddings involving the traces hold.

Theorem 1.10 (Trace embeddings in bounded domains). *Let Ω be a bounded Lipschitz domain. Then*

- *If $1 \le p < N$ and $p^\sharp = \frac{Np-p}{N-p}$, then $W^{1,p}(\Omega) \hookrightarrow L^{p^\sharp}(\partial\Omega)$, i.e., there exists a constant $C > 0$ depending on p, N, such that*

$$\|v\|_{L^{p^\sharp}(\partial\Omega)} \le C\|v\|_{W^{1,p}(\Omega)} \tag{1.22}$$

for any $v \in W^{1,p}(\Omega)$.

Moreover, if $1 < p < N$ and $1 \le q < p^\sharp$ then $W^{1,p}(\Omega)$ is continuously and compactly embedded in $L^q(\partial\Omega)$.

- *If $p = N$, then $W^{1,N}(\Omega)$ is continuously and compactly embedded in $L^q(\partial\Omega)$ for any $q \ge 1$.*
- *If $p > n$, then $W^{1,p}(\Omega)$ is continuously and compactly embedded in $C^0(\overline{\Omega})$ and, therefore, in $C^0(\partial\Omega)$.*

As recalled before, if Ω is a bounded domain the Poincaré's inequality and a pure Sobolev inequality are not true in the space $W^{1,p}(\Omega)$, but they do hold in subspaces bigger than $W_0^{1,p}(\Omega)$, in particular in the space of functions whose trace vanishes on a part of the boundary with positive $(N-1)$-dimensional measure.

Theorem 1.11 (More general Poincaré's, pure Sobolev and trace inequalities). *Let Ω be a bounded Lipschitz domain, $1 < p < \infty$, and let us denote by $[\mathrm{Trace}(v) = 0]$ the set $\{x \in \partial\Omega : \mathrm{Trace}(v)(x) = 0\}$.*

For any $a > 0$, there exists a constant $C > 0$, depending on p, N, a, such that for any $v \in W^{1,p}(\Omega)$ with $\mathrm{meas}_{N-1}([\mathrm{Trace}(v) = 0]) \geq a > 0$ the following statements hold:

i) *If $1 < p < N$,*

$$\|v\|_{L^{p^*}(\Omega)} \leq C\|\nabla v\|_{L^p(\Omega)} \tag{1.23}$$

$$\|v\|_{L^{p^\sharp}(\partial\Omega)} \leq C\|\nabla v\|_{L^p(\Omega)} \tag{1.24}$$

ii) *If $1 < p < \infty$,*

$$\|v\|_{L^p(\Omega)} \leq C|\Omega_v|^{\frac{1}{N}}\|\nabla v\|_{L^p(\Omega)} \tag{1.25}$$

where, as before, $\Omega_v = \{x \in \Omega : v(x) \neq 0\} \subset \Omega$ and $|\Omega_v| = \mathrm{meas}_N(\Omega_v)$.

The same inequalities hold
- *for any $v \in W^{1,p}(\Omega)$ with $\mathrm{meas}_N([v = 0]) \geq a > 0$ with a constant C depending on p, N, a.*
- *for any $v \in W^{1,p}(\Omega)$ with $\fint_\Omega v\, dx = 0$ with a constant C depending on p, N.*

Proof. Step 1. We start by proving for any $p > 1$ a first Poincaré's inequality with a constant depending on p and N. Namely, we prove that there exists a constant $C > 0$ such that

$$\|v\|_{L^p(\Omega)} \leq C\|\nabla v\|_{L^p(\Omega)} \tag{1.26}$$

for any $v \in W^{1,p}(\Omega)$ with $\mathrm{meas}_{N-1}([\mathrm{Trace}(v) = 0]) \geq a > 0$.

Suppose by contradiction that this is not true. Then there exists a sequence v_n with $\mathrm{meas}_{N-1}([\mathrm{Trace}(v_n) = 0]) \geq a > 0$ and

$$\|v_n\|_{L^p} \geq n\|\nabla v_n\|_{L^p} \tag{1.27}$$

and by homogeneity we can assume that $\|v_n\|_{L^p} = 1$.

If this is the case, then the sequence is bounded in $W^{1,p}(\Omega)$, so that, up to a subsequence, v_n converges weakly in $W^{1,p}(\Omega)$ to a function $v \in W^{1,p}(\Omega)$. Thanks to the compact embedding into $L^p(\Omega)$, a subsequence converges to v a. e. and strongly in $L^p(\Omega)$, so that $\|v\|_{L^p} = 1$, and $v \neq 0$.

On the other hand by (1.27) and by the weak semicontinuity of the norm, we have that $\|\nabla v\|_{L^p} \leq \liminf \|\nabla v_n\|_{L^p} = 0$, so that $\nabla v = 0$ a. e. in Ω, and v is equal to a constant c. The constant c cannot be zero, since $\|v\|_{L^p} = 1$ and this is a contradiction. Indeed by the compact trace embeddings $\mathrm{Trace}(v_n) \to \mathrm{Trace}(v)$ $(N-1)$-a. e. in $\partial\Omega$ and strongly in $L^p(\partial\Omega)$; if $\mathrm{Trace}(v_n) = 0$ on S_n, with $\mathrm{meas}_{N-1}(S_n) \geq a > 0$, then $v = 0$ a.e on $S = \bigcap_{j=1}^{\infty} \bigcup_{n=j}^{\infty} S_n$ with $\mathrm{meas}_{N-1}(S) \geq a$, which is impossible since $v \equiv c \neq 0$.

Step 2. By the Sobolev and Trace embeddings for $W^{1,p}(\Omega)$ and (1.26), it follows that the pure Sobolev inequality (1.23) and the pure Trace inequality (1.24) hold.

Step 3. Proceeding exactly as in the proof of Theorem 1.4 we get the Poincaré's inequality (1.25) with the dependence of the constant on the measure of Ω.

The proof for functions with vanishing mean or vanishing in a set of positive measure in Ω is exactly the same. $\qquad\square$

Remark 1.12. The best constant for Sobolev embedding has been studied; see [17, 36, 213] for the case of $\mathbb{R}^N$ and [164] for the case of functions vanishing on a part of the boundary, and the references therein.

The previous theorem applies in particular to the subspace of $W^{1,p}(\Omega)$ of the functions whose trace vanishes on a fixed part of the boundary having positive $(N-1)$-measure. This class of functions will be considered in the next chapters, therefore, we state explicitly the results relative to it.

Let Ω be a Lipschitz domain in $\mathbb{R}^N$, and let $\Gamma \subset \partial\Omega$ a relatively *open* subset of the boundary. We denote by

$$W_0^{1,p}(\Omega \cup \Gamma)$$

the closure in $W^{1,p}(\Omega)$ of the subspace $C_c^1(\Omega \cup \Gamma)$. It coincides with the subspace of the functions in $W^{1,p}(\Omega)$ whose trace vanishes on $\partial\Omega \setminus \Gamma$ and obviously $W_0^{1,p}(\Omega \cup \Gamma) = W_0^{1,p}(\Omega)$ if $\Gamma = \emptyset$, while $W_0^{1,p}(\Omega \cup \Gamma) = W^{1,p}(\Omega)$ if $\Gamma = \partial\Omega$. In particular, we will consider the space

$$H_0^1(\Omega \cup \Gamma) = W_0^{1,2}(\Omega \cup \Gamma)$$

which is the closure of $C_c^\infty(\Omega \cup \Gamma)$ in the Hilbert space $H^1(\Omega)$, and coincides with the space of functions $v \in H^1(\Omega)$ such that $\mathrm{Trace}(v) = 0$ on $\partial\Omega \setminus \Gamma$. Setting

$$\Gamma_0 = \partial\Omega \setminus \overline{\Gamma}$$

we have also

$$\Gamma = \partial\Omega \setminus \overline{\Gamma_0}$$

since, in the relative topology of $\partial\Omega$, Γ is open and Γ_0 is its exterior.

We require the following conditions:

$$\Gamma \text{ is either empty or a smooth } (N-1) - \text{submanifold} \qquad (1.28)$$

$$\partial\Omega \setminus (\Gamma_0 \cup \Gamma) = \overline{\Gamma_0} \cap \overline{\Gamma} = \partial\Gamma_0 \cap \partial\Gamma$$

$$\text{is either empty or a smooth } (N-2)-\text{submanifold} \tag{1.29}$$

Let us summarize the previous results in the following statement.

Theorem 1.13 (Sobolev, Trace and Poincaré's inequalities in $H_0^1(\Omega \cup \Gamma)$). *Let Ω be a bounded Lipschitz domain in $\mathbb{R}^N$, $N \geq 2$, Γ and Γ_0 subsets of the boundary, as described in (1.28), (1.29). Moreover, let us assume that Γ_0 has positive $(N-1)$-Hausdorff measure. If $N \geq 3$, then there exist constants $C_1, C_2 > 0$ such that*

$$\left(\int_\Omega |v|^{\frac{2N}{N-2}} \, dx \right)^{\frac{N-2}{2N}} \leq C_1 \left(\int_\Omega |\nabla v|^2 \, dx \right)^{\frac{1}{2}} \tag{1.30}$$

$$\left(\int_\Gamma |v|^{\frac{2N-2}{N-2}} \, d\sigma \right)^{\frac{N-2}{2N-2}} \leq C_2 \left(\int_\Omega |\nabla v|^2 \, dx \right)^{\frac{1}{2}} \tag{1.31}$$

for any $v \in H_0^1(\Omega \cup \Gamma)$.

If $N = 2$, the norm in the left-hand side can be replaced by any L^q norm, $q > 1$.

Moreover, if $N \geq 2$, then there exists a constant C depending on N and $\mathrm{meas}_{(N-1)}(\partial\Omega \setminus \Gamma)$ such that

$$\left(\int_\Omega |v|^2 \, dx \right)^{\frac{1}{2}} \leq C |\Omega_v|^{\frac{1}{N}} \left(\int_\Omega |\nabla v|^2 \, dx \right)^{\frac{1}{2}} \tag{1.32}$$

for any $v \in H_0^1(\Omega \cup \Gamma)$, where $\Omega_v = \{x \in \Omega : v(x) \neq 0\} \subset \Omega$ and $|\Omega_v|$ is its N-dimensional measure.

Remark 1.14. Note that the constants which appear in the previous theorem depend on the $(N-1)$-measure of the portion of the boundary where the functions involved vanish. In some cases, considering subsets D of a given domain Ω, it could be important to have the previous inequalities with constants independent of the specific subset D. When dealing with some domains with cylindrical symmetry, we can use the basic inequalities that hold in the half spaces and get further Poincaré's inequalities with fixed constants, as we show below.

Denoting by $x = (x', x_N)$ a point $x = (x_1, \ldots, x_{N-1}, x_N) \in \mathbb{R}^N$, the domains we consider will be subsets of the half-space $\mathbb{R}_+^N = \{x = (x_1, \ldots, x_N) \in \mathbb{R}^N : x_N > 0\}$ defined as follows.

Definition 1.15. We say that a bounded Lipschitz domain Ω in $\mathbb{R}^N$, $N \geq 2$, has cylindrical symmetry or is a cylindrically symmetric domain if assuming that

$$\inf\{t \in \mathbb{R} : (x', t) \in \Omega\} = 0, \quad \sup\{t \in \mathbb{R} : (x', t) \in \Omega\} = b > 0$$

then, for every $h \in [0, b)$, the set

$$\overline{\Omega^h} = \overline{\Omega} \cap \{x_N = h\}$$

is either a closed $(N-1)$-dimensional ball or a closed $(N-1)$-dimensional annulus with the center on the x_N axis.

For such domains, we will always set

$$\Gamma_0 = \partial\Omega \cap \mathbb{R}^N_+; \quad \Gamma = \partial\Omega \setminus \overline{\Gamma_0} = \Omega^0 = \mathrm{int}(\overline{\Omega} \cap \mathbb{R}^N_0) \tag{1.33}$$

Thus Γ is a relatively open flat part of the boundary at the height $x_N = 0$, which, by our assumptions, is either a $(N-1)$-dimensional ball or a $(N-1)$-dimensional annulus.

Examples 1.16. Cylindrically symmetric domains are the following:
- a half-ball

$$(B^N_R)_+ = B_R \cap \mathbb{R}^N_+ = \{x = (x_1, \ldots, x_N) \in \mathbb{R}^N : |x| < R; \, x_N > 0\},$$

- a half-annulus

$$(A^N_{R_1, R_2})_+ = \{x = (x_1, \ldots, x_N) \in \mathbb{R}^N : R_1 < |x| < R_2; \, x_N > 0\},$$

- a cylinder

$$C_{R,b} = \{x = (x', x_N) \in \mathbb{R}^N : |x'| < R; \, 0 < x_N < b\},$$

- an annular cylinder

$$C_{R_1, R_2, b} = \{x = (x', x_N) \in \mathbb{R}^N : R_1 < |x'| < R; \, 0 < x_N < b\},$$

- a cone

$$K_{R,b} = \left\{x = (x', x_N) \in \mathbb{R}^N : |x'| < \frac{R}{b}(b - x_N); \, 0 < x_N < b\right\} .$$

Note that Γ_0 is smooth in the first two examples, and is not in the other cases (the cone is not smooth at the vertex, and the cylinders at height b).

When considering cylindrically symmetric domains, we will take advantage of their simple geometry to prove all the relevant inequalities that we need, starting from (1.10), (1.14), no matter how big is the part of the boundary where the functions vanish.

Note that if $v \in H^1_0(\Omega \cup \Gamma)$ then the trivial extension of v to $\mathbb{R}^N_+$ belongs to $H^1(\mathbb{R}^N_+)$ and has vanishing trace on $\mathbb{R}^N_0 \setminus \Gamma$. As a consequence, using Hölder's inequality, we obtain Poincaré's type inequalities both in Ω and on the flat boundary Γ. More precisely, we have the following.

Theorem 1.17 (Poincaré's inequalities in cylindrically symmetric domains). *Let $N \geq 2$ and let Ω be a bounded cylindrically symmetric domain. There exist constants $C_1, C_2 > 0$, depending only on N, such that for any $v \in H_0^1(\Omega \cup \Gamma)$*

$$\int_\Omega |v|^2 \, dx \leq C_1 |[v \neq 0]|^{\frac{2}{N}} \int_\Omega |\nabla v|^2 \, dx \tag{1.34}$$

$$\int_\Gamma |v|^2 \, dx' \leq C_2 |[v \neq 0]|^{\frac{1}{N}} \int_\Omega |\nabla v|^2 \, dx \tag{1.35}$$

where $[v \neq 0] = \{x \in \Omega : v(x) \neq 0\}$ and $|[v \neq 0]|$ denotes its measure.

Proof. Let us first consider the case $N \geq 3$. By density, we can assume that $v \in C_c^\infty(\Omega \cup \Gamma)$ and we denote by v also the trivial extension to $\mathbb{R}_+^N$.

By Hölder's and Sobolev inequalities, we have that

$$\int_\Omega |v|^2 \, dx = \int_{[v \neq 0]} |v|^2 1 \, dx \leq \left(\int_{\mathbb{R}_+^N} |v|^{\frac{2N}{N-2}} \, dx \right)^{\frac{N-2}{N}} |[v \neq 0]|^{\frac{2}{N}}$$

$$\leq C_1 |[v \neq 0]|^{\frac{2}{N}} \int_{\mathbb{R}_+^N} |\nabla v|^2 \, dx = C_1 |[v \neq 0]|^{\frac{2}{N}} \int_\Omega |\nabla v|^2 \, dx$$

so that we get (1.34).

If instead $N = 2$ then $p^* = 2$ for $p = 1$ and, since $\nabla v = 0$ a. e. on $[v = 0]$, by Sobolev and Hölder's inequalities we get

$$\int_\Omega |v|^2 \, dx = \int_{[v \neq 0]} |v|^{1^*} \, dx \leq C \left(\int_{[v \neq 0]} |\nabla v| \, dx \right)^2 \leq C |[v \neq 0]| \int_{[v \neq 0]} |\nabla v|^2 \, dx$$

and (1.34) follows.

To get (1.35), we observe that for any $x' = (x_1, \ldots, x_{N-1}) \in \mathbb{R}^{N-1}$ we have that

$$v^2(x', 0) = -\int_0^{+\infty} \frac{\partial v^2}{\partial x_N}(x', t) \, dt = -2 \int_0^{+\infty} v(x', t) \frac{\partial v}{\partial x_N}(x', t) \, dt$$

Integrating over Γ and using the Poincaré's inequality (1.34) and Hölder's inequality, we get

$$\int_\Gamma v^2(x', 0) \, dx' \leq 2 \int_{\mathbb{R}_+^N} |v| |\nabla v| \, dx \leq 2 \left(\int_\Omega |v|^2 \, dx \right)^{\frac{1}{2}} \left(\int_\Omega |\nabla v|^2 \, dx \right)^{\frac{1}{2}} dx$$

$$\leq C_2 |[v \neq 0]|^{\frac{1}{N}} \int_\Omega |\nabla v|^2 \, dx \qquad \qquad \square$$

1.2 Maximum principles

Here, we consider standard semilinear elliptic equations of the type

$$- \Delta u = f(x, u) \quad \text{in } \Omega \tag{1.36}$$

where Ω is a Lipschitz domain in $\mathbb{R}^N$, $N \geq 2$ and f is a continuous function in $\overline{\Omega} \times \mathbb{R}$.

A weak solution of (1.36) is a function $u \in H^1_{\mathrm{loc}}(\Omega)$ such that the function $x \mapsto f(x, u(x))$ belongs to $L^1_{\mathrm{loc}}(\Omega)$ and

$$\int_\Omega \nabla u \cdot \nabla \varphi \, dx = \int_\Omega f(x, u) \varphi \, dx \tag{1.37}$$

for any $\varphi \in C_c^\infty(\Omega)$.

When $N \geq 3$ and $u \in H^1(\Omega)$, we assume that the function $x \mapsto f(x, u(x))$ belongs to $L^{\frac{2N}{N+2}}(\Omega)$ (note that $\frac{2N}{N+2}$ is the conjugate exponent of the critical Sobolev exponent $2^* = \frac{2N}{N-2}$). Then, by density, (1.37) is satisfied for any $\varphi \in H^1_0(\Omega)$.

We say that $u \in H^1(\Omega)$ (weakly) satisfies the inequality

$$u \leq 0 \text{ on } \partial\Omega \quad (u \geq 0 \text{ on } \partial\Omega)$$

if $u^+ \in H^1_0(\Omega)$ $(u^- \in H^1_0(\Omega))$. Moreover, we write

$$- \Delta u \geq \ (\leq) f(x, u) \quad \text{in } \Omega \tag{1.38}$$

(in a weak sense) if

$$\int_\Omega \nabla u \cdot \nabla \varphi \, dx \geq \ (\leq) \int_\Omega f(x, u) \varphi \, dx \tag{1.39}$$

for any $\varphi \in H^1_0(\Omega)$ with $\varphi \geq 0$ in Ω.

A weak solution of the Dirichlet problem

$$\begin{cases} -\Delta u = f(x, u) & \text{in } \Omega \\ u = 0 & \text{on } \partial\Omega \end{cases} \tag{1.40}$$

is a function $u \in H^1_0(\Omega)$ such that (1.37) holds for any $\varphi \in H^1_0(\Omega)$.

Let Ω be a bounded Lipschitz domain, Γ a relatively open portion of its boundary, $\Gamma_0 = \partial\Omega \setminus \overline{\Gamma}$, $f : \overline{\Omega} \times \mathbb{R}$ and $g : \Gamma \times \mathbb{R} \to \mathbb{R}$ continuous functions and let us denote by v the outer normal on $\partial\Omega$.

A weak solution of the Dirichlet–Neumann problem (with nonlinear boundary conditions)

$$\begin{cases} -\Delta u = f(x, u) & \text{in } \Omega \\ u = 0 & \text{on } \Gamma_0 \\ \dfrac{\partial u}{\partial v} = g(x, u) & \text{on } \Gamma \end{cases} \tag{1.41}$$

is a function $u \in H_0^1(\Omega \cup \Gamma)$ such that

$$\int_\Omega \nabla u \cdot \nabla \varphi \, dx = \int_\Omega f(x, u)\varphi \, dx + \int_\Gamma g(x', u|_\Gamma)\varphi|_\Gamma \, dx' \tag{1.42}$$

for any $\varphi \in H_0^1(\Omega \cup \Gamma)$.

Analogously, we say that $u \in H^1(\Omega)$ weakly satisfies the inequality

$$\begin{cases} -\Delta u \leq f(x, u) & \text{in } \Omega \\ u \leq 0 & \text{on } \Gamma_0 \\ \dfrac{\partial u}{\partial \nu} \leq g(x, u) & \text{on } \Gamma \end{cases} \tag{1.43}$$

if $u^+ \in H_0^1(\Omega \cup \Gamma)$ and

$$\int_\Omega \nabla u \cdot \nabla \varphi \, dx \leq \int_\Omega f(x, u)\varphi \, dx + \int_\Gamma g(x', u)\varphi \, dx' \tag{1.44}$$

for any $\varphi \in H_0^1(\Omega \cup \Gamma)$, $\varphi \geq 0$, with the obvious modification in the case of the reversed inequality.

1.2.1 Weak maximum principle

Definition 1.18. Let D be a bounded domain in $\mathbb{R}^N$ and $c \in L^\infty(D)$.

We say that the operator $L = -\Delta + c$ satisfies the maximum principle in D (or the maximum principle holds for the operator L in D) if for any $v \in H^1(D)$ such that

$$\begin{cases} -\Delta v + c(x)v \leq 0 & \text{in } D \\ v \leq 0 & \text{on } \partial D \end{cases}$$

it holds that $v \leq 0$ in D.

If $c \geq 0$, then the maximum principle holds in any bounded domain Ω, namely the following classical (weak) form of the weak maximum principle holds (see, e. g., [115, 126, 193, 194, 218] among others for many generalizations).

Theorem 1.19 (First case for weak maximum principle). *Let* $N \geq 2$, Ω *a bounded domain in* $\mathbb{R}^N$ *and let* $v \in H^1(\Omega)$ *weakly satisfy*

$$\begin{cases} -\Delta v + c(x)v \leq 0 & \text{in } \Omega \\ v \leq 0 & \text{on } \partial\Omega \end{cases} \tag{1.45}$$

with $c \in L^\infty(\Omega)$ *and* $c \geq 0$ *a. e. in* Ω.

Then $v \leq 0$ *in* Ω.

Proof. By hypothesis, the nonnegative function v^+ belongs to $H_0^1(\Omega)$ and can be used as a test function in (1.45), yielding

$$\int_\Omega |\nabla v^+|^2 \, dx \leq \int_\Omega |\nabla v^+|^2 + c(x)(v^+)^2 \, dx \leq 0$$

so that $v^+ = 0$, since $\int_\Omega |\nabla v^+|^2$, which by Poincaré's inequality is an equivalent norm on $H_0^1(\Omega)$, vanishes. $\qquad\square$

In general, the maximum principle holds in domains with small measure, as we prove now using the Poincaré inequality, regardless of the sign of $c(x)$.

Theorem 1.20 (Weak maximum principle in small domains). *Let $N \geq 2$, Ω a bounded domain in $\mathbb{R}^N$ and $M \geq 0$. Assume that $c \in L^\infty(\Omega)$, $\|c\|_{L^\infty(\Omega)} \leq M$.*

Then there exists $\delta > 0$, depending on M, such that the maximum principle holds for the operator $-\Delta + c$ in $\Omega' \subset \Omega$ provided $|\Omega'| = meas_N(\Omega') < \delta$.

More generally, there exists $\delta > 0$, depending on M, such that the following holds. If $\Omega' \subseteq \Omega$ is a subdomain of Ω and $v \in H^1(\Omega')$, then

$$\begin{cases} -\Delta v + c(x)v \leq 0 & \text{in } \Omega' \\ v \leq 0 & \text{on } \partial\Omega' \end{cases} \qquad \Longrightarrow \qquad v \leq 0 \text{ in } \Omega' \qquad (1.46)$$

provided $|[v > 0]| < \delta$, where $[v > 0] = \{x \in \Omega' : v(x) > 0\}$ and $|[v > 0]| = meas_N([v > 0])$. (In particular, this condition is satisfied if $|\Omega'| < \delta$.)

Proof. By hypothesis, the nonnegative function v^+ belongs to $H_0^1(\Omega')$ and can be used as a test function in (1.46), yielding

$$\int_{\Omega'} |\nabla v^+|^2 \leq M \int_{\Omega'} (v^+)^2$$

On the other hand by the Poincaré's inequality (1.8), we get

$$M \int_{\Omega'} (v^+)^2 \leq MC|[v > 0]|^{\frac{2}{N}} \int_{\Omega'} |\nabla v^+|^2$$

If the measure of $[v > 0] = \{x \in \Omega' : v(x) > 0\}$ is sufficiently small, precisely when $MC|[v > 0]|^{\frac{2}{N}} < 1$, then necessarily $\int_{\Omega'} |\nabla v^+|^2 \, dx = 0$, which implies that $v^+ \equiv 0$ in Ω' (since $(\int_{\Omega'} |\nabla v^+|^2 \, dx)^{\frac{1}{2}}$ is an equivalent norm on $H_0^1(\Omega')$). $\qquad\square$

As a corollary, we have the following.

Theorem 1.21 (Weak comparison principle in small domains). *Let Ω be a bounded domain in $\mathbb{R}^N$, $f = f(x,s) : \Omega \times \mathbb{R} \to \mathbb{R}$ a continuous function which is locally Lipschitz continuous in $s \in \mathbb{R}$ uniformly w. r. t. $x \in \Omega$, and $u, v \in H^1(\Omega) \cap L^\infty(\Omega)$ such that $\|u\|_{L^\infty(\Omega)}$, $\|v\|_{L^\infty(\Omega)} \leq A$.*

Then there exists $\delta > 0$, depending on f and A (more precisely on the Lipschitz constant of f for s in $[-A, A]$) such that the following holds:
if $\Omega' \subseteq \Omega$ is a bounded subdomain of Ω with $|[u > v] \cap \Omega'| < \delta$ and

$$\begin{cases} -\Delta u \leq f(x, u), \ -\Delta v \geq f(x, v) & \text{in } \Omega' \\ u \leq v & \text{on } \partial\Omega' \end{cases} \tag{1.47}$$

then $u \leq v$ in Ω'.

Proof. If $\Omega' \subseteq \Omega$ and u, v satisfy (1.47), the difference $u - v$ satisfies the linear inequality

$$\begin{cases} -\Delta(u - v) + c(x)(u - v) \leq 0 & \text{in } \Omega' \\ u - v \leq 0 & \text{on } \partial\Omega' \end{cases}$$

where $c : \Omega \to \mathbb{R}$ is defined by

$$-c(x) = \begin{cases} \frac{f(x,u(x)) - f(x,v(x))}{u(x) - v(x)} & \text{if } u(x) \neq v(x) \\ 0 & \text{if } u(x) = v(x) \end{cases}$$

Since f is Lipschitz continuous w. r. t. $s \in [-A, A]$, we get that $c \in L^\infty(\Omega)$ with $\|c\|_{L^\infty(\Omega)} \leq M$, where M is a Lipschitz constant of f for $s \in [-A, A]$.

So the result is a consequence of the previous maximum principle. $\qquad\square$

Remark 1.22. Of course, the choice of the value $c(x)$ in points where $u(x) = v(x)$ is arbitrary, but note that if f is a C^1-function, then we can write a linear inequality for the difference $u - v$ with a continuous coefficient, namely $u - v$ satisfies the inequality

$$-\Delta(u - v) + \tilde{c}(x)(u - v) \leq 0$$

where $\tilde{c}$ is the *continuous* function given by

$$-\tilde{c}(x) = \begin{cases} \frac{f(x,u(x)) - f(x,v(x))}{u(x) - v(x)} & \text{if } u(x) \neq v(x) \\ f'(x, u(x)) & \text{if } u(x) = v(x) \end{cases} \tag{1.48}$$

$$= \int_0^1 f'(x, tu(x) + (1 - t)v(x)) \, dt$$

For what concerns mixed boundary value problems, analogous properties hold.

Let us fix Ω and the subsets Γ and Γ_0 of $\partial\Omega$ that satisfy (1.28), (1.29). As before, we do not exclude the case of $\Gamma = \emptyset$ (i. e., when the Dirichlet problem is studied), in which case $\Gamma_0 = \partial\Omega$.

Definition 1.23. Let Ω be a bounded Lipschitz domain in $\mathbb{R}^N$, with $\Gamma, \Gamma_0 \subseteq \partial\Omega$ satisfying (1.28), (1.29), $c \in L^\infty(\Omega)$, $d \in L^\infty(\Gamma)$. We say that the operator $L = (-\Delta + c; \frac{\partial}{\partial v} + d)$ satisfies the maximum principle in $\Omega \cup \Gamma$ if for any $v \in H^1(\Omega)$ such that

$$\begin{cases} -\Delta v + cv \le 0 & \text{in } \Omega \\ v \le 0 & \text{on } \Gamma_0 \\ \dfrac{\partial v}{\partial \upsilon} + dv \le 0 & \text{on } \Gamma \end{cases}$$

it holds that $v \le 0$ in Ω.

As before, it is easy to see that for any bounded domain Ω:

if $c, d \ge 0$ then the maximum principle holds in $\Omega \cup \Gamma$.

Definition 1.24. Let Ω be a bounded Lipschitz domain in $\mathbb{R}^N$, with $\Gamma, \Gamma_0 \subseteq \partial\Omega$ satisfying (1.28), (1.29). If $D \subset \Omega$ is an open subdomain (i. e., it is open and connected), let us set $\Gamma_D = \Gamma \cap \partial D$, $(\Gamma_0)_D = \partial D \setminus \overline{\Gamma_D}$. In particular when $\Gamma = \emptyset$ then $\Gamma_D = \emptyset$ and $(\Gamma_0)_D = \partial D$.

We say that D is a *regular subdomain* of Ω if one of the following conditions holds:
1. $\Gamma = \emptyset$ (then D can be any open connected subset of Ω)
2. $\Gamma \ne \emptyset$ and $\Gamma_D = \Gamma \cap \partial D = \emptyset$
3. D is a Lipschitz domain, $\Gamma_D \ne \emptyset$, and both Γ_D and $(\Gamma_0)_D$ satisfy the conditions analogous to (1.28), (1.29).

In cylindrically symmetric domains, a maximum principle in regular subdomains with small measure, analogous to Theorem 1.20, holds.

Theorem 1.25 (Weak maximum principle in small cylindrical domains). *Let $N \ge 2$, Ω a cylindrically symmetric domain, $\alpha \in L^\infty(\Omega)$, $\beta \in L^\infty(\Gamma)$ with $\|\alpha\|_{L^\infty(\Omega)} \le M$, $\|\beta\|_{L^\infty(\Gamma)} \le M$. Then there exists $\delta > 0$, depending on M, such that the following holds:*
if $\Omega' \subseteq \Omega$ is a regular subdomain of Ω and $v \in H^1(\Omega')$ satisfies

$$\begin{cases} -\Delta v + \alpha(x)v \le 0 & \text{in } \Omega' \\ v \le 0 & \text{on } \Gamma'_0 = \partial\Omega' \cap \mathbb{R}^N_+ \\ \dfrac{\partial v}{\partial \upsilon} + \beta(x)v \le 0 & \text{on } \Gamma' = \partial\Omega' \setminus \overline{\Gamma'_0} \end{cases} \tag{1.49}$$

then $v \le 0$ in Ω', provided $|[v > 0]| < \delta$, where $[v > 0] = \{x \in \Omega' : v(x) > 0\}$ and $|[v > 0]| = meas_N([v > 0])$.

In particular, this condition is satisfied if $|\Omega'| < \delta$.

Proof. By hypothesis, the nonnegative function v^+ belongs to $H_0^1(\Omega' \cup \Gamma')$ and can be used as a test function, yielding

$$\int_{\Omega'} |\nabla v^+|^2 \le M\left(\int_{\Omega'} (v^+)^2 + \int_{\Gamma'} (v^+)^2 \right)$$

On the other hand, by the Poincaré's inequalities (1.34), (1.35), we get

$$M\left(\int_{\Omega'} (v^+)^2 + \int_{\Gamma'} (v^+)^2\right)$$

$$\leq MC(|[v > 0]|^{\frac{2}{N}} + |[v > 0]|^{\frac{1}{N}}) \int_{\Omega'} |\nabla v^+|^2$$

If the measure of $([v > 0])$ is sufficiently small, then $MC(|[v > 0]|^{\frac{2}{N}} + |[v > 0]|^{\frac{1}{N}}) < 1$, which implies that $v^+ \equiv 0$ in Ω'. $\qquad\square$

As before, a corollary is the following

Theorem 1.26 (Weak comparison principle in small cylindrical domains). *Let Ω be a cylindrically symmetric domain, $f(|x|, s)$ and $g(|x'|, s)$ two functions defined in $[0, +\infty) \times \mathbb{R}$ which are locally Lipschitz continuous functions w. r. t. the second variable, and $u, v \in H^1(\Omega) \cap L^\infty(\Omega)$ with their trace belonging to $L^\infty(\partial\Omega)$, such that $\|u\|_{L^\infty(\Omega)}, \|v\|_{L^\infty(\Omega)} \leq A$. Then there exists $\delta > 0$, depending on A, f, g such that the following holds: if $\Omega' \subseteq \Omega$ is a regular subdomain of Ω, $|[u > v] \cap \Omega'| < \delta$ and*

$$\begin{cases} -\Delta u \leq f(|x|, u), \ -\Delta v \geq f(|x|, v) & \text{in } \Omega' \\ u \leq v & \text{on } \Gamma'_0 = \partial\Omega' \cap \mathbb{R}^N_+ \\ \dfrac{\partial u}{\partial v} \leq g(|x|, u), \ \dfrac{\partial v}{\partial v} \geq g(|x|, v) & \text{on } \Gamma' = \partial\Omega' \setminus \overline{\Gamma'_0} \end{cases} \tag{1.50}$$

then $u \leq v$ in Ω'.

Proof. Since f is Lipschitz continuous, the difference $u - v$ satisfies a linear inequality, so the result is a consequence of the previous maximum principle, but we also give a direct short proof.

Since $u, v \in L^\infty$ and f, g are locally Lipschitz continuous in the second variable, there exists $\Lambda > 0$ such that the functions $\tilde{f}_\Lambda(|x|, s) = f(|x|, s) - \Lambda s$, $\tilde{g}_\Lambda(|x|, s) = g(|x|, s) - \Lambda s$, are nonincreasing in $s \in [-A, A]$ for any $x \in \Omega$ (so that if $u > v$ then $f(|x|, u) - \Lambda u \leq f(|x|, v), -\Lambda v$).

By hypothesis, the nonnegative function $(u - v)^+$ belongs to $H^1_0(\Omega \cup \Gamma)$ and can be used as a test function in the inequalities, yielding

$$\int_\Omega |\nabla(u - v)^+|^2 \, dx - \int_\Omega \Lambda((u - v)^+)^2 \, dx - \int_\Gamma \Lambda((u - v)^+)^2 \, dx' \leq 0$$

On the other hand, by Poincaré's inequalities (1.34), (1.35), we get that

$$\int_\Omega |\nabla(u - v)^+|^2 \, dx - \int_\Omega \Lambda((u - v)^+)^2 \, dx - \int_\Gamma \Lambda((u - v)^+)^2 \, dx'$$

$$\geq (1 - \Lambda C(|[u > v] \cap \Omega'|^{\frac{2}{N}} + |[u > v] \cap \Omega'|^{\frac{1}{N}})) \int_\Omega |\nabla(u - v)^+|^2 \, dx$$

and the last quantity is strictly positive if the measure of $([u > v])$ is sufficiently small and $(u - v)^+$ does not vanish. $\qquad\square$

Remark 1.27. For general elliptic operators in nondivergence form, a maximum principle in domains with small measure has been proved in [31, 32], using the Aleksandrov–Bakelman–Pucci inequality (see [126]).

1.2.2 Strong maximum principle

Let us recall the following versions of the strong maximum principle and Hopf's lemma (see, e. g., [126, 193, 194] and the references therein for many generalizations).

We say that Ω satisfies an interior sphere condition at $x_0 \in \partial\Omega$ if there exists a ball $B \subset \Omega$ which has x_0 on its boundary.

Theorem 1.28 (Strong maximum principle and Hopf's lemma). *Let Ω be a (bounded or unbounded) domain in $\mathbb{R}^N$, and let $v \in H^1_{loc}(\Omega) \cap C^1(\Omega)$ be a weak solution of*

$$\begin{cases} -\Delta v + c(x)v \geq 0 & \text{in } \Omega \\ v \geq 0 & \text{in } \Omega \end{cases} \tag{1.51}$$

with $c \in L^\infty_{loc}(\Omega)$.

Then either $v \equiv 0$ in Ω or $v > 0$ in Ω.

Moreover if $c \in L^\infty(\Omega)$, $v \in C^0(\Omega \cup \{x_0\})$, $v > 0$ in Ω and $x_0 \in \partial\Omega$ is a point where $v(x_0) = 0$ and the interior sphere condition is satisfied then

$$\frac{\partial v}{\partial s}(x_0) > 0 \tag{1.52}$$

for any inward directional derivative (if v does not belong to the class $C^1(\Omega \cup \{x_0\})$ this means that if s is a inward direction, i. e. there exists $\delta > 0$ such that $x_0 + ts \in \Omega$ for any $t \in (0, \delta)$, then $\liminf_{t\to 0^+} \frac{v(x_0+ts)-v(x_0)}{t} > 0$).

The proof of this theorem is based on the following lemma, which proves Hopf's lemma in a particular case.

Lemma 1.29. *Let $B = B(y, r_1)$ a ball, $x_0 \in \partial B$ and let $v \in H^1(B) \cap C^0(B \cup \{x_0\})$ be a weak solution of $-\Delta v + c(x)v \geq 0$ in B with $c \in L^\infty(B)$. Assume further that $v > 0$ in B and $v(x_0) = 0$.*

Then $\frac{\partial v}{\partial s}(x_0) > 0$ for any inner direction s (this means that if $(y - x_0) \cdot s > 0$ then $\liminf_{t\to 0^+} \frac{v(x_0+ts)-v(x_0)}{t} > 0$).

Proof. Since we know that $v \geq 0$, we can suppose that $c \geq 0$ by substituting c with its positive part c^+. In fact, by hypothesis, we have that

$$-\Delta v + c^+(x)v \geq -\Delta v + c(x)v \geq 0 \quad \text{in } B$$

Then we consider the annulus $A = B \setminus B'$ where $B = B(y, r_1)$, $B' = B(y, \frac{r_1}{2})$, and the function

$$z(x) = e^{-\alpha r_1^2} - e^{-\alpha |x-y|^2}$$

with $\alpha > 0$ to be determined.

The function z satisfies $z < 0$ in A, $z = 0$ on ∂B, $z < 0$ on $\partial B'$. Moreover, if $x \in A$ then

$$\Delta z = -e^{-\alpha|x-y|^2}\left(4\alpha^2|x-y|^2 - 2\alpha N\right)$$

so that

$$-\Delta z + c(x)z = e^{-\alpha|x-y|^2}\left(4\alpha^2|x - y1|^2 - 2\alpha N + c(x)\frac{e^{-\alpha r_1^2}}{e^{-\alpha|x-y|^2}} - c(x)\right)$$

$$\geq e^{-\alpha|x-y|^2}\left(4\alpha^2\frac{r_1^2}{4} - 2\alpha N - \|c\|_\infty\right) \geq e^{\frac{r_1^2}{4}}\left(4\alpha^2\left(\frac{r_1^2}{4}\right)^2 - 2\alpha N - \|c\|_\infty\right) > 0$$

if α is sufficiently large.

Since $v > 0$ in B, it has a positive miminum on $\partial B'$, and if ε is sufficiently small then $v + \varepsilon z > 0$ on $\partial B'$. By the weak maximum principle, since $w = v + \varepsilon z \geq 0$ on ∂A and $-\Delta w + c(x)w \geq \varepsilon(-\Delta z + c(x)z) \geq 0$ in A, we get that $w = v + \varepsilon z \geq 0$ in A, which in turn implies that

$$\frac{\partial w}{\partial s}(x_0) = \frac{\partial v}{\partial s}(x_0) + \varepsilon\frac{\partial z}{\partial s} \geq 0$$

Since $\frac{\partial z}{\partial s}(x_0) = 2\alpha(x_0 - y) \cdot se^{-\alpha|x_0-y|^2} < 0$, we get that $\frac{\partial v}{\partial s}(x_0) > 0$. $\qquad \square$

Proof of Theorem 1.28. Assume that $v \not\equiv 0$ in Ω and let $\Omega^+ = \{x \in \Omega : v(x) > 0\}$. Since $v \geq 0$ and $v \not\equiv 0$ in Ω, then $\Omega^+ \neq \emptyset$. Suppose by contradiction that there exist points in Ω where v vanishes, i. e., $\Omega \setminus \Omega^+ \neq \emptyset$.

Then $\partial\Omega^+ \cap \Omega \neq \emptyset$ and we can find a point $x_1 \in \Omega^+$ which is closer to $\partial\Omega^+$ than to $\partial\Omega$. If $r_1 = \sup\{r > 0 : B(x_1, r) \subset \Omega^+\}$, then $B = B(x_1, r_1) \subset \Omega^+$, $\overline{B} \subset \Omega$ and there exists $x_0 \in \partial\Omega^+ \cap \Omega$ such that $v(x_0) = 0$. By Lemma 1.29 $\frac{\partial v}{\partial s}(x_0) > 0$ for any inward directional derivative, and this contradicts the fact that $\nabla v(x_0) = 0$, since x_0 is a minimum point in Ω for the function v. So if $v \not\equiv 0$ in Ω necessarily $v > 0$ in Ω.

The last part of the assertion, regarding the sign of the inner derivative, has already been proved in Lemma 1.29. $\qquad \square$

Let us remark that a consequence of Hopf's lemma for Dirichlet–Neumann problems, under a global C^1 regularity hypothesis, is the following result.

Theorem 1.30 (Strong maximum principle in $H_0^1(\Omega \cup \Gamma)$). *Let Ω be a bounded domain in $\mathbb{R}^N$ with the subsets Γ, Γ_0 of $\partial\Omega$ satisfying (1.28), (1.29) and let $v \in H_0^1(\Omega \cup \Gamma) \cap C^1(\Omega \cup \Gamma)$ be a weak solution of*

$$\begin{cases} -\Delta v + c(x)v \geq 0 & \text{in } \Omega \\ v \geq 0 & \text{in } \Omega \\ v \geq 0 & \text{on } \Gamma_0 \\ \dfrac{\partial v}{\partial v} + d(x)v \geq 0 & \text{on } \Gamma \end{cases} \tag{1.53}$$

with $c \in L^\infty(\Omega)$, $d \in C^0(\Gamma)$. Assume further that Ω satisfies an interior sphere condition at every point in Γ.

Then either $v \equiv 0$ in Ω or $v > 0$ in $\Omega \cup \Gamma$.

Proof. By the previous strong maximum principle if $v \not\equiv 0$ in Ω, then $v > 0$ in Ω and hence, by continuity, $v \geq 0$ on Γ. Let $x_0 \in \Gamma$ and assume by contradiction that $v(x_0) = 0$. Then by Hopf's lemma $\frac{\partial v}{\partial v}(x_0) < 0$ where v is the exterior normal, since v is positive in Ω and vanishes in x_0. This contradicts the Neumann condition $\frac{\partial v}{\partial v}(x_0) + d(x)v(x_0) = \frac{\partial v}{\partial v}(x_0) \geq 0$. Hence v is positive on Γ. $\qquad\square$

The strong maximum principle implies the following.

Theorem 1.31 (Strong comparison principle). *Let Ω be a (bounded or unbounded) domain in $\mathbb{R}^N$, and let $u, v \in H_{loc}^1(\Omega) \cap C^1(\Omega)$ satisfy weakly*

$$\begin{cases} -\Delta u \leq f(x, u); \quad -\Delta v \geq f(x, v) & \text{in } \Omega \\ u \leq v & \text{in } \Omega \end{cases} \tag{1.54}$$

where $f = f(x, u) : \Omega \times \mathbb{R} \to \mathbb{R}$ is a continuous function which is locally Lipschitz continuous in u, uniformly w.r.t $x \in \Omega$.

Then either $u \equiv v$ in Ω or $u < v$ in Ω. In the latter case, let $x_0 \in \partial\Omega$ a point where $u(x_0) = v(x_0)$ and the interior sphere condition is satisfied. Assume that $B \subset \Omega$ is a ball which has x_0 on its boundary and $u, v \in C^1(\overline{B})$. Then $\frac{\partial u}{\partial s}(x_0) < \frac{\partial v}{\partial s}(x_0)$ for any inward directional derivative.

Proof. It is a consequence of the strong maximum principle (Theorem 1.28). Indeed, as in Theorem 1.21, it suffices to observe that by hypothesis the difference $w = u - v$ is nonpositive in Ω, and satisfies the linear inequality

$$-\Delta(u - v) + c(x)(u - v) \leq 0$$

where

$$-c(x) = \begin{cases} \dfrac{f(x, u(x)) - f(x, v(x))}{u(x) - v(x)} & \text{if } u(x) \neq v(x) \\ 0 & \text{if } u(x) = v(x) \end{cases} \in L_{loc}^\infty(\Omega)$$

Moreover, if $B \subset \Omega$ is a ball which has x_0 on its boundary and $u, v \in C^1(\overline{B})$, then $c \in L^\infty(B)$. $\qquad\square$

1.3 Compact self-adjoint operators

Here, we recall some basic properties of compact self-adjoint operators in Hilbert spaces and the related spectral theory (see [41, 151] or any text in linear functional analysis for more details).

Let H be an Hilbert space on the field $\mathbb{K}$ (= $\mathbb{R}$ or $\mathbb{C}$) with scalar product $(.,.)$ and let us denote by $\mathcal{B}(H)$ the space of bounded linear operators $A : H \to H$ with the norm

$$\|A\|_{\mathcal{B}(H)} = \sup_{x \neq 0,\, x \in H} \frac{\|Ax\|_H}{\|x\|_H} = \sup_{x \in H,\, \|x\|=1} \|Ax\|$$

If $h \in H$, let us denote by $h^\perp = \{k \in H : (k,h) = 0\}$ the set of the vectors which are orthogonal to h, which is a closed subspace of H. Analogously, if $S \subset H$ then $S^\perp$ is the subspace of the vectors $k \in H$ orthogonal to every $h \in S$.

Definition 1.32. An operator $A \in B(H)$ is **self-adjoint** if

$$(Ax, y) = (x, Ay) \quad \text{for any } x, y \in H$$

Some properties of self-adjoint operators are summarized in the following proposition.

Proposition 1.33. *Let H be an Hilbert space and $A \in \mathcal{B}(H)$ a self-adjoint operator.*
1. *(Au, u) is real for any $u \in H$.*
2. *Let us define*

$$m = m_A = \inf_{\|u\|=1} (Au, u) = \inf_{u \neq 0} \frac{(Au, u)}{\|u\|^2}$$

$$M = M_A = \sup_{\|u\|=1} (Au, u) = \sup_{u \neq 0} \frac{(Au, u)}{\|u\|^2}$$

so that $m\|u\|^2 \leq (Au, u) \leq M\|u\|^2 \ \forall u \in H$, and let $\lambda \in \{m_A, M_A\}$ be such that $|\lambda| = \max\{|m|, M\}$. Then

$$|\lambda| = \sup_{\|u\|=1} |(Au, u)| = \sup_{u \neq 0} \frac{|(Au, u)|}{\|u\|^2} = \|A\|$$

3. *$A = 0$ if and only if $m = M = 0$.*

Proof. The proof of statement 1. is straightforward, and (3) follows immediately from (1) and (2). Let us prove statement 2.

It is immediate to see that $|\lambda| = \sup_{\|u\|=1} |(Au, u)| = \sup_{u \neq 0} \frac{|(Au,u)|}{\|u\|^2}$ and we have to prove that $|\lambda| = \|A\|$. If $u \in H$ with $\|u\| = 1$, then

$$|(Au, u)| \leq \|Au\|\|u\| \leq \|A\|\|u\|^2 = \|A\|$$

so that $|\lambda| \leq \|A\|$.

Let us show the opposite inequality, namely $\|Au\| \le |\lambda| = \sup_{\|u\|=1} |(Au, u)|$ for any $u \in H$ with $\|u\| = 1$. This is obvious if $Au = 0$, while if $Au \ne 0$ it derives from the following identity:

$$\|Au\| = \frac{(Au, Au)}{\|Au\|} = \frac{1}{4}\left(A\left(u + \frac{Au}{\|Au\|} \right), u + \frac{Au}{\|Au\|} \right) - \frac{1}{4}\left(A\left(u - \frac{Au}{\|Au\|} \right), u - \frac{Au}{\|Au\|} \right)$$

which follows easily using the bilinearity of the scalar product and the self-adjointness of A. From this identity, the parallelogram identity and the fact that $\|u\| = 1$ we get

$$\|Au\| \le \frac{1}{4}|\lambda|\left\| u + \frac{Au}{\|Au\|} \right\|^2 + \frac{1}{4}|\lambda|\left\| u - \frac{Au}{\|Au\|} \right\|^2 = \frac{1}{4}|\lambda|\left(2\|u\|^2 + 2\frac{\|Au\|^2}{\|Au\|^2} \right) = |\lambda| \qquad \square$$

Let $A \in \mathcal{B}(H)$ and $\mu \in \mathbb{K}$. We consider the operator $A_\mu = (A - \mu I) : H \to H$ where $I : x \in H \mapsto x$ is the identity map.

Definition 1.34. The **spectrum** $\sigma(A)$ of A is the set of $\mu \in \mathbb{K}$ such that $A_\mu = (A - \mu I)$ **does not have** a continuous inverse $(A - \mu I)^{-1} : H \to H$.

The **resolvent set** of A is $\rho(A) = \mathbb{K} \setminus \sigma(A)$, i.e., it is the set of $\mu \in \mathbb{K}$ such that $A_\mu = (A - \mu I)$ is bijective with bounded inverse $(A - \mu I)^{-1} : H \to H$.

In particular, the **point spectrum** of A is the set of $\mu \in \sigma(A)$ such that $\mathrm{Ker}(A_\mu) \ne \{0\}$. If μ belongs to the point spectrum, then it is called an **eigenvalue** of A and the subspace $\mathrm{Ker}(A_\mu)$ is called the **eigenspace** corresponding to the eigenvalue μ. Any nonzero element of the eigenspace, namely any $x \ne 0$ such that $Ax = \mu x$, is called an **eigenvector** of A corresponding to the eigenvalue μ.

The dimension of the eigenspace $\mathrm{Ker}(A_\mu)$, if finite, is called the **multiplicity** of the eigenvalue μ, otherwise we say that μ has infinite multiplicity. If an eigenvalue μ has multiplicity 1, we will call μ a **simple eigenvalue**.

It is easy to see that the eigenvalues of a self-adjoint operators are real. Moreover, eigenvectors u_1, u_2 corresponding to different eigenvalues $\mu_1 \ne \mu_2$ are orthogonal, since $\mu_1(u_1, u_2) = (Au_1, u_2) = (u_1, Au_2) = \mu_2(u_1, u_2)$.

Definition 1.35. An operator $A \in \mathcal{B}(H)$ is said to be **compact** if it maps bounded sets into precompact sets, i.e., any *bounded* sequence $\{x_n\}$ in H has a subsequence $\{x_{k_n}\}$ such that $y_{k_n} = A(x_{k_n})$ converges in H to some $y \in H$.

The collection of compact linear operators on H will be denoted by $\mathcal{K}(H)$.

It is easy to see from the definition that if $A \in \mathcal{K}(H)$ and $B \in \mathcal{B}(H)$, then the compositions AB and BA are compact. As a consequence, if A is a compact operator on an infinite dimensional Hilbert space H, then $0 \in \sigma(A)$ (for if A is invertible in $\mathcal{B}(H)$ then $I = AA^{-1}$ is compact, but the unit sphere is never compact if H is infinite dimensional).

The following proposition shows a fundamental property of compact self-adjoint operators.

Proposition 1.36. *Let $A : H \to H$ be a nonzero compact self-adjoint operator and let $\mu \in \{m_A, M_A\}$ be such that $|\mu| = \max\{|m|, M\}$. Then μ is an eigenvalue of A.*

Proof. By Proposition 1.33, $|\mu| = \sup_{\|u\|=1} |(Au, u)| = \|A\| \neq 0$. Let u_n be a sequence, with $\|u_n\| = 1$, such that $(Au_n, u_n) \to \mu$. Then

$$\|\mu u_n - Au_n\|^2 = (\mu u_n - Au_n, \mu u_n - Au_n) = \mu^2 \|u_n\|^2 + \|Au_n\|^2 - 2\mu(Au_n, u_n)$$

$$\leq \mu^2 + \|A\|^2 - 2\mu(Au_n, u_n) \to \mu^2 + \mu^2 - 2\mu^2 = 0 \quad \text{as } n \to \infty$$

Hence $\mu u_n - Au_n \to 0$. By the compactness of A, there exists a subsequence u_{k_n} such that $Au_{k_n} \to z \in H$ as $n \to \infty$. This implies that $\mu u_{k_n} = \mu u_{k_n} - Au_{k_n} + Au_{k_n} \to 0 + z = z$. Setting $v = \frac{1}{\mu}z$ we deduce that $u_{k_n} \to v$. By the continuity of the norm and of the operator A, it follows that $\|v\| = \lim \|u_{k_n}\| = 1$, in particular $v \neq 0$, and $Av = \lim Au_{k_n} = z = \mu v$. Thus v is an eigenvector corresponding to the nonzero eigenvalue μ. $\qquad\square$

By iterating Proposition 1.36, we get a fundamental result about compact selfadjoint operators in Hilbert spaces.

Theorem 1.37 (Spectral theorem for compact self-adjoint operators). *Let H be a Hilbert space and $A : H \to H$ a nonzero compact self-adjoint operator.*

Then there exists a (finite or infinite) sequence of eigenvalues $\{\mu_n\}_{n \in J}$, $J \subseteq \mathbb{N}^+$, and a corresponding sequence of eigenvectors $\{u_n\}$ which form an orthonormal basis of $(\operatorname{Ker} A)^\perp$. Moreover:

1. *$Ax = \sum_{n \in J} \mu_n(x, u_n)u_n$ for any $x \in H$.*
2. *Any nonzero eigenvalue μ of A belongs to $\{\mu_n\}$, and the dimension of the eigenspace of μ is finite and it is the cardinality of the set $\{i \in \mathbb{N} : \mu_i = \mu\}$.*
3. *$|\mu_1| = \|A\|$, $|\mu_1| \geq |\mu_2| \geq \dots$. If the sequence is infinite, then $\lim_{n \to \infty} \mu_n = 0$.*

Proof. The proof is based on an iteration of the previous proposition. We know that if $\mu_1 \in \{m_A, M_A\}$ with $|\mu_1| = \max\{|m|, M\}$ then $|\mu_1| = \|A\| \neq 0$, μ_1 is an eigenvalue of A, and there exists a corresponding eigenvector u_1 with $\|u_1\| = 1$.

Let us set $X_1 = H$, $A_1 = A$, $m_1 = m_A$, $M_1 = M_A$, $X_2 = [u_1]^\perp = \{x \in H : (x, u_1) = 0\}$, and $A_2 = A|_{X_2} : X_2 \to H$. Since X_2 is a closed subspace of H, it is a complete Hilbert space. Moreover, X_2 is invariant with respect to A, i. e., $A(X_2) \subseteq X_2$. Indeed, if $x \perp u_1$, then $(Ax, u_1) = (x, Au_1) = \mu_1(x, u_1) = 0$.

Hence $A_2 : X_2 \to X_2$ is a compact self-adjoint operator on the Hilbert space X_2. If $A_2 = 0$, we stop and obtain that $Ax = \mu_1(x, u_1)u_1$ for any $x \in H$, because any $x \in H$ can be written as $x = (x, u_1)u_1 + y$ with $y \in X_2 = [u_1]^\perp$, so that $Ax = (x, u_1)Au_1 + Ay = \mu_1(x, u_1)u_1 + 0$.

If instead A_2 is a *nonzero* compact self-adjoint operator in X_2, by the previous Proposition 1.36 we obtain that if we set $m_2 = m_{A_2}$, $M_2 = M_{A_2}$ and $\mu_2 \in \{m_2, M_2\}$ with $|\mu_2| = \max\{|m_2|, M_2\}$, then μ_2 is an eigenvalue of A_2.

Hence there exists a corresponding eigenvector u_2 with $\|u_2\| = 1$, and it is orthogonal to u_1 because $u_2 \in X_2$. Note that $m_1 \leq m_2 \leq 0 \leq M_2 \leq M_1$, so that $|\mu_2| \leq |\mu_1|$.

Continuing in this way, we have two alternatives. Either we stop at the nth step with $X_n = \{u_1, \ldots, u_{n-1}\}^\perp$, i. e., the operator is zero on $X_{n+1} = \{u_1, \ldots, u_n\}^\perp$, or we construct an infinite sequence of subspaces X_n and operators A_n with the above properties.

In the first case, since any $x \in H$ can be written as $x = (x, u_1)u_1 + \cdots + (x, u_n)u_n + y$ with $y \in X_{n+1}$, we have that (1) holds.

In the second alternative, we obtain infinite sequences of subspaces $X_1 = H \supset X_2 \supset \ldots X_n \supset \ldots$ with $X_n = \{u_1, \ldots, u_{n-1}\}^\perp$, of operators $A_n = A_n = A]_{X_n} : X_n \to X_n$, of eigenvalues μ_n with $|\mu_1| \geq |\mu_2| \geq \ldots$ and of mutually orthogonal corresponding unit eigenvectors u_n.

Let us show that $\lim_{n\to\infty} \mu_n = 0$. Since $|\mu_n|$ is nonincreasing, there exists $\lim_{n\to\infty} |\mu_n| = \varepsilon \geq 0$. Suppose by contradiction that $\varepsilon > 0$. Then $\left\| \frac{u_n}{\mu_n} \right\| = \frac{\|u_n\|}{|\mu_n|} \leq \frac{1}{\varepsilon}$, and the sequence $\frac{u_n}{\mu_n}$ is bounded. By the compactness of A, there exists a subsequence u_{k_n} with $A\left(\frac{u_{k_n}}{\mu_{k_n}}\right)$ converging in H.

This is impossible, since $A\left(\frac{u_{k_n}}{\mu_{k_n}}\right) = u_{k_n}$, and u_{k_n} is an orthonormal sequence. So $\lim_{n\to\infty} \mu_n = 0$ if the sequence is infinite.

If $x \in H$ and we define $y_n = x - \sum_{i=1}^n (x, u_i)u_i$, then y_n is orthogonal to $u_1, \ldots, u_n$, so that $y_n \in X_{n+1}$, and by the Pythagorean theorem, $\|x\|^2 = \|y_n\|^2 + \sum_{i=1}^n |x, u_i|^2$. It follows that $\|y_n\| \leq \|x\|$, hence the sequence y_n is bounded. Moreover, $Ax = Ay_n + \sum_{i=1}^n \mu_i(x, u_i)u_i$, and (1) is equivalent to show that $Ay_n \to 0$ as $n \to \infty$. Since $Ay_n = A_{n+1}y_n$ we have that $\|Ay_n\| \leq \|A_{n+1}\|\|y_n\| \leq |\mu_{n+1}|\|x\| \to 0$ as $n \to \infty$. Thus (1) is proved.

The sequence of the eigenvectors (u_i) form an orthonormal basis of the subspace $(\operatorname{Ker} A)^\perp$. Indeed, if $y \in \operatorname{Ker} A$, then $\mu_i(u_i, y) = (Au_i, y) = (u_i, Ay) = 0$, so that $(u_i, y) = 0$, and every u_i belongs to $(\operatorname{Ker} A)^\perp$. Moreover, for any $x \in H$ we have that $x - \sum_{i\in J}(x, u_i)u_i \in \operatorname{Ker} A$ by (1), since $A(x - \sum_{i\in J}(x, u_i)u_i) = Ax - \sum_{i\in J}(x, u_i)Au_i = Ax - \sum_{i\in J}(x, u_i)\mu_i u_i = 0$.

It follows that if $x \in (\operatorname{Ker} A)^\perp$ then $x - \sum_{i\in J}(x, u_i)u_i \in \operatorname{Ker} A \cap (\operatorname{Ker} A)^\perp = \{0\}$. Therefore, $x = \sum_{i\in J}(x, u_i)u_i$ and the sequence (u_i) form an orthonormal basis of the closed subspace $(\operatorname{Ker} A)^\perp$.

Let μ be a nonzero eigenvalue, and suppose by contradiction that it is different from any μ_n previously obtained. Then there exists a corresponding eigenvector $u \neq 0$ orthogonal to all the eigenvectors u_n, and $Au = \mu u \neq 0$. On the other hand, $Au = \sum_{j\in J}\mu_j(u, u_j)u_j = 0$, which gives a contradiction.

A nonzero eigenvalue μ can appear only a finite number of times in the sequence, since $\lim \mu_n = 0$. If it appears p times with p associate mutually orthogonal unit eigenvectors, these eigenvectors generate its eigenspace. Indeed, if there exists a unit eigenvector u corresponding to the eigenvalue μ and orthogonal to the previous ones, it is also orthogonal to all the eigenvectors u_n obtained in the previous construction, and as before we would obtain that $0 \neq \mu u = Au = \sum_{j\in J}(u, u_j)u_j = 0$. $\qquad\square$

Corollary 1.38 (Spectral theorem for positive compact self-adjoint operators). *Let H be a Hilbert space and $A : H \to H$ a nonzero compact self-adjoint operator which is* positive definite, *i. e.*,

$$(Af, f) > 0 \quad \text{for any } f \in H, f \neq 0$$

Then the previous sequence of eigenvectors $\{u_n\}$ form an orthonormal basis of H, and all the eigenvalues are positive and form a nonincreasing sequence $\mu_1 = \|A\| \geq \mu_2 \geq \dots$.

Proof. It is enough to observe that since A is positive definite then $\text{Ker}(A) = \{0\}$. $\square$

Let H be an infinite dimensional Hilbert space and $A \in \mathcal{B}(H)$ a nonzero self-adjoint compact operator. We know that the value 0 belongs to the spectrum of A and it can be an eigenvalue (if A is not injective) or not (in that case A is injective but either it is not surjective or the inverse is not continuous).

The previous theorem gives the sequence of nonzero eigenvalues satisfying the properties 1–3. We now show that any other number $\mu \neq 0$, $\mu \neq \mu_k$, belongs to the resolvent set $\rho(A) = \mathbb{K} \setminus \sigma(A)$, i. e., there exists the continuous inverse $(A - \mu I)^{-1}$: $H \to H$.

Theorem 1.39 (Fredholm alternative). *Let $A : H \to H$ a nonzero compact self-adjoint operator on the Hilbert space H, and $\{\mu_n\}_{n\in J \subseteq \mathbb{N}^+}$, $\{u_n\}_{n\in J \subseteq \mathbb{N}^+}$, the sequences of nonzero eigenvalues and corresponding mutually orthogonal eigenvectors of A.*

1. *If $\mu \neq 0$ and $\mu \neq \mu_k$ $\forall k \in J$, then $\mu \in \rho(A) = \mathbb{K} \setminus \sigma(A)$. More precisely, for any $h \in H$ there exists a unique $f \in H$ such that $Af = \mu f + h$, and*

$$f = (A - \mu I)^{-1}h = \frac{-1}{\mu}\left[h + \sum_{n\in J} \frac{\mu_n}{\mu - \mu_n}(h, u_n)u_n\right] \tag{1.55}$$

   (the series converges and defines a continuous linear operator).
2. *If $\mu = \mu_k$ for some $k \in J$, then $\text{Im}(A - \mu I) = [\text{Ker}(A - \mu I)]^{\perp}$, i. e., the equation (in the unknown f)*

$$Af = \mu_k f + h$$

   has a solution if and only if $(h, u) = 0$ $\forall u \in \text{Ker}(A - \mu I)$.
   If this is the case, then the solutions of the equation are given by $f + \text{Ker}(A - \mu I) = \{f + u : u \in \text{Ker}(A - \mu I)\}$, with

$$f = \frac{-1}{\mu}\left[h + \sum_{n\in J, \mu_n \neq \mu} \frac{\mu_n}{\mu - \mu_n}(h, u_n)u_n\right] \tag{1.56}$$

Proof.
1. Let us first show that given $h \in H$, if $\mu \neq \mu_k$ $\forall k \in J$ and there exists $f \in H$ such that $Af = \mu f + h$, then f is necessarily given by (1.55) where the series converges. In this way, as we show below we obtain the continuous inverse of $A - \mu I$.

If $\mu f + h = Af = \sum_{n \in J} \mu_n (f, u_n) u_n$, then $\mu f = -h + \sum_{n \in J} \mu_n (f, u_n) u_n$. Taking the scalar product with u_i, $i \in J$, we get $(\mu f, u_i) = -(h, u_i) + \mu_i (f, u_i)$, so that $(\mu - \mu_i)(f, u_i) = -(h, u_i)$ and $(f, u_i) = \frac{1}{(\mu - \mu_i)}(-h, u_i)$. Inserting this value in the previous equality, we get that $\mu f = -h + \sum_{n \in J} \mu_n \frac{1}{(\mu - \mu_n)}(-h, u_n) u_n$ and (1.55) follows. Let us show that the series converges for any $h \in H$ and the convergence is uniform with respect to any unit vector h. This will show that $(A - \mu I)^{-1}$ is bounded, and inserting the value of f given by (1.55) in the formula $(A - \mu I)f = \sum_{n \in J} \mu_n (f, u_n) u_n - \mu f$ it is easy to see that one gets h.

If we set $\alpha = \sup_{k \in J} \frac{|\mu_k|}{|\mu - \mu_k|}$, then $\alpha < \infty$ because $\mu \neq \mu_k$ for any $k \in J$ and $\mu_k \to 0$ as $k \to \infty$.

If $m < n$, thanks to the Pythagorean theorem, we have that

$$\|s_n - s_m\|^2 = \left\| \sum_{k=m+1}^{n} \frac{\mu_k}{\mu - \mu_k}(h, u_k) u_k \right\|^2$$

$$= \sum_{k=m+1}^{n} \left| \frac{\mu_k}{\mu - \mu_k} \right|^2 |(h, u_k)|^2 \leq \alpha^2 \sum_{k=m+1}^{\infty} |(h, u_k)|^2$$

and the last term tends to 0 as $m \to \infty$ because the series $\sum_{k=1}^{\infty} |(h, u_k)|^2 \leq \|h\|^2$ converges. Moreover,

$$(A - \mu I)^{-1} h = \frac{-1}{\mu}\left[h + \sum_{n \in J} \frac{\mu_n}{\mu - \mu_n}(h, u_n) u_n \right] = \frac{-1}{\mu} Ih + Bh$$

with $Bh = \frac{-1}{\mu} \sum_{n \in J} \frac{\mu_n}{\mu - \mu_n}$ bounded, since an analogous computation shows that $\|Bh\|^2 \leq \frac{1}{|\mu|} \alpha^2 \|h\|^2$.

2. Considering the case when $\mu = \mu_k$ for some $k \in J$, the previous computations show that the series (1.56) converges and defines a bounded linear operator, because the terms with $\mu_n = \mu_k$ are omitted.

    Finally, if $(A - \mu_k I)f = h$ and u is an eigenvector corresponding to the eigenvalue μ_k, then $(h, u) = ((A - \mu_k I)f, u) = (f, (A - \mu_k I)u) = 0$ so that $(h, u) = 0$. Vice versa, it is easy to verify that if $(h, u) = 0$ for any $u \in \mathrm{Ker}(A - \mu I)$, then the series defines a vector f such that $Af = h$, and any other vector g satisfying $Ag = h$ has the form $g = f + u : u \in \mathrm{Ker}(A - \mu I)$. $\qquad\square$

Remark 1.40. The name Fredholm alternative depends on the fact that by the previous theorem the following alternative holds.

Either

(i) the homogeneous equation

$$Af - \mu f = 0 \tag{1.57}$$

admits only the trivial solution $f = 0$ and then the inhomogeneous equation

$$Af - \mu f = h \tag{1.58}$$

has a unique solution for any $h \in H$

or

(ii) the homogeneous equation (1.57) has nontrivial solutions, and then the nonhomogeneous equation (1.58) has a solution if and only if $h \in (\mathrm{Ker}(A - \mu I))^{\perp}$.

1.4 Eigenvalues of elliptic problems

In this section, we review the classical theory of eigenvalues for boundary value problems involving elliptic operators. We consider linear elliptic mixed boundary value problems in a general bounded Lipschitz domain $\Omega \subset \mathbb{R}^N$, $N \geq 2$, and we denote by Γ_0, Γ relatively open disjoint subsets of the boundary $\partial\Omega$ that satisfy (1.28), (1.29).

We construct (as in [78]) a sequence of eigenvalues and prove their variational characterization adapting the standard methods used for Dirichlet problems (based on the theory of positive compact self-adjoint operators) by working in the product space $\mathbf{V} = L^2(\Omega) \times L^2(\Gamma)$. Of course, the case of a Dirichlet boundary problem is a particular case of Dirichlet–Neumann problems when $\Gamma = \emptyset$. We only observe that in the case of Dirichlet problems obviously we do not work in the product space $L^2(\Omega) \times L^2(\Gamma)$ but instead only in $L^2(\Omega)$.

As in Section 1.1, we denote by the same symbol a function belonging to $H_0^1(\Omega \cup \Gamma)$ and its trace on the boundary.

Let us consider the bilinear form in $H_0^1(\Omega \cup \Gamma)$ defined by

$$B(u, \varphi) = \int_{\Omega} (\nabla u \cdot \nabla \varphi + cu\varphi)\, dx + \int_{\Gamma} du\, \varphi\, d\sigma \tag{1.59}$$

with the associated quadratic form

$$Q(u) = B(u, u) = \int_{\Omega} (|\nabla u|^2 + c|u|^2)\, dx + \int_{\Gamma} d|u|^2\, d\sigma \tag{1.60}$$

where we assume that

$$c \in L^\infty(\Omega), \quad d \in L^\infty(\Gamma) \tag{1.61}$$

Remark 1.41. In the spectral theory that follows, it would be enough to assume the following hypotheses on c, d:

$$c^- \in L^\infty(\Omega), \quad d^- \in L^\infty(\Gamma) \tag{1.62}$$

and

$$c \in L^{\frac{N}{2}}(\Omega), \quad d \in L^{N-1}(\Gamma) \tag{1.63}$$

if $N \geq 3$, while c and d can belong to any L^q space with $q > 1$, if $N = 2$.

However, in order to make the presentation not too heavy, in the sequel we will consider only the case when c and d are bounded.

If $\Lambda \geq 0$ let us define, together with (1.59), the bilinear form

$$B^\Lambda(u, \varphi) = \int_\Omega (\nabla u \cdot \nabla \varphi + (c + \Lambda)u\varphi)\, dx + \int_{\partial\Omega} (d + \Lambda)u\varphi\, d\sigma \tag{1.64}$$

Since (1.61) holds, B and B^Λ are continuous symmetric bilinear forms on $H_0^1(\Omega \cup \Gamma)$. Moreover, again by (1.61), a number $\Lambda_0 \geq 0 \in \mathbb{R}$ can be chosen in such a way that for any $\Lambda \geq \Lambda_0$ the bilinear form B^Λ is coercive in $H_0^1(\Omega \cup \Gamma)$, i. e., it gives an equivalent scalar product in $H_0^1(\Omega \cup \Gamma)$. Hence we fix any $\Lambda \geq \Lambda_0$.

Let us consider the Hilbert space

$$\mathbf{V} = L^2(\Omega) \times L^2(\Gamma) \tag{1.65}$$

with the scalar product

$$((f_1, f_2) \cdot (g_1, g_2)) = \int_\Omega f_1 g_1\, dx + \int_\Gamma f_2 g_2\, d\sigma \tag{1.66}$$

We can identify $f = (f_1, f_2) \in \mathbf{V}$ with the continuous linear functional

$$\varphi \in H_0^1(\Omega \cup \Gamma) \mapsto (f_1, \varphi)_{L^2(\Omega)} + (f_2, \varphi)_{L^2(\Gamma)} \tag{1.67}$$

By the Riesz representation theorem, for any $f = (f_1, f_2) \in \mathbf{V}$ there exists a unique $u =: T(f) \in H_0^1(\Omega \cup \Gamma)$ such that

$$B^\Lambda(u, \varphi) = (f, \varphi)_\mathbf{V} = (f_1, \varphi)_{L^2(\Omega)} + (f_2, \varphi)_{L^2(\Gamma)} \quad \forall \varphi \in H_0^1(\Omega \cup \Gamma)$$

i. e., $u = T(f)$ is the unique weak solution of the problem

$$\begin{cases} -\Delta u + (c(x) + \Lambda)u = f_1 & \text{in } \Omega \\ u = 0 & \text{on } \Gamma_0 \\ \dfrac{\partial u}{\partial v} + (d(x) + \Lambda)u = f_2 & \text{on } \Gamma \end{cases} \tag{1.68}$$

The solution u belongs to $H_0^1(\Omega \cup \Gamma)$ and

$$\|u\|_{H_0^1(\Omega \cup \Gamma)} \leq C\|f\|_\mathbf{V} \quad \text{for some } C > 0$$

If we identify a function $u \in H_0^1(\Omega \cup \Gamma)$ with the pair $(u, \text{Trace}(u)) \in H_0^1(\Omega \cup \Gamma) \times L^2(\Gamma) \subset \mathbf{V}$, we can consider T as a continuous linear operator $T : \mathbf{V} \to \mathbf{V}$ defined by $f = (f_1, f_2) \mapsto (u, \text{Trace}(u))$, which maps $\mathbf{V}$ into $\mathbf{V}$. We observe that T is *compact* because of the compact embedding of $H_0^1(\Omega \cup \Gamma)$ in $\mathbf{V}$. Moreover, T is a *positive self-adjoint* operator. Indeed (recall that B^Λ is an equivalent scalar product in $H_0^1(\Omega \cup \Gamma)$) if $f = (f_1, f_2) \neq 0$, and hence $u \neq 0$, we have that

$$(T(f), f)_\mathbf{V} = (u, f)_\mathbf{V} = B^\Lambda(u, u) > 0$$

so that T is positive. If $T(f) = u$, $T(g) = v$, i.e., $B^\Lambda(u, \phi) = (f, \phi)_\mathbf{V}$, $B^\Lambda(v, \phi) = (g, \phi)_\mathbf{V}$, then

$$(T(f), g)_\mathbf{V} = (u, g)_\mathbf{V} = (g, u)_\mathbf{V} = B^\Lambda(v, u) = B^\Lambda(u, v) = (f, v)_\mathbf{V} = (f, T(g))_\mathbf{V}$$

which means that T is self-adjoint.

Thus, by the spectral theory of positive compact self-adjoint operators in Hilbert spaces there exist a nonincreasing sequence $\{\mu_j^\Lambda\}$ of positive eigenvalues with $\lim_{j\to\infty} \mu_j^\Lambda = 0$ and a corresponding sequence $\{w_j\} \subset H_0^1(\Omega \cup \Gamma)$ of eigenvectors such that $T(w_j) = \mu_j^\Lambda w_j$ and w_j is an orthonormal basis of $\mathbf{V}$.

If we set $\lambda_j^\Lambda = \frac{1}{\mu_j^\Lambda}$, then w_j solve the problem

$$\begin{cases} -\Delta w_j + (c(x) + \Lambda)w_j = \lambda_j^\Lambda w_j & \text{in } \Omega \\ w_j = 0 & \text{on } \Gamma_0 \\ \dfrac{\partial w_j}{\partial v} + (d(x) + \Lambda)w_j = \lambda_j^\Lambda w_j & \text{on } \Gamma \end{cases} \tag{1.69}$$

Translating, and denoting by λ_j the differences $\lambda_j = \lambda_j^\Lambda - \Lambda$, we get the existence of a sequence $\{\lambda_j\}$ of eigenvalues, with $-\infty < \lambda_1 \le \lambda_2 \le \dots$, $\lim_{j\to+\infty} \lambda_j = +\infty$, and of a corresponding sequence of eigenfunctions $\{w_j\}$ that weakly solve the following *mixed boundary eigenvalue problem*:

$$\begin{cases} -\Delta w + c(x)w = \lambda w & \text{in } \Omega \\ w = 0 & \text{on } \Gamma_0 \\ \dfrac{\partial w}{\partial v} + d(x)w = \lambda w & \text{on } \Gamma \end{cases} \tag{1.70}$$

Moreover, since $c \in L^\infty(\Omega)$, by standard elliptic regularity theory the eigenfunctions w_j belong at least to $C^1(\Omega)$ (and they are far more regular if the coefficients and the domain are smooth).

In the next theorem, we show the variational characterization of the eigenvalues λ_j of (1.70) as well as the main properties of the first eigenfunctions, i.e., the eigenfunctions corresponding to the first eigenvalue.

Theorem 1.42. *Assume that Ω is a bounded Lipschitz domain in $\mathbb{R}^N$, $N \ge 2$, Γ_0 and Γ are relatively open disjoint subsets of the boundary $\partial\Omega$ that satisfy (1.28), (1.29) and (1.61) holds.*

Then there exist sequences of eigenvalues $\{\lambda_j\}_{j\in\mathbb{N}} \subset \mathbb{R}$, with $\lim_{j\to\infty} \lambda_j = +\infty$, and eigenfunctions $\{w_j\}_{j\in\mathbb{N}} \subset H_0^1(\Omega \cup \Gamma)$ which satisfy (1.70) and such that the sequence $\{(w_j, \mathrm{Trace}(w_j))\}$ is an orthonormal basis of the space $\mathbf{V} = L^2(\Omega) \times L^2(\Gamma)$.

Moreover, defining for $v \in H_0^1(\Omega \cup \Gamma)$, $v \ne 0$, the Rayleigh quotient

$$R(v) = \frac{Q(v)}{(v, v)_{L^2(\Omega)} + (v, v)_{L^2(\Gamma)}} = \frac{Q(v)}{\|v\|_\mathbf{V}^2} \tag{1.71}$$

with Q as in (1.60), and denoting by $\boldsymbol{H}_k$ a k-dimensional subspace of $H_0^1(\Omega \cup \Gamma)$, the following properties hold:

(i)

$$\lambda_1 = \min_{\substack{v \in H_0^1(\Omega\cup\Gamma) \\ v \neq 0}} R(v) = \min_{\substack{v \in H_0^1(\Omega\cup\Gamma) \\ (v,v)_{\mathbf{V}}=1}} Q(v)$$

(ii) If $m \geq 2$, then

$$\lambda_m = \min_{\substack{v \in H_0^1(\Omega\cup\Gamma) \\ v \neq 0 \\ v \perp w_1,\ldots,v \perp w_{m-1}}} R(v) = \min_{\substack{v \in H_0^1(\Omega\cup\Gamma) \\ (v,v)_{\mathbf{V}}=1 \\ v \perp w_1,\ldots,v \perp w_{m-1}}} Q(v)$$

where the orthogonality conditions are meant in the space $\boldsymbol{V}$.

(iii) If $m \geq 2$, then

$$\lambda_m = \min_{\boldsymbol{H}_m} \max_{\substack{v \in \boldsymbol{H}_m \\ v \neq 0}} R(v) = \min_{\boldsymbol{H}_m} \max_{\substack{v \in \boldsymbol{H}_m \\ \|v\|_{\mathbf{V}}=1}} Q(v)$$

(iv) If $m \geq 2$, then[2]

$$\lambda_m = \max_{\boldsymbol{H}_{m-1}} \min_{\substack{v \perp \boldsymbol{H}_{m-1} \\ v \neq 0}} R(v) = \max_{\boldsymbol{H}_{m-1}} \min_{\substack{v \perp \boldsymbol{H}_{m-1} \\ \|v\|_{\mathbf{V}}=1}} Q(v)$$

where the orthogonality condition is meant in the space $\boldsymbol{V}$.

(v) If $w \in H_0^1(\Omega \cup \Gamma)$, $w \neq 0$ and $R(w) = \lambda_1$, then w is an eigenfunction corresponding to λ_1.

(vi) Any first eigenfunction does not change sign in Ω and the first eigenvalue is simple, i. e., up to scalar multiplication there is only one eigenfunction corresponding to the first eigenvalue.

Proof. We have already proved the existence of the sequences of eigenvalues and eigenfunctions. Moreover, we observe that the second equality in i)–iv) is a consequence of the homogeneity of R: for any $k \in \mathbb{R} \setminus \{0\}$, $v \in H_0^1(\Omega \cup \Gamma) \setminus \{0\}$ it holds $R(kv) = R(v) = Q(\frac{v}{\|v\|_{\mathbf{V}}})$.

As in (1.64), we define $R^\Lambda(v) = \frac{B^\Lambda(v,v)}{(v,v)_{\mathbf{V}}} = \frac{B^\Lambda(v,v)}{(v,v)_{L^2(\Omega)}+(v,v)_{L^2(\Gamma)}}$ for $v \in H_0^1(\Omega \cup \Gamma)$ $v \neq 0$. Since $R^\Lambda(v) = R(v) + \Lambda$, once the properties are proved for the eigenvalues λ_j^Λ of (1.69) we recover the results for λ_j by translation. Therefore, for simplicity of notation, we make the proof for $\Lambda = 0$, and we assume that $B = B^0$ is coercive, and hence gives an equivalent scalar product in $H_0^1(\Omega \cup \Gamma)$.

---

[2] In the sequel, we will refer to this result also when $m = 1$, meaning that $\boldsymbol{H}_0 = \{0\}$, and $\lambda_1 = \min_{v \in H_0^1(\Omega\cup\Gamma), v \neq 0} R(v)$, proved in (i).

(i) The sequence $\{w_j\}$ is an orthonormal basis of $\mathbf{V}$. By definition for every $u \in H_0^1(\Omega \cup \Gamma)$, we have

$$B(w_k, u) = \lambda_k (w_k, u)_{\mathbf{V}} \quad \forall k \in \mathbb{N}^+ \tag{1.72}$$

In particular, $B(w_k, w_j) = \lambda_k (w_k, w_j)_{\mathbf{V}} \ \forall k, j \in \mathbb{N}^+$, hence the sequence $\{(\lambda_j)^{-\frac{1}{2}} w_j\}$ is an orthonormal basis of $H_0^1(\Omega \cup \Gamma)$ with respect to the scalar product given by the bilinear form B.

If $v \in H_0^1(\Omega \cup \Gamma)$ and $(v, v)_V = 1$, using the Fourier expansions in $\mathbf{V}$ and in $H_0^1(\Omega \cup \Gamma)$ we have

$$v = \sum_{k \in J} \beta_k w_k, \quad (v, v) = \sum_{k \in J} |\beta_k|^2 = 1$$

and

$$v = \sum_{k=1}^{+\infty} \alpha_k ((\lambda_k)^{-\frac{1}{2}} w_k), \quad B(v, v) = \sum_{k=1}^{+\infty} |\alpha_k|^2$$

It is easy to see, using that $\beta_k = (v, w_k)_{\mathbf{V}}$, $\alpha_k = B(v, (\lambda_k)^{-\frac{1}{2}} w_k)$ and (1.72) that $\alpha_k = (\lambda_k)^{\frac{1}{2}} \beta_k$. Thus

$$B(v, v) = \sum_{k=1}^{+\infty} \lambda_k \beta_k^2 \geq \lambda_1 \sum_{k=1}^{+\infty} \beta_k^2 = \lambda_1 = B(w_1, w_1)$$

and (i) follows.

(ii) If $v \perp \{w_1, \ldots, w_{m-1}\}$ and $(v, v)_{\mathbf{V}} = 1$, then $v = \sum_{k=m}^{\infty} \beta_k w_k$ and as before $B(v, v) \geq \lambda_m$. Since $B(w_m, w_m) = \lambda_m$, the claim (ii) follows.

(iii) If $\dim(\mathbf{H}_m) = m$ and $\{v_1, \ldots, v_m\}$ is a basis of $\mathbf{H}_m$, there exists a linear combination $0 \neq \bar{v} = \sum_{i=1}^{m} a_i v_i$ which is orthogonal to all $w_1, \ldots w_{m-1}$ (m unknown coefficients and $(m-1)$ equations), so that by (ii) we obtain that $\max_{\substack{v \in \mathbf{H}_m \\ v \neq 0}} R(v) \geq \lambda_m$. On the other hand, if $\mathbf{H}_m = \operatorname{span}(w_1, \ldots, w_m)$ then $\max_{\substack{v \in \mathbf{H}_m \\ v \neq 0}} R(v) = \lambda_m$. Indeed if $v = \sum_{k=1}^{m} \beta_k w_k$, as in the proof of (i) we have $R(v) = \frac{\sum_{k=1}^{m} \lambda_k \beta_k^2}{\sum_{k=1}^{m} \beta_k^2} \leq \lambda_m$ and $R(w_m) = \lambda_m$. Hence (iii) follows.

(iv) The proof is similar to the previous one. If $\{v_1, \ldots, v_{m-1}\}$ is a basis of an $m-1$-dimensional subspace $\mathbf{H}_{m-1}$, there exists a linear combination $0 \neq \bar{w} = \sum_{i=1}^{m} a_i w_i$ of the first m eigenfunctions which is orthogonal to $\mathbf{H}_{m-1}$, and $R(w) \leq \lambda_m$. Hence $\min_{\substack{v \perp \mathbf{H}_{m-1} \\ v \neq 0}} R(v) \leq \lambda_m$ for every $(m-1)$-dimensional subspace $\mathbf{H}_{m-1}$. On the other hand, by taking $\mathbf{H}_{m-1} = \operatorname{span}(w_1, \ldots, w_{m-1})$ then $\min_{\substack{v \perp \mathbf{H}_{m-1} \\ v \neq 0}} R(v) = \lambda_m$ by (ii), so that (iv) follows.

(v) By normalizing, we can suppose that $(w, w)_{\mathbf{V}} = 1$. Let $v \in H_0^1(\Omega \cup \Gamma)$, $t > 0$. Then by (i) $R(w + tv) = \frac{B(w+tv, w+tv)}{(w+tv, w+tv)_{\mathbf{V}}} \geq \lambda_1$, i. e., $B(w, w) + t^2 B(v, v) + 2tB(w, v) \geq \lambda_1[(w, w)_{\mathbf{V}} + t^2(v, v)_{\mathbf{V}} + 2t(w, v)_{\mathbf{V}}] = \lambda_1 + \lambda_1 t^2(v, v) + 2t\lambda_1(w, v)$. Since $B(w, w) = \lambda_1$, dividing by t and letting $t \to 0$ we obtain that $B(w, v) \geq \lambda_1(w, v)_{\mathbf{V}}$ and changing v with $-v$ we deduce that $B(w, v) = \lambda_1(w, v)_{\mathbf{V}}$ for any $v \in H_0^1(\Omega \cup \Gamma)$, i. e., w is a first eigenfunction.

(vi) Let w_1 be a first eigenfunction. Multiplying (1.70) by w_1^+ and integrating we deduce that $B(w_1^+, w_1^+) = \lambda_1(w_1^+, w_1^+)_{\mathbf{V}}$, so that by (v) w_1^+ is a first eigenfunction if it is not identically zero. The same applies to w_1^-. Suppose now that $w_1^+ \not\equiv 0$. Then it is a first eigenfunction and solves (1.70) and, by the strong maximum principle (Theorem 1.28), $w_1^+ = w_1 > 0$ in Ω.

If w_1, w_2 are two eigenfunctions corresponding to λ_1, they do not change sign in Ω, so that they cannot be orthogonal in $\mathbf{V}$. This implies that the first eigenvalue is simple. $\qquad\square$

Remark 1.43.

1. In statement (iii) of Theorem 1.42, we write max instead of sup since the unit sphere in a finite dimensional space is compact and, therefore, $\sup_{\substack{v \in \mathbf{H}_m \\ \|v\|_{\mathbf{V}}=1}} Q(v) = \max_{\substack{v \in \mathbf{H}_m \\ \|v\|_{\mathbf{V}}=1}} Q(v)$. Instead, the fact that

$$\inf_{\mathbf{H}_m} \max_{\substack{v \in \mathbf{H}_m \\ v \neq 0}} R(v) = \min_{\mathbf{H}_m} \max_{\substack{v \in \mathbf{H}_m \\ v \neq 0}} R(v)$$

is part of the proof of statement (iii).

2. Let us explain why we write min instead of inf in statement (iv) of Theorem 1.42. If $\mathbf{H}_{m-1}$ is a $(m-1)$ dimensional subspace of $H_0^1(\Omega \cup \Gamma)$, then $S = \mathbf{H}_{m-1}^{\perp}$ (where the orthogonality is in $\mathbf{V} = L^2(\Omega) \times L^2(\Gamma)$) is a subspace of $H_0^1(\Omega \cup \Gamma)$ which is closed in $\mathbf{V}$ and, therefore, in $H_0^1(\Omega \cup \Gamma)$. So S is a Hilbert space, and thanks to the compact embedding of $H_0^1(\Omega \cup \Gamma)$ into $\mathbf{V}$ the unit sphere of $\mathbf{V} \cap S$ is weakly closed in S. Moreover, $Q(v)$ (which by the hypothesis made at the beginning of the proof is in fact the square of a norm in $H_0^1(\Omega \cup \Gamma)$) is weakly lower semicontinuous. Therefore, by standard arguments of calculus of variations we have that $\inf_{\substack{v \perp \mathbf{H}_{m-1} \\ \|v\|_{\mathbf{V}}=1}} Q(v)$ is attained and it is in fact a minimum (note that the same argument can be applied to say that in (i) and (ii) the minimum is achieved, though it has been proved directly). Instead, the fact that

$$\sup_{\mathbf{H}_{m-1}} \min_{\substack{v \perp \mathbf{H}_{m-1} \\ v \neq 0}} R(v) = \max_{\mathbf{H}_{m-1}} \min_{\substack{v \perp \mathbf{H}_{m-1} \\ v \neq 0}} R(v)$$

is part of the proof of statement (iv).

Next, we prove some **monotonicity and continuity results on the eigenvalues** that will be needed in the sequel.

For any regular subdomain D of Ω (see Definition 1.24), consider the eigenvalue problem

$$\begin{cases} -\Delta w + c'(x)w = \lambda w & \text{in } D \\ w = 0 & \text{on } (\Gamma_0)_D = \partial D \setminus \overline{\Gamma_D} \\ \dfrac{\partial w}{\partial v} + d'(x)w = \lambda w & \text{on } \Gamma_D = \Gamma \cap \partial D \end{cases} \qquad (1.73)$$

with $c' \in L^\infty(D)$, $d' \in L^\infty(\Gamma_D)$. We use the notation

$$\lambda_k\left(-\Delta + c'; \frac{\partial}{\partial v} + d'; D \cup \Gamma_D\right) \qquad (1.74)$$

for the eigenvalues of problem (1.73).

In particular, when $\Gamma_D = \emptyset$, we write

$$\lambda_k(-\Delta + c'; D) \qquad (1.75)$$

for the eigenvalues for the Dirichlet problem, with corresponding eigenfunctions w_k in $H_0^1(D)$ that solve the problem

$$\begin{cases} -\Delta w + c'(x)w = \lambda w & \text{in } D \\ w = 0 & \text{on } \partial D \end{cases}$$

We keep the simple notation λ_k for the eigenvalues of (1.70), namely

$$\lambda_k = \lambda_k\left(-\Delta + c; \frac{\partial}{\partial v} + d; \Omega \cup \Gamma\right)$$

Finally, if $v \in H_0^1(D \cup \Gamma_D)$ we write

$$Q(u; D \cup \Gamma_D) = \int_D \left(|\nabla u|^2 + c'|u|^2\right) dx + \int_{\Gamma_D} d'\, u^2 \delta\sigma \qquad (1.76)$$

$$R(v; D \cup \Gamma_D) = \frac{Q(u; D \cup \Gamma_D)}{(v, v)_{L^2(\Omega)} + (v, v)_{L^2(\Gamma)}} = \frac{Q(u; D \cup \Gamma_D)}{\|v\|_{\mathbf{V}}^2} \qquad (1.77)$$

for the quadratic form and the Rayleigh quotient corresponding to the choices of D, c', d'. We will write $Q(v)$ and $R(v)$ for the quadratic form and the Rayleigh quotient given by (1.76), (1.77) when D, c', d' are fixed.

Let us recall that $\mathbf{V} = L^2(\Omega) \times L^2(\Gamma)$ and denote by $\mathbf{H}_k$ a k-dimensional subspace of $H_0^1(\Omega \cup \Gamma)$. Moreover, we denote by X a dense subspace of $H_0^1(\Omega \cup \Gamma)$ and by X_k a k-dimensional subspace of X.

In particular, we will consider spaces of smooth functions *with compact support* in $\Omega \cup \Gamma$, e. g., $X = C_c^1(\Omega \cup \Gamma)$ (or $X = C_c^\infty(\Omega \cup \Gamma)$), which is a dense subspace of $H_0^1(\Omega \cup \Gamma)$.

Theorem 1.44. *Assume that Ω is a bounded Lipschitz domain in $\mathbb{R}^N$, $N \geq 2$, Γ_0 and Γ are relatively open disjoint subsets of the boundary $\partial\Omega$ that satisfy (1.28), (1.29) and that (1.61) holds.*

(i)   *If X is a dense subspace of the space $H_0^1(\Omega \cup \Gamma)$ (in particular if $X = C_c^1(\Omega \cup \Gamma)$) and we denote by X_k a k-dimensional subspace of X then for any $m \in \mathbb{N}^+$*

$$\lambda_m = \inf_{X_m} \max_{\substack{v \in X_m \\ v \neq 0}} R(v)$$

(ii)   *(Monotonicity w. r. t. subdomains) If $\Omega^1 \subset \Omega^2 \subset \Omega$ are regular subdomains of the domain Ω and $\Gamma^i = \Gamma_{\Omega^i} = \Gamma \cap \Omega^i$, $i = 1, 2$, then*

$$\lambda_m^1 \geq \lambda_m^2 \quad \text{for any } m \in \mathbb{N}^+$$

where $\lambda_m^i = \lambda_m(-\Delta + c; \frac{\partial}{\partial v} + d; \Omega^i \cup \Gamma^i)$, $i = 1, 2$.
Moreover, if $\Omega^1 \subset \Omega^2 \subset \Omega$ and $\partial\Omega^1 \cap \Omega^2 \neq \emptyset$ then the strict inequality $\lambda_1^1 > \lambda_1^2$ holds for the first eigenvalue.

(iii)   *(Continuity w. r. t. subdomains) For any $m \in \mathbb{N}^+$, it holds*

$$\lambda_m = \inf\left\{\lambda_m\left(-\Delta + c; \ \frac{\partial}{\partial v} + d; \ D \cup \Gamma_D\right)\right\}$$

the infimum being among the regular subdomains D such that $\overline{D \cup \Gamma_D}$ is a compact subset of $\Omega \cup \Gamma$.

(iv)   *(Monotonicity w. r. t. the coefficients) If $c' \in L^\infty(\Omega)$, $d' \in L^\infty(\Gamma)$ and $c \geq c'$ a. e. in Ω, $d \geq d'$ a. e. on Γ, then $\lambda_m \geq \lambda'_m$ for any $m \in \mathbb{N}^+$, where $\lambda'_m = \lambda_m(-\Delta + c'; \frac{\partial}{\partial v} + d'; \Omega \cup \Gamma)$.*

(v)   *(Continuity w. r. t. the coefficients) If c^n is a sequence in $L^\infty(\Omega)$, d^n is a sequence in $L^\infty(\Gamma)$ and $c^n \to c$ in $L^\infty(\Omega)$, $d^n \to d$ in $L^\infty(\Gamma)$, then $\lim_{n\to\infty} \lambda_m^n = \lambda_m$ $\forall m \in \mathbb{N}^+$, where $\lambda_m^n = \lambda_m(-\Delta + c^n; \frac{\partial}{\partial v} + d^n; \Omega \cup \Gamma)$*

(vi)   *Let $\Gamma = \emptyset$ and set $\lambda_1(D) = \lambda_1(-\Delta + c; \ D)$ if D is an open subset of Ω. Then*

$$\lim_{meas(D)\to 0} \lambda_1(D) = +\infty$$

(vii)   *Let Ω be a cylindrically symmetric domain (see Definition 1.15), and let us consider a regular subdomain $\Omega' \subset \Omega$ (see Definition 1.24). Setting $\Gamma'_0 = \partial\Omega' \cap \mathbb{R}_+^N$, $\Gamma' = \partial\Omega' \setminus \overline{\Gamma'_0}$, $\lambda_1(\Omega' \cup \Gamma') = \lambda_1(-\Delta + c; \frac{\partial}{\partial v} + d; \Omega' \cup \Gamma')$, we have*

$$\lim_{meas(\Omega')\to 0} \lambda_1(\Omega' \cup \Gamma') = +\infty$$

In the proof, we will exploit the following simple result.

Lemma 1.45. *Let X be a normed space, and let us consider for $m \in \mathbb{N}^+$ a set of linearly independent vectors $z_1, \ldots, z_m$. Assume that φ_i^n are sequences of vectors converging to z_i in X for $i = 1, \ldots, m$. Then the vectors $\varphi_1^n, \ldots, \varphi_m^n$ are linearly independent if n is sufficiently large.*

Proof. Suppose by contradiction that (up to a subsequence) for any $n \in \mathbb{N}$ there exist $t_1^n, \ldots, t_m^n$ not all vanishing such that $\sum_{k=1}^{m} t_k^n \varphi_k^n = 0$. Setting $s_k^n = \dfrac{t_k^n}{(\sum_{i=1}^{m} |t_i^n|^2)^{\frac{1}{2}}}$ we have that $\sum_{k=1}^{m} s_k^n \varphi_k^n = 0$, $\sum_{k=1}^{m} |s_k^n|^2 = 1$ and the sequences s_k^n are bounded for any $k = 1, \ldots, m$. Then (up to a subsequence) the coefficients $s_k^n \to s_k \in \mathbb{R}$ as $n \to \infty$, with $\sum_{k=1}^{m} |s_k|^2 = 1$, so that letting $n \to \infty$ we get that $0 = \sum_{k=1}^{m} s_k^n \varphi_k^n \to \sum_{k=1}^{m} s_k z_k$, i. e., the vectors $z_1, \ldots, z_m$ are linearly dependent, which is a contradiction. $\qquad\square$

Proof of Theorem 1.44.

(i) Let us recall that $\lambda_m = \min_{\mathbf{H}_m} \max_{\substack{v \in \mathbf{H}_m \\ v \neq 0}} R(v)$ and set

$$\tilde{\lambda}_m = \inf_{X_m} \max_{\substack{v \in X_m \\ v \neq 0}} R(v)$$

Obviously, $\lambda_m \leq \tilde{\lambda}_m$, since $X \subset H_0^1(\Omega \cup \Gamma)$, so we have to show the opposite inequality.

Let $w_1, \ldots, w_m$ the eigenfunctions corresponding to the eigenvalues $\lambda_k = \lambda_k(-\Delta + c; \frac{\partial}{\partial v} + d; \Omega \cup \Gamma)$, $k = 1, \ldots, m$, normalized in $\mathbf{V} = L^2(\Omega) \times L^2(\Gamma)$, i. e., such that $\|w_i\|_{\mathbf{V}} = 1$, $i = 1, \ldots, m$.

By the density of X in $H_0^1(\Omega \cup \Gamma)$ and by Lemma 1.45 there exist sequences φ_i^n, $i = 1, \ldots, m$, such that $\varphi_1^n, \ldots, \varphi_m^n$ are linearly independent and φ_i^n converges to w_i in $H_0^1(\Omega \cup \Gamma)$, as $n \to \infty$.

Let us set $X^n = \mathrm{span}(\varphi_1^n, \ldots, \varphi_m^n)$, and denote by ψ^n the point of X^n such that $\max_{v \in X^n, v \neq 0} R(v) = R(\psi^n)$. Then

$$\psi^n = \sum_{k=1}^{m} s_k^n \varphi_k^n$$

for some $s_1^n, \ldots, s_m^n \in \mathbb{R}$. Since the function $R(v)$ is homogeneous of degree zero, without loss of generality and arguing as in Lemma 1.45, we can assume that $\sum_{k=1}^{m} |s_k^n|^2 = 1$ and the sequences s_k^n are bounded. Hence, up to a subsequence $s_k^n \to s_k \in \mathbb{R}$ as $n \to \infty$, with $\sum_{k=1}^{m} |s_k|^2 = 1$ and setting

$$\psi = \sum_{k=1}^{m} s_k w_k$$

we have

$$\psi^n \to \psi$$

Therefore, $R(\psi^n) \to R(\psi)$.

Since $w_1, \ldots, w_m$ are the first m eigenfunctions, $\|w_k\|_{\mathbf{V}} = 1$, $k = 1, \ldots, m$ and they are orthogonal in $\mathbf{V}$ and $H_0^1(\Omega \cup \Gamma)$ as well, we have that

$$\|\psi\|_{\mathbf{V}}^2 = \sum_{k=1}^{m} |s_k|^2 = 1$$

$$R(\psi) = Q(\psi) = \sum_{k=1}^{m} |s_k|^2 \lambda_k \le \sum_{k=1}^{m} |s_k|^2 \lambda_m = \lambda_m$$

Hence $R(\psi^n) \to R(\psi) \le \lambda_m$, which implies that $\tilde{\lambda}_m \le \lambda_m$, and we conclude that $\lambda_m = \tilde{\lambda}_m$.

(ii) Since $C_c^1(\Omega^1 \cup \Gamma^1) \subset C_c^1(\Omega^2 \cup \Gamma^2)$ any m-dimensional subspace X_m of $C_c^1(\Omega^1 \cup \Gamma^1)$ is an m-dimensional subspace of $C_c^1(\Omega^2 \cup \Gamma^2)$, $m \ge 1$. Hence the monotonicity of the eigenvalues follows by the variational characterization of the eigenvalues proved in (i).

Moreover, if $\partial\Omega^1 \cap \Omega^2 \ne \emptyset$ and $m = 1$ then the equality $\lambda_1^1 = \lambda_1^2$ cannot hold. Indeed if $\lambda_1^1 = \lambda_1^2$ then the trivial extension of the first eigenfunction in $\Omega^1 \cup \Gamma^1$ to $\Omega^2 \cup \Gamma^2$ would be a first eigenfunction in Ω^2 by Theorem 1.42 (v), vanishing in some point of Ω^2, contradicting the property (vi) stated in the same theorem.

(iii) By (i), for any $\varepsilon > 0$ there exists an m-dimensional subspace $X_m \subset C_c^1(\Omega \cup \Gamma)$ such that $\lambda_m \le \max_{\substack{v \in X_m \\ v \ne 0}} R(v) < \lambda_m + \varepsilon$.

If we consider a regular subdomain D such that $D \cup \Gamma_D$ contains the support of m functions which span the subspace X_m, using (ii) we get that $\lambda_m \le \lambda_m(-\Delta + c(x);\ \frac{\partial}{\partial v} + d(x);\ D \cup \Gamma_D) \le \max_{v \in X_m, v \ne 0} R(v) \le \lambda_m + \varepsilon$.

(iv) If $\mathbf{H}_{m-1}$ is a $(m-1)$-dimensional subspace of $H_0^1(\Omega \cup \Gamma)$ and $v \perp \mathbf{H}_{m-1}$, we have that

$$Q(v) = \int_{\Omega} \left[|\nabla v|^2 + c|v|^2 \right] dx + \int_{\Gamma} d|v|^2 \, d\sigma$$

$$\ge \int_{\Omega} \left[|\nabla v|^2 + c'|v|^2 \, dx \right] + \int_{\Gamma} d'|v|^2 = Q'(v) \tag{1.78}$$

Hence, denoting by $R'(v)$ the Rayleigh quotient with coefficients c', d', we have that $R(v) \ge R'(v)$. By the arbitrarity of v, we get

$$\min_{v \perp \mathbf{H}_{m-1}, v \ne 0} R(v) \ge \min_{v \perp \mathbf{H}_{m-1}, v \ne 0} R'(v)$$

Then we easily conclude recalling that $\lambda_m = \max_{\mathbf{H}_{m-1}} \min_{\substack{v \perp \mathbf{H}_{m-1} \\ v \ne 0}} R(v)$ (and $\lambda_m' = \max_{\mathbf{H}_{m-1}} \min_{\substack{v \perp \mathbf{H}_{m-1} \\ v \ne 0}} R'(v)$) by Theorem 1.42(iv).

(v) Let $\mathbf{H}_{m-1}$ be an $(m-1)$-dimensional subspace of $H_0^1(\Omega \cup \Gamma)$ and $v \perp \mathbf{H}_{m-1}$. Denoting by λ_m^n, Q^n and R^n the eigenvalues, the quadratic form and the Raylegh quotient associated to the coefficients c^n and d^n, we have

$$|Q(v) - Q^n(v)| = \int_\Omega |c - c^n||v|^2\, dx + \int_\Gamma |d - d^n||v|^2$$

$$\leq \left(\|c - c^n\|_{L^\infty(\Omega)} + \|d - d^n\|_{L^\infty(\Gamma)}\right)\left(\int_\Omega |v|^2 + \int_\Gamma |v|^2\right)$$

It follows that

$$|R(v) - R^n(v)| \leq \|c - c^n\|_{L^\infty(\Omega)} + \|d - d^n\|_{L^\infty(\Gamma)}$$

From this inequality, it is not difficult to deduce that

$$\left|\min_{\substack{v\perp\mathbf{H}_{m-1}\\ v\neq 0}} R(v) - \min_{\substack{v\perp\mathbf{H}_{m-1}\\ v\neq 0}} R^n(v)\right|$$

$$\leq \|c - c^n\|_{L^\infty(\Omega)} + \|d - d^n\|_{L^\infty(\Gamma)}$$

and finally

$$|\lambda_m - \lambda_m^n| = \left|\max_{\mathbf{H}_{m-1}} \min_{\substack{v\perp\mathbf{H}_{m-1}\\ v\neq 0}} R(v) - \max_{\mathbf{H}_{m-1}} \min_{\substack{v\perp\mathbf{H}_{m-1}\\ v\neq 0}} R^n(v)\right|$$

$$\leq \|c - c^n\|_{L^\infty(\Omega)} + \|d - d^n\|_{L^\infty(\Gamma)} \to 0 \text{ as } n \to \infty$$

(vi) If $v \in H_0^1(D)$, since $c \in L^\infty(\Omega)$, there exists $C \geq 0$ such that

$$Q(v; D) \geq \int_D |\nabla v|^2\, dx - C \int_D |v|^2\, dx,$$

while by Poincaré's inequality

$$\int_D |v|^2\, dx \leq C' |D|^{\frac{2}{N}} \int_D |\nabla v|^2\, dx.$$

It follows that

$$R(v; D) = \frac{Q(v; D)}{\int_D |v|^2} \geq \frac{\int_D |\nabla v|^2\, dx}{\int_D |v|^2\, dx} - C \geq \frac{1}{C'|D|^{\frac{2}{N}}} - C \to +\infty \text{ if } |D| \to 0.$$

(vii) If $v \in H_0^1(\Omega' \cup \Gamma')$, since $c, d \in L^\infty$ there exists $C \geq 0$ such that

$$Q(v; \Omega' \cup \Gamma') \geq \int_{\Omega'} |\nabla v|^2\, dx - C\left(\int_{\Omega'} |v|^2\, dx + \int_{\Gamma'} |v|^2\, d\sigma\right),$$

while by Poincaré's inequalities (Theorem 1.17) we have that there exists a constant C' such that

$$\int_{\Omega'} |v|^2 \, dx + \int_{\Gamma'} |v|^2 \, d\sigma \leq C'(|\Omega'|^{\frac{2}{N}} + |\Omega'|^{\frac{1}{N}}) \int_{\Omega'} |\nabla v|^2 \, dx.$$

It follows that

$$R(v; \Omega' \cup \Gamma') = \frac{Q(v; \Omega' \cup \Gamma')}{(\int_{\Omega'} |v|^2 \, dx + \int_{\Gamma'} |v|^2 \, d\sigma)} \geq \frac{\int_{\Omega'} |\nabla v|^2 \, dx}{(\int_{\Omega'} |v|^2 \, dx + \int_{\Gamma'} |v|^2 \, d\sigma)} - C$$

$$\geq \frac{1}{C'(|\Omega'|^{\frac{2}{N}} + |\Omega'|^{\frac{1}{N}})} - C \to +\infty \text{ if } |\Omega'| \to 0$$

$\square$

In Theorem 1.44, the space X can be any dense subspace of $H_0^1(\Omega \cup \Gamma)$, and in particular we consider the space $X = C_c^1(\Omega \cup \Gamma)$.

In some applications (see Chapter 4), when $\Omega = B_R$ is a ball and we consider the Dirichlet problem (i. e., when $\Gamma = \emptyset$), we will need to approximate the eigenvalues in the ball with the eigenvalues in the annulus $B_R \setminus B_{\frac{1}{h}}$ with $h \in \mathbb{N}$ large. This can be done using Theorem 1.44 and the following result.

Proposition 1.46. *Let $N \geq 2$ and $B_R = B_R(0)$ a ball in $\mathbb{R}^N$ centered at the origin, with radius $R > 0$. Then $C_c^\infty(B_R \setminus \{0\})$ is dense in $H_0^1(B_R)$, i. e., for any function $v \in H_0^1(B_R)$ there exists a sequence $\{w_n\} \subset C_c^\infty(B_R \setminus \{0\})$ that converges to v in $H^1(B_R)$.*

Proof. Let us denote by $H_0^1(B_R \setminus \{0\})$ the closure of the subspace $C_c^\infty(B_R \setminus \{0\})$ in $H^1(B_R \setminus \{0\})$ (or in $H^1(B_R)$ which is the same space). Since by definition $C_c^\infty(B_R)$ is dense in $H_0^1(B_R)$ and $C_c^\infty(B_R \setminus \{0\})$ is dense in $H_0^1(B_R \setminus \{0\})$, it suffices to prove that for any $v \in C_c^\infty(B_R)$ there exists a sequence $v_n \in H_0^1(B_R \setminus \{0\})$ that converges to v in $H^1(B_R)$. In particular, we will consider functions $v_n \in H^1(B_R)$ with compact support in $B_R \setminus \{0\}$, therefore belonging to $H_0^1(B_R \setminus \{0\})$ by (ii) of Theorem 1.3.

Let us first assume that $N \geq 3$, $v \in C_c^\infty(B_R)$ and consider the functions:

$$g_n(t) = \begin{cases} 0 & \text{if } 0 \leq t \leq \frac{1}{2n} \\ 2nt - 1 & \text{if } \frac{1}{2n} < t < \frac{1}{n} \,, \\ 1 & \text{if } t \geq \frac{1}{n} \end{cases} \quad t \in [0, +\infty), \ n \in \mathbb{N}^+$$

Then $g_n \in C^{0,1}([0, +\infty))$ satisfies $g_n(t) = 0$ for $0 \leq t \leq \frac{1}{2n}$, $g_n(t) = 1$ for $t \geq \frac{1}{n}$, $0 \leq g_n(t) \leq 1 \ \forall t \in [0, +\infty)$ and $|g_n'(t)| \leq 2n$.

Let us set $v_n(x) = g_n(|x|)v(x)$. Then, since the support of v_n is a compact subset of $B_R \setminus \{0\}$, it follows that $v_n \in H_0^1(B_R \setminus \{0\})$ with $\nabla v_n = g_n(|x|)\nabla v(x) + \nabla g_n(|x|)v(x)$, and we claim that $v_n \to v$ in $H^1(B_R)$. Indeed, it is immediate to see that as $n \to \infty$ we have

$$\int_{B_R} |v_n - v|^2 \, dx = \int_{B_R} (1 - g_n(|x|))^2 |v|^2 \, dx \le \int_{B_{\frac{1}{n}}} |v|^2 \, dx \to 0$$

$$\int_{B_R} |\nabla v_n - \nabla v|^2 \, dx \le C \left(\int_{B_R} |1 - g_n(||x|)|^2 |\nabla v|^2 \, dx + \int_{B_R} |g_n'(|x|)|^2 |v|^2 \, dx \right)$$

$$\le C \left(\int_{B_{\frac{1}{n}}} |\nabla v|^2 \, dx + n^2 \int_{B_{\frac{1}{n}}} |v|^2 \, dx \right)$$

The last term converges to 0 as $n \to \infty$ since $\int_{B_{\frac{1}{n}}} |\nabla v|^2 \, dx \to 0$ and

$$\int_{B_R} |g_n'(|x|)|^2 |v|^2 \, dx \le Cn^2 \int_{B_{\frac{1}{n}}} |v|^2 \, dx$$

$$\le n^2 (\text{meas}_N(B_{\frac{1}{n}}))^{\frac{2}{N}} \left(\int_{B_{\frac{1}{n}}} |v|^{2^*} \right)^{\frac{N-2}{N}} \le C \left(\int_{B_{\frac{1}{n}}} |v|^{2^*} \right)^{\frac{N-2}{N}} \to 0 \qquad (1.79)$$

If instead $N = 2$ and $v \in C_c^1(B_R)$, we consider the functions

$$g_n(t) = \begin{cases} 0 & \text{if } 0 \le t \le \frac{1}{n^2} \\ 2 + \dfrac{\log(t)}{\log(n)} & \text{if } \frac{1}{n^2} < t < \frac{1}{n} \\ 1 & \text{if } t \ge \frac{1}{n} \end{cases}$$

Then $g_n(t) = 0$ if $0 \le t \le \frac{1}{2n}$, $g_n(t) = 1$ if $t \ge \frac{1}{n}$, $0 \le g_n(t) \le 1 \ \forall t \in [0, +\infty)$, and $g_n'(t) = \frac{1}{t \log(n)}$ if $\frac{1}{n^2} < t < \frac{1}{n}$.

Setting $v_n(x) = g_n(|x|)v(x)$, we have that $v_n \in H_0^1(B_R \setminus \{0\})$ and it converges to v in $H^1(B_R)$.

Indeed, it is enough to proceed as before substituting the estimate (1.79) by

$$\int_{B_R} |g_n'(|x|)|^2 |v|^2 \, dx \le C\|v\|_{L^\infty}^2 \int_{\frac{1}{n^2}}^{\frac{1}{n}} \frac{1}{\rho^2 \log^2(n)} \rho \, d\rho$$

$$= \frac{\log(\frac{1}{n}) - \log(\frac{1}{n^2})}{\log^2(n)} = \frac{\log(n)}{\log^2(n)} = \frac{1}{\log(n)} \to 0 \qquad (1.80)$$

$$\square$$

Corollary 1.47. *Let B_R be the ball centered at the origin of radius $R > 0$ and $c \in L^\infty(B_R)$. Let us denote, for $h \in \mathbb{N}^+$, by $A_h = \{x \in \mathbb{R}^N : \frac{1}{h} < |x| < R\}$ the annulus obtained by removing a small ball around the origin, and let $\lambda_m = \lambda_m(-\Delta + c; B_R)$ be the eigenvalues of the Dirichlet problem in B_R. Then*

$$\lambda_m = \inf_{h \in \mathbb{N}} \{\lambda_m(-\Delta + c; A_h)\} = \lim_{h \to \infty} \lambda_m(-\Delta + c; A_h)$$

Proof. By Proposition 1.46, we know that $C_c^1(B_R \setminus \{0\})$ is dense in $H_0^1(B_R)$. Then by (i) of Theorem 1.44, for any $\varepsilon > 0$ there exists an m dimensional subspace $X_m \subset C_c^1(B_R \setminus \{0\})$ such that $\max_{v \in X_m, v \neq 0} R(v) < \lambda_m + \varepsilon$. If h is large, then A_h will contain all supports of m functions spanning X_m. Hence, using also the monotonicity of the eigenvalues given by (ii) of Theorem 1.44, we get

$$\lambda_m \leq \lambda_m(-\Delta + c; A_h) \leq \max_{v \in X_m, v \neq 0} R(v) \leq \lambda_m + \varepsilon.$$

By the arbitrarity of $\varepsilon > 0$, it follows that $\lambda_m = \lim_{h \to \infty} \{\lambda_m(-\Delta + c; A_h)\}$ and since the sequence $\alpha_h = \lambda_m(-\Delta + c; A_h)$ is nonincreasing, we also get $\lambda_m = \inf_{h \in \mathbb{N}} \{\lambda_m(-\Delta + c; A_h)\}$. $\qquad\square$

Remark 1.48. In Proposition 1.46 and Corollary 1.47, we can substitute the ball with any bounded domain Ω and the origin 0 with any interior point.

Remark 1.49. If $N \geq 3$ adapting the proof of Theorem 1.46 for the two-dimensional case, and using cylindrical coordinates, it is not difficult to show that $C_c^\infty(B_R \setminus F)$ is dense in $H_0^1(B_R)$ if $F = \{x = (x_1, \ldots, x_n) \in \mathbb{R}^N : x_1 = x_2 = 0\}$.

Let us end the section with a result that gives a *necessary and sufficient condition for the validity of the maximum principle* that depends on the spectrum of the corresponding operator. Recalling Definition 1.23 for an operator satisfying the maximum principle, we have the following.

Theorem 1.50. *Let Ω be a bounded domain in $\mathbb{R}^N$, with $\Gamma, \Gamma_0 \subseteq \partial\Omega$ satisfying (1.28), (1.29), $c \in L^\infty(\Omega)$, $d \in L^\infty(\Gamma)$.*

The operator $L = (-\Delta + c; \frac{\partial}{\partial v} + d)$ satisfies the maximum principle in $\Omega \cup \Gamma$ if and only if $\lambda_1(-\Delta + c; \frac{\partial}{\partial v} + d; \Omega) > 0$.

Proof. If $\lambda_1(\Omega) \leq 0$, then a positive first eigenfunction φ_1 satisfies

$$\begin{cases} -\Delta\varphi_1 + c(x)\varphi_1 = \lambda_1\varphi_1 \leq 0 & \text{in } \Omega \\ \varphi_1 = 0 & \text{on } \Gamma_0 \\ \dfrac{\partial\varphi_1}{\partial v} + d\varphi_1 = \lambda_1\varphi_1 & \text{on } \Gamma \end{cases} \tag{1.81}$$

and hence the maximum principle cannot hold, otherwise φ_1 should be nonpositive.

If instead $\lambda_1(-\Delta + c; \frac{\partial}{\partial v} + d; \Omega) > 0$, by the variational characterization of the first eigenvalue it holds $\lambda_1 = \min_{v \in H^1(\Omega \cup \Gamma), v \neq 0} R(v) > 0$, so that $Q(v) > 0$ for any $v \neq 0$ in $H_0^1(\Omega \cup \Gamma)$.

Suppose that $v \in H^1(\Omega)$ satisfies

$$\begin{cases} -\Delta v + cv \leq 0 & \text{in } \Omega \\ v \leq 0 & \text{on } \Gamma_0 \\ \dfrac{\partial v}{\partial v} + dv \leq 0 & \text{on } \Gamma \end{cases}$$

Then $v^+ \in H_0^1(\Omega \cup \Gamma)$ can be used as a test function, and we obtain that $Q(v^+) \le 0$, which implies $v^+ \equiv 0$ in Ω. $\qquad\square$

Remark 1.51. The maximum principle in small domains (Theorem 1.20) can be seen also as a consequence of Theorem 1.50 and (vi) in Theorem 1.44.

Analogously, if Ω is a cylindrically symmetric domain, Theorem 1.25 follows from Theorem 1.50 and (vii) in Theorem 1.44.

1.5 Systems of elliptic equations

In this section, we extend some results stated for elliptic equations in previous sections to the case of elliptic linear and semilinear systems. For simplicity, only the case of Dirichlet problems is considered.

Let Ω be any bounded domain in $\mathbb{R}^N$, $N \ge 2$, and D a $m \times m$ matrix with bounded entries:

$$D = (d_{ij})_{i,j=1}^m, \quad d_{ij} \in L^\infty(\Omega) \tag{1.82}$$

We consider the linear elliptic system

$$\begin{cases} -\Delta U + D(x)U = F & \text{in } \Omega \\ U = 0 & \text{on } \partial\Omega \end{cases} \tag{1.83}$$

i. e.,

$$\begin{cases} -\Delta u_1 + d_{11}u_1 + \cdots + d_{1m}u_m = f_1 & \text{in } \Omega \\ \dotsc\dotsc & \dotsc \\ -\Delta u_m + d_{m1}u_1 + \cdots + d_{mm}u_m = f_m & \text{in } \Omega \\ u_1 = \cdots = u_m = 0 & \text{on } \partial\Omega \end{cases}$$

where $F = (f_1, \ldots, f_m) \in (L^2(\Omega))^m$, $U = (U_1, \ldots, U_m)$.

Definition 1.52. The matrix D or the associated system (1.83) is said to be
- *cooperative* or *weakly coupled* in Ω if

$$d_{ij} \le 0 \text{ a. e. in } \Omega, \quad \text{whenever } i \ne j \tag{1.84}$$

- *fully coupled* in Ω if it is weakly coupled in Ω and the following condition holds:

$$\forall I, J \subset \{1, \ldots, m\}, I, J \ne \emptyset, I \cap J = \emptyset, I \cup J = \{1, \ldots, m\}$$
$$\exists i_0 \in I, j_0 \in J : \text{meas}(\{x \in \Omega : d_{i_0 j_0} < 0\}) > 0 \tag{1.85}$$

Before going on we fix some **notation and definitions.**

- Inequalities involving vectors should be understood to hold componentwise, e. g., if $\Psi = (\psi_1, \ldots, \psi_m)$, Ψ nonnegative means that $\psi_j \geq 0$ for any index $j = 1, \ldots, m$.
- If $m \geq 2$ and $1 \leq p \leq \infty$, we will consider the Banach spaces

$$\mathbf{L}^p(\Omega) = \left(L^p(\Omega)\right)^m, \quad \mathbf{W}^{1,p}(\Omega) = \left(W^{1,p}(\Omega)\right)^m$$

If $p = 2$, we consider in particular the Hilbert spaces

$$\mathbf{L}^2(\Omega) = \left(L^2(\Omega)\right)^m, \quad \mathbf{H}^1(\Omega) = \left(H^1(\Omega)\right)^m$$

and the space $\mathbf{H}_0^1(\Omega) = (H_0^1(\Omega))^m$, the closure in $\mathbf{H}^1(\Omega)$ of the subspace $C_c^1(\Omega; \mathbb{R}^m)$. If $f = (f_1, \ldots, f_m)$, $g = (g_1, \ldots, g_m)$, the scalar products are defined by

$$(f, g)_{\mathbf{L}^2(\Omega)} = \sum_{i=1}^m (f_i, g_i)_{L^2(\Omega)} = \sum_{i=1}^m \int_\Omega f_i g_i \, dx$$

$$(f, g)_{\mathbf{H}_0^1(\Omega)} = \sum_{i=1}^m (f_i, g_i)_{H_0^1(\Omega)} = \sum_{i=1}^m \int_\Omega \nabla f_i \cdot \nabla g_i \, dx \tag{1.86}$$

- If $U = (u_1, \ldots, u_m)$, $\Psi = (\psi_1, \ldots, \psi_m) \in \mathbf{H}_0^1(\Omega)$, and the matrix D satisfies (1.82), we set

$$\nabla U \cdot \nabla \Psi = \sum_{i=1}^m \nabla u_i \cdot \nabla \psi_i \tag{1.87}$$

$$D(x)(U, \Psi) = \sum_{i,j=1}^m d_{ij}(x) u_i \psi_j \tag{1.88}$$

$$B(U, \Phi) = \int_\Omega \left[\nabla U \cdot \nabla \Phi + D(U, \Phi) \right] dx$$

$$= \int_\Omega \left[\sum_{i=1}^m \nabla u_i \cdot \nabla \phi_i + \sum_{i,j=1}^m d_{ij} u_i \phi_j \right] dx \tag{1.89}$$

i. e., $D(x)(U, \Psi)$ is the action of the bilinear form associated to the matrix D on the pair (U, Ψ), and B is the bilinear form in $\mathbf{H}_0^1(\Omega)$ associated to the operator $-\Delta + D$ and to system (1.83).

- If $U = (u_1, \ldots, u_m) \in \mathbf{H}^1(\Omega)$, we say that U weakly satisfies

$$U \leq 0 \text{ on } \partial\Omega \quad (U \geq 0 \text{ on } \partial\Omega)$$

if $U^+ \in \mathbf{H}_0^1(\Omega)$ $(U^- \in \mathbf{H}_0^1(\Omega))$, i. e., if $u_i^+ \in H_0^1(\Omega)$ $(u_i^- \in H_0^1(\Omega))$ for any $i = 1, \ldots, m$.
- If $U = (u_1, \ldots, u_m) \in \mathbf{H}^1(\Omega)$ and D satisfies (1.82), we say that U weakly satisfies the inequality

$$-\Delta U + D(x)U \geq 0 \quad \text{in } \Omega \tag{1.90}$$

if for any $i = 1, \ldots, m$ the inequalities

$$-\Delta u_i + \sum_{j=1}^{m} d_{ij} u_j \geq 0 \quad \text{in } \Omega$$

are satisfied in a weak sense, i. e.,

$$\int_{\Omega} \left(\nabla u_i \cdot \nabla \psi + \sum_{j=1}^{m} d_{ij} u_j \psi \right) dx \geq 0 \tag{1.91}$$

for any $\psi \in H_0^1(\Omega)$ with $\psi \geq 0$ and any $i = 1, \ldots, m$.
This implies that

$$\int_{\Omega} \nabla U \cdot \nabla \Psi + D(x)(U, \Psi)$$

$$= \int_{\Omega} \left[\sum_{i=1}^{m} \nabla u_i \cdot \nabla \psi_i + \sum_{i,j=1}^{m} d_{ij}(x) u_i \psi_j \right] dx \geq 0 \tag{1.92}$$

for any nonnegative $\Psi = (\psi_1, \ldots, \psi_m) \in \mathbf{H}_0^1(\Omega)$. As a matter of fact, (1.92) is equivalent to (1.90), as it is easy to see considering test functions Ψ with all components vanishing except one.

It is well known that either condition (1.84) or conditions (1.84) and (1.85) together are needed in the proofs of maximum principles for systems (see [94, 97, 201] and the references therein). In particular, if both are fulfilled the strong maximum principle holds as it is shown in the next theorem.

Theorem 1.53 (Strong maximum principle and Hopf's lemma). *Suppose that* (1.82) *and* (1.84) *hold and* $U = (u_1, \ldots, u_m) \in C^1(\Omega; \mathbb{R}^m)$ *is a weak solution of the inequalities*

$$-\Delta U + D(x)U \geq 0 \text{ in } \Omega \quad \text{and} \quad U \geq 0 \text{ in } \Omega$$

Then:
1. *for any $k \in \{1, \ldots, m\}$ either $u_k \equiv 0$ or $u_k > 0$ in Ω; in the latter case if $u_k \in C^1(\overline{\Omega}; \mathbb{R}^m)$, Ω satisfies the interior sphere condition at $P \in \partial\Omega$ and $u_k(P) = 0$ then $\frac{\partial u_k}{\partial v}(P) < 0$, where v is the unit exterior normal vector at P;*
2. *if in addition (1.85) holds, then the same alternative holds for all $k = 1, \ldots, m$, i. e. either $U \equiv 0$ in Ω or $U > 0$ in Ω. In the latter case if $U \in C^1(\overline{\Omega}; \mathbb{R}^m)$, Ω satisfies the interior sphere condition at $P \in \partial\Omega$ and $U(P) = 0$ then $\frac{\partial U}{\partial v}(P) < 0$, where v is the unit exterior normal vector at P.*

Proof. It is enough to apply the scalar strong maximum principle using (1.84) to obtain that each component u_k is either identically equal to zero or positive. Moreover, if (1.85) holds, then the same alternative holds for all the components.

Indeed, since $d_{ij} \leq 0$ if $i \neq j$ and $u_i \geq 0$, for any equation we have

$$-\Delta u_j + d_{jj}u_j \geq \sum_{i \neq j} -d_{ij}u_i \geq 0$$

This implies that, for any $j = 1, \ldots, m$, either $u_j \equiv 0$ or $u_j > 0$ and in the latter case by Hopf's lemma we have the sign of the normal derivative on a point of the boundary where the function u_j vanishes.

Suppose now that (1.85) holds and U does not vanish identically in Ω and let $J \subset \{1, \ldots, m\}$ the set of indexes j such that $u_j > 0$ in Ω. Assume by contradiction that J is a proper subset of $\{1, \ldots, m\}$, and let $I = \{1, \ldots, m\} \setminus J$ the set of indexes i such that $u_i \equiv 0$. If i_0, j_0 are as in (1.85) then

$$-\Delta u_{i_0} + d_{i_0 i_0} u_{i_0} \geq -\sum_{j \neq i_0} -d_{i_0 j}u_j \geq -d_{i_0 j_0}u_{j_0} \neq 0$$

Hence, by the scalar strong maximum principle, $u_{i_0} > 0$, which is a contradiction. □

Remark 1.54. Maximum principles, Harnack inequalities and other estimates for systems have been studied in many papers with general conditions and also for elliptic operators not in divergence form (see [201] and the references therein).

We consider now the bilinear form and the quadratic form associated to the system (1.83), namely

$$B(U, \Phi) = \int_\Omega [\nabla U \cdot \nabla \Phi + C(U, \Phi)] = \int_\Omega \left[\sum_{i=1}^m \nabla u_i \cdot \nabla \phi_i + \sum_{i,j=1}^m c_{ij}u_i\phi_j \right] \tag{1.93}$$

for $U = (u_1, \ldots, u_m)$, $\Phi = (\phi_1, \ldots, \phi_m) \in \mathbf{H}_0^1(\Omega)$ and

$$Q(\Psi) = B(\Psi, \Psi) = \int_\Omega (|\nabla \Psi|^2 + D(x)(\Psi, \Psi))dx$$

$$= \int_\Omega \left(\sum_{i=1}^m |\nabla \psi_i|^2 + \sum_{i,j=1}^m d_{ij}(x)\psi_i\psi_j \right) dx \tag{1.94}$$

for $\Psi = (\psi_1, \ldots, \psi_m) \in \mathbf{H}_0^1(\Omega)$.

Sometimes we will also write $Q(\Psi; \Omega)$ instead of $Q(\Psi)$ specifying the domain.

It is easy to see that this quadratic form coincides with the quadratic form associated to the symmetric system

$$\begin{cases} -\Delta U + C(x)U = F & \text{in } \Omega \\ U = 0 & \text{on } \partial\Omega \end{cases} \tag{1.95}$$

i. e.,

$$\begin{cases} -\Delta u_1 + c_{11}u_1 + \cdots + c_{1m}u_m & = f_1 \\ \cdots\cdots & \cdots \\ -\Delta u_m + c_{m1}u_1 + \cdots + c_{mm}u_m & = f_m \end{cases}$$

where $C = \frac{1}{2}(D + D^t)$ and D^t is the transpose of D, i. e.,

$$C = (c_{ij}), \quad c_{ij} = \frac{1}{2}(d_{ij} + d_{ji}) \tag{1.96}$$

So, to study the sign of the quadratic form Q, we can also use the properties of the symmetric system and this will be crucial when studying the Morse index of a solution of a semilinear system in Chapter 7. Therefore, we review briefly the spectral theory for these kinds of symmetric systems, and use it to prove some results that we need for the possible nonsymmetric system (1.83).

Remark 1.55. If system (1.83) is cooperative, respectively fully coupled, so is the associate symmetric system (1.95).

1.5.1 Spectral theory for symmetric systems

Let Ω be a bounded domain in $\mathbb{R}^N$, $N \geq 2$, and let $C = C(x) = (c_{ij}(x))_{i,j=1}^{m}$ be a symmetric matrix whose elements are bounded functions:

$$c_{ij} \in L^\infty(\Omega), \quad c_{ij} = c_{ji} \quad \text{a. e. in } \Omega \tag{1.97}$$

Let us consider the bilinear forms

$$B(U, \Phi) = \int_\Omega [\nabla U \cdot \nabla \Phi + C(U, \Phi)] = \int_\Omega \left[\sum_{i=1}^{m} \nabla u_i \cdot \nabla \phi_i + \sum_{i,j=1}^{m} c_{ij} u_i \phi_j \right] \tag{1.98}$$

and, for $\Lambda > 0$

$$\begin{aligned} B^\Lambda(U, \Phi) &= \int_\Omega [\nabla U \cdot \nabla \Phi + (C + \Lambda I)(U, \Phi)] \\ &= \int_\Omega \left[\sum_{i=1}^{m} (\nabla u_i \cdot \nabla \phi_i + \Lambda u_i \phi_i) + \sum_{i,j=1}^{m} c_{ij} u_i \phi_j \right] \end{aligned} \tag{1.99}$$

Since $c_{ij} \in L^\infty$ and $c_{ij} = c_{ji}$, B and B^Λ are continuous symmetric bilinear forms, and, since $|\int_\Omega c_{ij} u_i \phi_j| \leq C \int_\Omega (u_i^2 + \phi_j^2)$, there exists $\Lambda \geq 0$ such that B^Λ is coercive in $\mathbf{H}_0^1(\Omega)$, i. e., it is an equivalent scalar product in $\mathbf{H}_0^1(\Omega)$.

By the Riesz representation theorem, identifying $F \in \mathbf{L}^2(\Omega)$ with the linear functional $U \in \mathbf{H}_0^1(\Omega) \mapsto (U, F)_{\mathbf{L}^2(\Omega)}$, for any $F \in L^2$ there exists a unique $U =: T(F) \in \mathbf{H}_0^1(\Omega)$ such that $B^\Lambda(U, \Phi) = (F, U)_{\mathbf{L}^2(\Omega)}$ for any $\Phi \in \mathbf{H}_0^1(\Omega)$ and $\|U\|_{\mathbf{H}_0^1(\Omega)} \leq C\|F\|_{\mathbf{L}^2(\Omega)}$ for some constant $C > 0$.

In other words, U is the unique weak solution of the system

$$\begin{cases} -\Delta U + (C(x) + \Lambda I)U = F & \text{in } \Omega \\ U = 0 & \text{on } \partial\Omega \end{cases} \tag{1.100}$$

i. e.,

$$\begin{cases} -\Delta u_1 + (c_{11} + \Lambda)u_1 + \cdots + c_{1m}u_m = f_1 & \text{in } \Omega \\ \cdots & \cdots \\ -\Delta u_m + c_{m1}u_1 + \cdots + (c_{mm} + \Lambda)u_m = f_m & \text{in } \Omega \\ u_1 = \cdots = u_m = 0 & \text{on } \partial\Omega \end{cases}$$

Moreover, $T : F \mapsto U$, maps $\mathbf{L}^2(\Omega)$ into $\mathbf{L}^2(\Omega)$ and is *compact* because of the compact embedding of $\mathbf{H}_0^1(\Omega)$ in $\mathbf{L}^2(\Omega)$. It is also a *positive* operator, since $(T(F), F)_{\mathbf{L}^2(\Omega)} = (U, F)_{\mathbf{L}^2(\Omega)} = B^\Lambda(U, U) > 0$ if $F \neq 0$ which implies $U \neq 0$ (recall that B^Λ is an equivalent scalar product in $\mathbf{H}_0^1(\Omega)$), and since C is symmetric, it is also *self-adjoint*.

Indeed, if $T(F) = U$, $T(G) = V$, i. e., $B^\Lambda(U, \Phi) = (F, U)_{\mathbf{L}^2(\Omega)}$, $B^\Lambda(V, \Phi) = (G, V)_{\mathbf{L}^2(\Omega)}$ for any $\Phi \in \mathbf{H}_0^1(\Omega)$, then $(T(F), G)_{\mathbf{L}^2(\Omega)} = (U, G)_{\mathbf{L}^2(\Omega)} = (G, U)_{\mathbf{L}^2(\Omega)} = B^\Lambda(V, U) = B^\Lambda(U, V) = (F, V)_{\mathbf{L}^2(\Omega)} = (F, T(G))_{\mathbf{L}^2(\Omega)}$.

Thus, by the spectral theory of positive compact self-adjoint operators in Hilbert spaces there exists a nonincreasing sequence $\{\mu_j^\Lambda\}$ of eigenvalues with $\lim_{j\to\infty} \mu_j^\Lambda = 0$ and a corresponding sequence $\{W^j\} \subset \mathbf{H}_0^1(\Omega)$ of eigenvectors such that $G(W^j) = \mu_j^\Lambda W^j$. Setting $\lambda_j^\Lambda = \frac{1}{\mu_j^\Lambda}$, then W^j solves the system $-\Delta W^j + (C + \Lambda I)W^j = \lambda_j^\Lambda W^j$ and $W^j = 0$ on $\partial\Omega$. Translating, and denoting by λ_j the differences $\lambda_j = \lambda_j^\Lambda - \Lambda$, we conclude that there exists a sequence $\{\lambda_j\}$ of eigenvalues, with $-\infty < \lambda_1 \leq \lambda_2 \leq \ldots$, $\lim_{j\to+\infty} \lambda_j = +\infty$, and a corresponding sequence of eigenfunctions $\{W^j\}$ which weakly solve the systems

$$\begin{cases} -\Delta W^j + CW^j = \lambda_j W^j & \text{in } \Omega \\ W^j = 0 & \text{on } \partial\Omega \end{cases} \tag{1.101}$$

i. e., if $W^j = (w_1, \ldots, w_m)$

$$\begin{cases} -\Delta w_1 + c_{11}w_1 + \cdots + c_{1m}w_m & = \lambda_j w_1 \\ \cdots\cdots & \cdots \\ -\Delta w_m + c_{m1}w_1 + \cdots + c_{mm}w_m & = \lambda_j w_m \end{cases}$$

Moreover, by (scalar) elliptic regularity theory applied iteratively to each equation, the eigenfunctions W^j belong at least to $C^1(\Omega; \mathbb{R}^m)$. We now collect the variational formulation and some properties of eigenvalues and eigenfunctions.

Theorem 1.56. *Let Ω be a bounded domain in $\mathbb{R}^N$, $N \geq 2$. Suppose that $C = (c_{ij})_{i,j=1}^{m}$ satisfies (1.97), and let $\{\lambda_j\}$, $\{W^j\}$ be the sequences of eigenvalues and eigenfunctions satisfying (1.101).*

Define the Rayleigh quotient

$$R(V) = \frac{Q(V)}{(V,V)_{L^2(\Omega)}} \quad \text{for } V \in H_0^1(\Omega), \quad V \neq 0 \tag{1.102}$$

with $Q(V) = B(V,V)$ and B as in (1.98). Then the following properties hold, where H_k denotes a k-dimensional subspace of $H_0^1(\Omega)$ and the orthogonality conditions $V \perp W^k$ or $V \perp H_k$ stand for the orthogonality in $L^2(\Omega)$.

(i)

$$\lambda_1 = \min_{\substack{V \in H_0^1(\Omega) \\ V \neq 0}} R(V) = \min_{\substack{V \in H_0^1(\Omega) \\ (V,V)_{L^2}=1}} Q(V)$$

(ii) *If $k \geq 2$, then*

$$\lambda_k = \min_{\substack{V \in H_0^1(\Omega) \\ V \neq 0 \\ V \perp W^1,\ldots,V \perp W^{k-1}}} R(V) = \min_{\substack{V \in H_0^1(\Omega) \\ (V,V)_{L^2(\Omega)}=1 \\ V \perp W^1,\ldots,V \perp W^{k-1}}} Q(V)$$

(iii) *If $k \geq 2$, then*

$$\lambda_k = \min_{H_k} \max_{\substack{V \in H_k \\ V \neq 0}} R(V)$$

(iv) *If $k \geq 2$, then*

$$\lambda_k = \max_{H_{k-1}} \min_{\substack{V \perp H_{k-1} \\ V \neq 0}} R(V)$$

(v) *If $W \in H_0^1(\Omega)$, $W \neq 0$ and $R(W) = \lambda_1$, then W is an eigenfunction corresponding to λ_1.*

(vi) *If the system is fully coupled in Ω, then the first eigenfunction does not change sign in Ω and the first eigenvalue is simple, i. e., up to scalar multiplication there is only one eigenfunction corresponding to the first eigenvalue.*

Proof. The proofs of (i), $\ldots$, (v) do not depend on cooperativeness and they are exactly the same as in Theorem 1.42, so we omit them.

To prove (vi), let us first show that if the system is cooperative and W is a first eigenfunction, then W^+ and W^- are eigenfunctions, if they do not vanish.

We multiply (1.101) by $W^+ = (w_1^+,\ldots,w_m^+)$ and integrate.

If in the ith equation, multiplied by w_i^+ we write $w_j = w_j^+ - w_j^-$ for $j \neq i$ and recall that by cooperativeness $-c_{ij} w_j^- w_i^+ \geq 0$, we deduce that

$$B(W^+, W^+) \leq \lambda_1(W^+, W^+)$$

so that, by (v), W^+ is a first eigenfunction if it does not vanish. The same argument applies to W^-. If the system is also fully coupled, then the assertion (vi) follows from the strong maximum principle, which is valid under the fully coupling hypothesis. In fact if, e. g., W^+ does not vanish, it is a nonnegative first eigenfunction and by the strong maximum principle (Theorem 1.53) it is strictly positive in Ω, i. e., $W^- \equiv 0$ and $W > 0$ in Ω if W it is positive somewhere.

If W^1, W^2 are two eigenfunctions corresponding to λ_1, they do not change sign in Ω, so that they cannot be orthogonal in $\mathbf{L}^2(\Omega)$. This implies that the first eigenvalue is simple. $\qquad\square$

When Ω' is a subdomain of Ω and $C' = C'(x) = (c'_{ij}(x))_{i,j=1}^m$ satisfies the analogous of (1.97) in Ω' (i. e., $c'_{ij} = c'_{ji}$ and they are bounded), we will denote by

$$\lambda_k(-\Delta + C'; \Omega') \tag{1.103}$$

the eigenvalues of the Dirichlet problem, with corresponding eigenfunctions W^k in $H_0^1(\Omega')$ which solve the problem

$$\begin{cases} -\Delta w + C'(x)w = \lambda W & \text{in } \Omega' \\ W = 0 & \text{on } \partial\Omega' \end{cases}$$

We keep the simple notation λ_k for the eigenvalues $\lambda_k = \lambda_k(-\Delta + C; \Omega)$.

Properties analogous to those of Theorem 1.44 also hold for systems.

Theorem 1.57. *Assume that Ω is a bounded domain in $\mathbb{R}^N$, $N \geq 2$ and $C = (c_{ij})_{i,j=1}^m$ satisfies (1.97).*

(i) *If X is a dense subspace of the space $\mathbf{H}_0^1(\Omega)$ (in particular, if $X = (C_c^1(\Omega))^m$) and we denote by X_k a k-dimensional subspace of X then*

$$\lambda_k = \inf_{X_k} \max_{\substack{v \in X_k \\ v \neq 0}} R(v)$$

(ii) *(Monotonicity w. r. t. subdomains) If $\Omega^1 \subset \Omega^2 \subset \Omega$ are subdomains of the domain Ω, then*

$$\lambda_k(-\Delta + C; \Omega^1) \geq \lambda_k(-\Delta + C; \Omega^2)$$

Moreover, if $\Omega^1 \subset \Omega^2 \subset \Omega$ and $\partial\Omega^1 \cap \Omega^2 \neq \emptyset$, then the strict inequality

$$\lambda_1(-\Delta + C; \Omega^1) > \lambda_1(-\Delta + C; \Omega^2)$$

holds (i. e., only for $k = 1$).

(iii) (*Continuity w. r. t. subdomains*) *For any* $k \in \mathbb{N}^+$, *it holds*

$$\lambda_k = \inf\{\lambda_k(-\Delta + C; \ \Omega')\}$$

where the infimum is taken among the subdomains Ω' *such that* $\overline{\Omega'}$ *is a compact subset of* Ω.

(iv) (*Monotonicity w. r. t. the coefficients*)

a) *Let* $C'(x)$ *be a symmetric matrix in* $\mathbf{L}^\infty(\Omega)$, *and assume that* $C \geq C'$, *i. e., the symmetric matrix* $C(x) - C'(x)$ *is positive semidefinite a. e. in* Ω *or equivalently*

$$\sum_{i,j=1}^m c_{ij}\xi_i\xi_j \geq \sum_{i,j=1}^m c'_{ij}\xi_i\xi_j \quad \forall \xi \in \mathbb{R}^m, \ a.e.\ in\ \Omega$$

Then, setting $\lambda'_k = \lambda_k(-\Delta + C'; \ \Omega)$, *for any* $k \geq 1$, *we have*

$$\lambda_k \geq \lambda'_k \tag{1.104}$$

b) *Assume that the system associated to the matrix* C *is fully coupled and let* $C''(x)$ *be a symmetric matrix in* $\mathbf{L}^\infty(\Omega)$ *such that* $c_{ij} \geq c''_{ij}$ *a. e. in* Ω, *for any* $i, j \in \{1, \dots, m\}$. *Then, setting* $\lambda''_1 = \lambda_1(-\Delta + C''; \ \Omega)$, *the following inequality holds for the first eigenvalues:*

$$\lambda_1 \geq \lambda''_1 \tag{1.105}$$

(v) (*Continuity w. r. t. the coefficients*) *If* C^n *is a sequence of symmetric matrices in* $\mathbf{L}^\infty(\Omega)$, *and* $C^n \to C$ *in* $\mathbf{L}^\infty(\Omega)$, *then the corresponding eigenvalues converge, i. e., for any* $k \in \mathbb{N}^+$ *we have*

$$\lambda_k(-\Delta + C^n; \ \Omega) \to \lambda_k(-\Delta + C; \ \Omega)$$

(vi) *Let us consider an open subset* $\Omega' \subset \Omega$ *and set* $\lambda_1(\Omega') = \lambda_1(-\Delta + C; \ \Omega')$. *Then*

$$\lim_{meas(\Omega') \to 0} \lambda_1(\Omega') = +\infty$$

Proof. The proofs of all the claims, except for part b) of claim (iv), which depends on the fully coupling of the system, are exactly the same as those in Theorem 1.44 and we omit them. Note that the analogous of claim (iv) in Theorem 1.44 is just part a) of claim (iv) of Theorem 1.57, which can be proved in the same way.

To prove part b) of claim (iv), let us observe that if $W^1 = (w_1, \dots, w_m)$ is a first eigenfunction for the system (1.101), then it does not change sign in Ω, since the system is fully coupled. By normalizing it, we can assume that $(W^1, W^1)_{\mathbf{L}^2(\Omega)} = 1$ and $W^1 > 0$ in Ω. Denoting by Q'' the bilinear form corresponding to the matrix C'', since $w_i w_j \geq 0$ and $c_{ij} \geq c''_{ij}$ we get that

$$\lambda_1 = Q(W^1) = \int_\Omega \left[\sum_{i=1}^m |\nabla w_i|^2 + \sum_{i,j=1}^m c_{ij} w_i w_j \right] dx$$

$$\geq \int_{\Omega} \left[\sum_{i=1}^{m} |\nabla w_i|^2 + \sum_{i,j=1}^{m} c_{ij}'' w_i w_j \right] dx = Q''(W^1) \geq \lambda_1'' \tag{1.106}$$

$\square$

Remark 1.58. Note that (1.105) only holds for the first eigenvalue, unlike (1.104) which requires the ordering of the matrices and not of the single entries.

1.5.2 Weak maximum principle for cooperative systems

Let us turn back to the (possibly) nonsymmetric cooperative system (1.83):

$$\begin{cases} -\Delta U + D(x)U = F & \text{in } \Omega \\ U = 0 & \text{on } \partial\Omega \end{cases}$$

and assume that the matrix $D = (d_{ij})_{i,j=1}^{m}$ satisfies

$$d_{ij} \in L^{\infty}(\Omega), \quad d_{ij} \leq 0, \quad \text{whenever } i \neq j \tag{1.107}$$

Then we consider the symmetric system (1.95) associated to (1.83) and we denote by $\lambda_j^{(s)} = \lambda_j^{(s)}(-\Delta + D; \Omega)$ the eigenvalues of the corresponding linear operator $-\Delta + C$ (where $C = \frac{1}{2}(D + D^t)$ i. e. $c_{ij} = \frac{1}{2}(d_{ij} + d_{ji})$) and by $W_j^{(s)}$ the corresponding eigenfunctions.

The eigenvalues $\lambda_j^{(s)}$ will be called **symmetric eigenvalues** of the (nonsymmetric) operator $-\Delta + D$.

The bilinear form corresponding to the symmetric operator will be denoted by $B^s(U, \Phi)$, i. e., if $U, \Phi \in \mathbf{H}_0^1(\Omega)$ we set

$$B^s(U, \Phi) = \int_{\Omega} \left[\nabla U \cdot \nabla \Phi + C(U, \Phi) \right] = \int_{\Omega} \left[\sum_{i=1}^{m} \nabla u_i \cdot \nabla \phi_i + \sum_{i,j=1}^{m} c_{ij} u_i \phi_j \right]$$

As already remarked, the quadratic form (1.94) associated to the system (1.83) coincides with that associated to the system (1.95), i. e.,

$$Q(\Psi; \Omega) = \int_{\Omega} |\nabla\Psi|^2 + D(x)(\Psi, \Psi) = \int_{\Omega} |\nabla\Psi|^2 + C(x)(\Psi, \Psi) = B^s(\Psi, \Psi)$$

for $\Psi \in H_0^1(\Omega; \mathbb{R}^m)$.

Definition 1.59. We say that the maximum principle holds for the operator $-\Delta + D$ in an open set $\Omega' \subseteq \Omega$ if for any $U \in \mathbf{H}^1(\Omega')$ such that $U \leq 0$ on $\partial\Omega'$ (i. e., $U^+ \in \mathbf{H}_0^1(\Omega')$) and $-\Delta U + D(x)U \leq 0$ in Ω' (i. e., $\int \nabla U \cdot \nabla\Phi + D(x)(U, \Phi) \leq 0$ for any nonnegative $\Phi \in \mathbf{H}_0^1(\Omega')$) it holds that $U \leq 0$ a. e. in Ω.

Let us denote by $\lambda_j^{(s)}(\Omega'), j \in \mathbb{N}^+$, the sequence of the eigenvalues of the symmetric system (1.95) in an open set $\Omega' \subseteq \Omega$.

Theorem 1.60 (Sufficient condition for weak maximum principle). *Under the hypothesis* (1.107), *if* $\lambda_1^{(s)}(\Omega') > 0$, *then the maximum principle holds for the operator* $-\Delta + D$ *in* $\Omega' \subseteq \Omega$.

Proof. By the variational characterization of the first eigenvalue, we have

$$\lambda_1^{(s)} = \min_{\substack{V \in \mathbf{H}^1(\Omega') \\ V \neq 0}} R(V) > 0$$

so that $Q(V) = B^s(V, V) > 0$ for any $V \neq 0$ in $\mathbf{H}_0^1(\Omega')$.

Assume that $U \leq 0$ on $\partial\Omega'$ and $-\Delta U + D(x)U \leq 0$ in Ω'. Then, testing the equation with $U^+ = (u_1^+, \ldots, u_m^+)$, writing in the ith equation $u_j = u_j^+ - u_j^-$ for $i \neq j$, and recalling that $-c_{ij}u_i^+ u_j^- \geq 0$ if $i \neq j$, we obtain that $B^s(U^+, U^+) \leq 0$, which implies $U^+ \equiv 0$ in Ω'. $\qquad\square$

As an almost immediate consequence we get the following "classical" and "small measure" forms of the weak maximum principle (see [54, 97, 193, 201]).

Theorem 1.61. *The following statements hold:*
(i) *If* (1.84) *holds and D is a. e. nonnegative definite in* Ω' *then the maximum principle holds for* $-\Delta + D$ *in* Ω'.
(ii) *There exists* $\delta > 0$, *depending on D, such that for any subdomain* $\Omega' \subseteq \Omega$ *the maximum principle holds for* $-\Delta + D$ *in* $\Omega' \subseteq \Omega$ *provided* $|\Omega'| \leq \delta$.

Proof. (i) If the matrix D is nonnegative definite, then

$$Q(\Psi) = B^s(\Psi, \Psi) \geq \int_\Omega |\nabla\Psi|^2 > 0 \quad \text{for any } \Psi \in \mathbf{H}_0^1(\Omega') \setminus \{0\}.$$

Hence the first symmetric eigenvalue is positive, and by Theorem 1.60 we get (i).

(ii) It is a consequence of Theorem 1.60 and the property (vi) in Theorem 1.57. $\quad\square$

Remark 1.62. Obviously, the converse of Theorem 1.60 holds if $D = C$, i. e., if D is symmetric: if the maximum principle holds for $-\Delta + C$ in Ω' then $\lambda_1^{(s)}(\Omega') > 0$. Indeed if $\lambda_1^{(s)}(\Omega') \leq 0$, since the system is cooperative (and symmetric), there exists a corresponding nontrivial nonnegative first eigenfunction $\Phi_1 \geq 0$, $\Phi \neq 0$, and the maximum principle does not hold, since $-\Delta\Phi_1 + C\Phi_1 = \lambda_1\Phi_1 \leq 0$ in Ω', $\Phi_1 = 0$ on $\partial\Omega'$, while $\Phi_1 \geq 0$ and $\Phi_1 \neq 0$.

However, the converse of Theorem 1.60 is not true for general nonsymmetric systems. Roughly speaking, the reason is that there is an equivalence between the validity of the maximum principle for the operator $-\Delta + D$ and the positivity of another eigenvalue, the *principal eigenvalue* $\tilde{\lambda}_1$, whose definition we recall below, and the inequality $\tilde{\lambda}_1(\Omega') \geq \lambda_1^{(s)}(\Omega')$, which can be strict, holds.

Definition 1.63. The **principal eigenvalue** of the operator $-\Delta + D$ in an open set $\Omega' \subseteq \Omega$ is defined as

$$\tilde{\lambda}_1(\Omega') = \sup\{\lambda \in \mathbb{R} : \exists \Psi \in W^{2,N}_{\mathrm{loc}}(\Omega'; \mathbb{R}^m) \text{ s.t.}$$
$$\Psi > 0, -\Delta\Psi + D(x)\Psi - \lambda\Psi \geq 0 \text{ in } \Omega'\} \tag{1.108}$$

Let us recall some of the properties of the principal eigenvalue (see [54] and the references therein, and also [32] for the case of scalar nonlinear equations for a discussion and a more general framework).

Theorem 1.64. *Assume that the system* (1.83) *is fully coupled in an open set* $\Omega' \subseteq \Omega$. *Then:*

(i) *there exists a positive eigenfunction* $\Psi_1 \in W^{2,N}_{\mathrm{loc}}(\Omega'; \mathbb{R}^m)$ *which satisfies*

$$\begin{cases} -\Delta\Psi_1 + D(x)\Psi_1 = \tilde{\lambda}_1(\Omega')\Psi_1 & \text{in } \Omega' \\ \Psi_1 > 0 & \text{in } \Omega' \\ \Psi_1 = 0 & \text{on } \partial\Omega' \end{cases} \tag{1.109}$$

Moreover, the principal eigenvalue is simple, i. e., any function that satisfy (1.109) *must be a multiple of* Ψ_1

(ii) *the maximum principle holds for the operator* $-\Delta + D$ *in* Ω' *if and only if* $\tilde{\lambda}_1(\Omega') > 0$

(iii) *if there exists* $\Psi \in W^{2,N}_{\mathrm{loc}}(\Omega'; \mathbb{R}^m)$ *such that* $\Psi > 0$ *and* $-\Delta\Psi + D(x)\Psi \geq 0$ *in* Ω', *then either* $\tilde{\lambda}_1(\Omega') > 0$ *or* $\tilde{\lambda}_1(\Omega') = 0$ *and* $\Psi = c\Psi_1$ *for some* $c > 0$

(iv) $\tilde{\lambda}_1(\Omega') \geq \lambda_1^{(s)}(\Omega')$, *with equality if and only if* Ψ_1 *is also the first eigenfunction of the symmetric operator* $-\Delta + C$ *in* Ω', $C = \frac{1}{2}(D + D^t)$.

If this is the case, the equality $C(x)\Psi_1 = D(x)\Psi_1$ *holds and, if* $m = 2$, *this implies that* $d_{12} = d_{21}$.

Proof. We refer to [54] for the proofs of (i)–(iii).

For what concerns (iv), we observe that testing (1.109) with Ψ_1 and recalling that the quadratic form associated to the operator $-\Delta + D$ coincides with the quadratic form associated to the symmetric operator $-\Delta + C$, $C = \frac{1}{2}(D + D^t)$, we obtain that

$$\tilde{\lambda}_1(\Omega') = \frac{Q(\Psi_1)}{(\Psi_1, \Psi_1)_{\mathbf{L}^2(\Omega)}} \geq \lambda_1^{(s)}(\Omega')$$

with equality if and only if Ψ_1 is the first symmetric eigenfunction, by Proposition 1.56(v). If this is the case, since Ψ_1 satisfies the system (1.109) and the system (1.83), the equality $D(x)\Psi_1 = C(x)\Psi_1$ follows. Since Ψ_1 is positive, if $m = 2$ we deduce that $d_{12} = d_{21}$. $\qquad\square$

1.5.3 Comparison principles for semilinear elliptic systems

We discuss now some weak and strong comparison principles for semilinear elliptic systems, analogous to those of Theorem 1.21 and Theorem 1.31 holding for equations. Let us consider a semilinear elliptic system of the type

$$\begin{cases} -\Delta U = F(x, U) & \text{in } \Omega \\ U = 0 & \text{on } \partial\Omega \end{cases} \tag{1.110}$$

i. e.

$$\begin{cases} -\Delta u_1 = f_1(x, u_1, \dots, u_m) & \text{in } \Omega \\ \quad \dots \dots & \quad \dots \\ -\Delta u_m = f_m(x, u_1, \dots, u_m) & \text{in } \Omega \\ u_1 = \cdots = u_m = 0 & \text{on } \partial\Omega \end{cases}$$

for the unknown vector valued function $U = (u_1, \dots, u_m) : \Omega \to \mathbb{R}^m$, where Ω is a bounded domain in $\mathbb{R}^N$ and $F = (f_1, \dots, f_m) : \overline{\Omega} \times \mathbb{R}^m \to \mathbb{R}^m$ is a C^1 function.

A weak solution of (1.110) is a function $U \in \mathbf{H}_0^1(\Omega)$ such that the function $x \mapsto F(x, U(x))$ belongs to $\mathbf{L}^q(\Omega)$, with $q > 1$ if $N = 2$, $q = \frac{2N}{N+2}$ if $N \geq 3$ (note that $\frac{2N}{N+2}$ is the conjugate exponent of the critical Sobolev exponent $2^* = \frac{2N}{N-2}$) and (see the notation about systems at the beginning of the section)

$$\int_\Omega \nabla U \cdot \nabla \Phi \, dx = \int_\Omega F(x, U) \cdot \Phi \, dx \tag{1.111}$$

for any $\Phi \in \mathbf{H}_0^1(\Omega)$.

If $U, V \in \mathbf{H}^1(\Omega)$, we write $U \leq V$ on $\partial\Omega$, if the difference $U - V$ weakly satisfies the inequality $U - V \leq 0$ on $\partial\Omega$, i. e., if $(U - V)^+ \in \mathbf{H}_0^1(\Omega)$.

Moreover, we say that U satisfies in a weak sense the inequality

$$-\Delta U \geq (\leq) \, F(x, U) \quad \text{in } \Omega \tag{1.112}$$

if for any $i = 1, \dots, m$ the component u_i of U weakly satisfies

$$-\Delta u_i \geq (\leq) \, f_i(x, U) \quad \text{in } \Omega, \quad \text{i. e.,}$$

$$\int_\Omega \nabla u_i \cdot \nabla \varphi \, dx \geq (\leq) \int_\Omega f_i(x, U)\varphi \, dx \tag{1.113}$$

for any $\varphi \in H_0^1(\Omega)$ with $\varphi \geq 0$ in Ω. This is equivalent to require that

$$\int_\Omega \nabla U \cdot \nabla \Phi \, dx \geq (\leq) \int_\Omega F(x, U) \cdot \Phi \, dx \tag{1.114}$$

for any $\Phi \in \mathbf{H}_0^1(\Omega)$ with $\Phi \geq 0$ in Ω.

Definition 1.65. We say that the system (1.110) is

- **cooperative or weakly coupled** in an open set $\Omega' \subseteq \Omega$ if

$$\frac{\partial f_i}{\partial s_j}(x, s_1, \ldots, s_m) \geq 0 \quad \text{for every } (x, s_1, \ldots, s_m) \in \Omega' \times \mathbb{R}^m \tag{1.115}$$

and every $i, j = 1, \ldots, m$ with $i \neq j$.

- **fully coupled** in an open set $\Omega' \subseteq \Omega$ along $U \in \mathbf{H}_0^1(\Omega) \cap C^0(\Omega; \mathbb{R}^m)$ if it is cooperative in Ω' and in addition $\forall I, J \subset \{1, \ldots, m\}$ such that $I \neq \emptyset, J \neq \emptyset, I \cap J = \emptyset, I \cup J = \{1, \ldots, m\}$ there exist $i_0 \in I, j_0 \in J$ such that

$$\text{meas}\left(\left\{x \in \Omega' : \frac{\partial f_{i_0}}{\partial s_{j_0}}(x, U(x)) > 0\right\}\right) > 0 \tag{1.116}$$

Theorem 1.66 (Weak comparison principle in small domains for systems). *Let Ω be a domain in $\mathbb{R}^N$, $F : \Omega \times \mathbb{R}^m \to \mathbb{R}^m$ a C^1 function and assume that (1.115) holds. Let $A > 0$ and $U, V \in \mathbf{H}^1(\Omega) \cap L^\infty(\Omega)$ such that*

$$\|U\|_{L^\infty(\Omega)} \leq A, \|V\|_{L^\infty(\Omega)} \leq A$$

Then there exists $\delta > 0$, depending on F and A such that the following holds: if $\Omega' \subseteq \Omega$ is a bounded subdomain of Ω, $\text{meas}_N([u > v] \cap \Omega') < \delta$ and

$$\begin{cases} -\Delta U \leq F(x, U), -\Delta V \geq F(x, V) & \text{in } \Omega' \\ U \leq V & \text{on } \partial\Omega' \end{cases} \tag{1.117}$$

then $U \leq V$ in Ω'.

Proof. Assume that $\Omega' \subseteq \Omega$ and U, V satisfy (1.117). Then, for any $i = 1, \ldots, m$:

$$f_i(x, U(x)) - f_i(x, V(x))$$

$$= \sum_{j=1}^m \int_0^1 \frac{\partial f_i}{\partial s_j}(|x|, tU(x) + (1-t)V(x)) \, dt (u_j(x) - v_j(x))$$

Let us define the matrix $B(x) = (b_{ij}(x))_{i,j=1}^m$, where

$$b_{ij}(x) = -\int_0^1 \frac{\partial f_i}{\partial s_j}[x, tU(x) + (1-t)V(x)] \, dt \tag{1.118}$$

Then we have that the function $W = U - V = (w_1, \ldots, w_m)$ satisfies

$$\begin{cases} -\Delta W + B(x)W \leq 0 & \text{in } \Omega' \\ W \leq 0 & \text{on } \partial\Omega' \end{cases} \tag{1.119}$$

and if $i \neq j$ then $b_{ij}(x) \leq 0$ by (1.115).

Hence the result is a consequence of Theorem 1.61. $\qquad\square$

Theorem 1.67 (Strong comparison principle for systems). *Let Ω be a (bounded or unbounded) domain in $\mathbb{R}^N$, and let $U, V \in C^1(\Omega)$ weakly satisfy*

$$\begin{cases} -\Delta U \leq F(x, U); \ -\Delta v \geq F(x, V) & \text{in } \Omega \\ U \leq V & \text{in } \Omega \end{cases} \tag{1.120}$$

where $F(x, U) : \Omega \times \mathbb{R}^m \to \mathbb{R}^m$ is a C^1 function and (1.115) holds.

1. *For every $i \in \{1, \ldots, m\}$ the following holds: either $u_i \equiv v_i$ in Ω or $u_i < v_i$ in Ω and, in the latter case, if $u_i, v_i \in C^1(\overline{\Omega})$, $u_i(x_0) = v_i(x_0)$ at a point $x_0 \in \partial\Omega$ where the interior sphere condition is satisfied then $\frac{\partial u_i}{\partial s}(x_0) < \frac{\partial v_i}{\partial s}(x_0)$ for any inward directional derivative.*

2. *If moreover $U \in C^1(\Omega; \mathbb{R}^m)$ is a solution of (1.110) and the system is fully coupled along U in Ω (i. e., also (1.116) with $\Omega' = \Omega$ holds) then either $U \equiv V$ in Ω or $U < V$ in Ω (i. e., the same alternative holds for any component u_i).*
   In the latter case, assume that $U, V \in C^1(\overline{\Omega})$ and let $x_0 \in \partial\Omega$ a point where $U(x_0) = V(x_0)$ and the interior sphere condition is satisfied. Then $\frac{\partial U}{\partial s}(x_0) < \frac{\partial V}{\partial s}(x_0)$ for any inward directional derivative.

Proof. The function $W = U - V = (w_1, \ldots, w_m)$ satisfies

$$\begin{cases} -\Delta W + B(x)W \leq 0 & \text{in } \Omega \\ W \leq 0 & \text{on } \partial\Omega \end{cases} \tag{1.121}$$

where

$$b_{ij}(x) = -\int_0^1 \frac{\partial f_i}{\partial s_j}\left[x, tU(x) + (1 - t)V(x)\right] dt$$

as in the previous theorem. Moreover, the linear system associated to the matrix B is cooperative because if $i \neq j$ then $b_{ij}(x) \leq 0$ by (1.115).

If $U \in C^1(\overline{\Omega}; \mathbb{R}^m)$ is a solution of (1.110) and the system is fully coupled along U, then the linear system associated to the matrix B is fully coupled as well. Indeed, if $x \in \Omega$, $i_0 \neq j_0$ and $\frac{\partial f_{i_0}}{\partial s_{j_0}}(x, U(x)) > 0$ then since $\frac{\partial f_{i_0}}{\partial s_{j_0}}(x, S) \geq 0$ for every $S \in \mathbb{R}^m$, we get that $b_{i_0 j_0}(x) = -\int_0^1 \frac{\partial f_{i_0}}{\partial s_{j_0}}[x, tU(x) + (1 - t)V(x)] dt < 0$.

Thus it is enough to apply Theorem 1.53. $\qquad\square$

2 Introduction to Morse theory

In this chapter, we introduce Morse theory both in finite and infinite dimension.

In the first section, we consider the case of functions defined on finite dimensional manifolds for which the main references are the books of M. Morse, J. Milnor and R. Bott [39, 177, 180]. We present in detail the Morse lemma, the deformation theorems and the Morse inequalities. The chief example of the height function on the torus will illustrate the meaning of the main results.

In the second section, we will describe the basic Morse theory on infinite dimensional manifolds following the book of K. C. Chang [60]. We first indicate how to study the local behavior of a functional f near a critical point and then how to study its global behavior, obtaining some kind of Morse inequalities. Finally, we show that a mountain pass critical point of a functional f in a Banach space has Morse index less than or equal to one.

2.1 Morse theory on finite dimensional manifolds

2.1.1 Introduction

Let M be a smooth, N-dimensional manifold and $f: M \longrightarrow \mathbb{R}$ a C^∞-function. A point $\bar{x} \in M$ is called a **critical point** of f if, for a given local coordinate system $(x_1, \ldots, x_N)$ in a neighborhood U of $\bar{x}$, it is:

$$\frac{\partial f}{\partial x_i}(\bar{x}) = 0, \quad \forall i = 1, \ldots, N.$$

In other words, this means that the induced derivative map

$$f_*: T_{\bar{x}} M \longrightarrow T_{f(\bar{x})} \mathbb{R}$$

on the tangent space at a point $\bar{x}$ is zero.

A number $c \in \mathbb{R}$ is said to be a **critical value** of f if the set $f^{-1}(c)$ contains at least one critical point; if this is not the case, we say that c is a **regular value.**

Morse theory aims to establish a relationship between the critical points of f and the topology of the manifold M, as expressed, for example, by its homology and cohomology groups. One way of doing this is to study the sublevels of f to detect a change of their topology when a critical value is crossed.

Let us illustrate this idea by a few examples.

Example 2.1. Let $f: \mathbb{R} \longrightarrow \mathbb{R}$ be a smooth function with only two critical values: $a_1 < a_2$, corresponding to a local minimum and a local maximum, respectively, as in Figure 2.1.

https://doi.org/10.1515/9783110538243-002

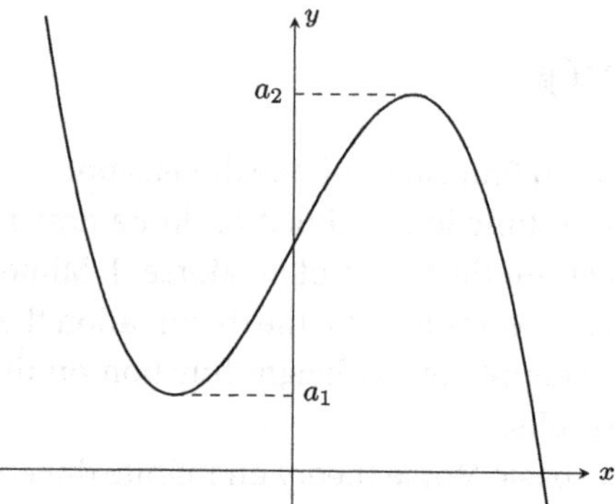

Figure 2.1: A function with two critical values.

Consider the sublevels:

$$f_a = \{\, x \in \mathbb{R} : f(x) \leq a \,\}, \quad a \in \mathbb{R}.$$

If $a \leq a_1$, then f_a is a half-line; if $a \in (a_1, a_2)$, then f_a is the union of a segment and a half-line; if $a > a_2$, then f_a is again a half-line. This shows that the topology of f_a changes while crossing the critical values.

Example 2.2. Let $f : \mathbb{R}^2 \longrightarrow \mathbb{R}$ the function defined as $f(x,y) = y^2 - x^2$. Obviously, it has $(0,0)$ as only critical point whose corresponding critical value is zero.

Let us consider the sublevels:

$$f_{-\epsilon} = \left\{\, (x,y) : y^2 - x^2 \leq -\epsilon \,\right\}, \quad f_{\epsilon} = \left\{\, (x,y) : y^2 - x^2 \leq \epsilon \,\right\}, \quad \epsilon > 0.$$

As we can see in Figure 2.2, they have different topological properties. Indeed, f_ϵ is a connected set, while $f_{-\epsilon}$ is not.

Example 2.3. Let $T \subset \mathbb{R}^3$ be a torus laying on the plane (x,y), using a (x,y,z) system of coordinates as below.

We consider the height function: $f(x,y,z) = z$, which has four critical points P_1, P_2, P_3, P_4 with corresponding critical values $0 = z_1 < z_2 < z_3 < z_4$ (see Figure 2.3).

The sublevels:

$$f_i = \{\, (x,y,z) \in T : z \leq a < z_i \,\}, \quad i = 1, \dots, 4$$

and

$$f_5 = \{\, (x,y,z) \in T : z \leq a, \, a > z_4 \,\}$$

are described in Figure 2.4.

It is clear that all these sets do not have the same homotopy type. Indeed f_1 is the empty set, f_2 is homeomorphic to a disk, f_3 is homeomorphic to a cylinder, f_4 is homeomorphic to a compact manifold of genus one having a circle as boundary and f_5 is the full torus.

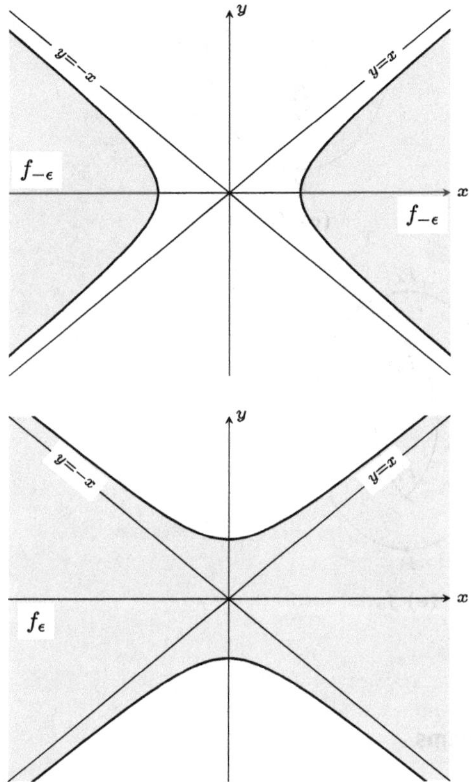

Figure 2.2: Sublevels of the function $f(x,y) = y^2 - x^2$.

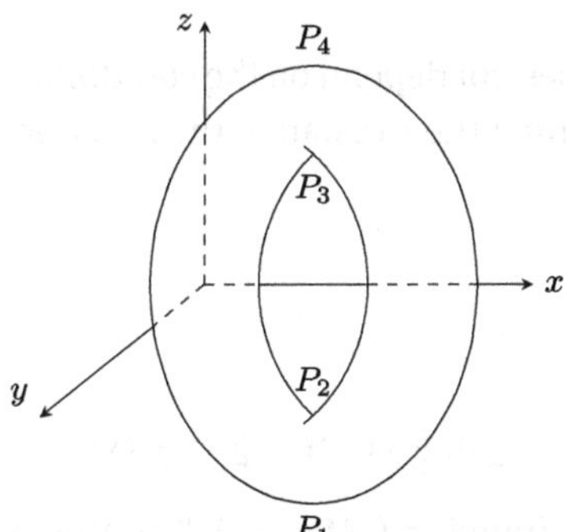

Figure 2.3: Critical points of the function $f(x,y,z) = z$ on the torus.

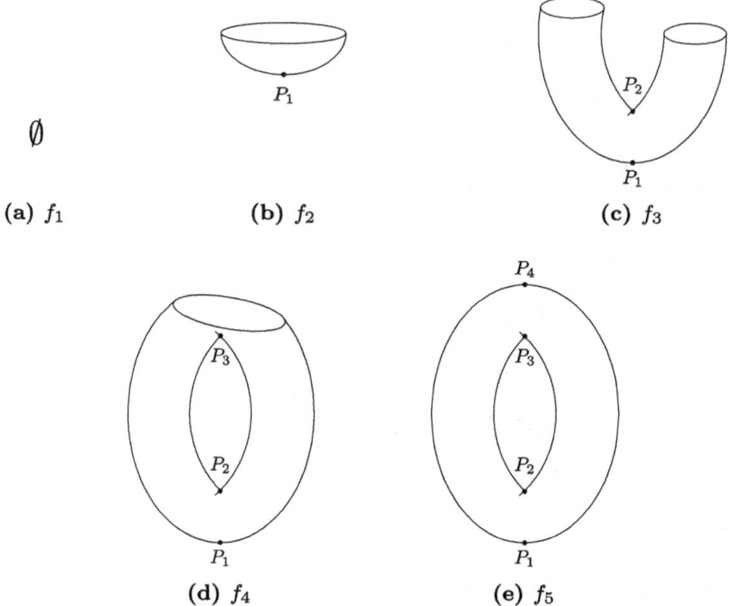

Figure 2.4: Sublevels of the function $f(x, y, z) = z$.

2.1.2 Morse lemma and deformation theorems

A critical point $\bar{x}$ of a function f on a manifold M, as in the previous section, is called **nondegenerate** if the Hessian matrix

$$\left(\frac{\partial^2 f}{\partial x_i \partial x_j}(\bar{x}) \right), \quad i, j = 1, \ldots, N,$$

in local coordinates, is nonsingular. This definition does not depend on the coordinate system, as follows also from the intrinsic way of defining the Hessian of the function f as a symmetric bilinear form:

$$f_{**}(v, w) = v(w(f)) = \sum_{i,j} a_i b_j \frac{\partial^2 f}{\partial x_i \partial x_j}(\bar{x})$$

where $v, w \in TM_{\bar{x}}$ and, in a local coordinate system, $v = \sum a_i \frac{\partial}{\partial x_i}(\bar{x})$, $w = \sum b_j \frac{\partial}{\partial x_j}(\bar{x})$.

Definition 2.4. Let $\bar{x}$ be a critical point of the smooth function $f: M \longrightarrow \mathbb{R}$. The Morse index of $\bar{x}$ is the maximal dimension of a subspace of $T_{\bar{x}}M$ on which f_{**} is negative definite, i. e., it is the number of the negative eigenvalues of the hessian matrix $\frac{\partial^2 f}{\partial x_i \partial x_j}$.

We are going to show that the behavior of f at the critical point $\bar{x}$ can be described by its Morse index.

Lemma 2.5 (Morse lemma). *Let M be a smooth, N-dimensional manifold and let $\bar{x}$ be a nondegenerate critical point of a smooth function $f\colon M \longrightarrow \mathbb{R}$. If the Morse index of $\bar{x}$ is k, then there exists a local coordinate system $(y_1,\dots,y_N)$ in a neighborhood U of $\bar{x}$, i.e., a local chart $\varphi\colon U \longrightarrow V$, where V is a neighborhood of $0 \in \mathbb{R}^N$, such that*

$$\begin{cases} \varphi^{-1}(0) = \bar{x} \\ f(\varphi^{-1}(y_1,\dots,y_n)) = f(\bar{x}) - \sum_{i=1}^{k} y_i^2 + \sum_{i=k+1}^{N} y_i^2 \end{cases} \tag{2.1}$$

Proof. Without loss of generality, we can assume that $\bar{x}$ is the origin $0 \in \mathbb{R}^n$ and that $f(\bar{x}) = f(0) = 0$. We have

$$f(\bar{x}) = \int_0^1 \sum_{i=1}^{N} \frac{\partial f}{\partial x_i}(tx_1,\dots,tx_n)x_i \, dt$$

$$= \sum_{i,j=1}^{N} \left[\int_0^1 \int_0^1 \frac{\partial^2 f}{\partial x_i \partial x_j}(stx) \, ds dt \right] x_i x_j$$

$$= \sum_{i,j=1}^{N} h_{ij}(x)x_i x_j.$$

We can assume that $h_{ij} = h_{ji}$, otherwise we take $\bar{h}_{ij} = \frac{1}{2}(h_{ij} + h_{ji})$. Then, using the diagonalization of quadratic forms and the fact that the matrix $(\frac{\partial^2 f}{\partial x_i \partial x_j}(0))$ is nonsingular, we can construct a nonsingular change of coordinates which gives (2.1). Indeed, after a linear change of coordinates which makes $(\frac{\partial^2 f}{\partial x_i \partial x_j}(0))$ diagonal, by the nondegeneracy hypothesis, we may assume that $h_{11}(0) \neq 0$ so that, in a neighborhood of the origin

$$f(x) = h_{11}\left[x_1^2 + \sum_{(i,j)\neq(1,1)} \frac{h_{ij}(x)}{h_{11}} x_i x_j \right].$$

Then, introducing the coordinates:

$$y_1 = \sqrt{|h_{11}|}\left[x_1 + \sum_{r\neq 1} \frac{h_{1r}(x)}{h_{11}} x_r \right], \qquad y_r = x_r \text{ for } r \neq 1$$

we have that, by the inverse function theorem, this is a regular change of coordinates and

$$f = y_1^2 + \sum_{i,j>1} y_i y_j H'_{i,j}(y_1,\dots,y_n).$$

Repeating this procedure, or arguing by induction, we get (2.1). $\qquad\square$

Remark 2.6. The proof can be done for functions $f \in C^2(M)$ (see [191]). Hence, as a consequence of (2.1) we get that the nondegenerate critical points of a C^2-function f are isolated. This can also be deduced directly by the nonsingularity of the matrix $\frac{\partial^2 f}{\partial x_i \partial x_j}$ which implies that the vector-valued map $\nabla f = (\frac{\partial f}{\partial x_1}, \ldots, \frac{\partial f}{\partial x_N})$ is locally invertible at $\bar{x}$, and hence vanishes only in $\bar{x}$.

The functions $f \in C^2(M)$ which have only nondegenerate critical points are usually called **Morse functions** and it can be proved that they are an open dense set in the space $C^2(M)$ with respect to the uniform topology (see [40]).

Next, we introduce some deformation theorems suitable to study the sublevels of a function $f: M \longrightarrow \mathbb{R}$.

A 1-**parameter group of diffeomorphisms** of a manifold M is a C^1-map $\varphi: \mathbb{R} \times M \longrightarrow M$ such that:

(i) $\forall t \in M$ the map $\varphi_t: M \longrightarrow M$ defined by $\varphi_t(x) = \varphi(t, x)$ is a diffeomorphism of M onto itself;

(ii) $\varphi_{t+s} = \varphi_t \circ \varphi_s \ \forall t, s \in \mathbb{R}$.

We recall that a vector field on M is a map $\underline{v}: M \longrightarrow TM$ such that $\pi \circ \underline{v}: M \longrightarrow M$ is the identity where $\pi: TM \longrightarrow M$ is the projection of the tangent bundle TM.

For a given vector field $\underline{v}$ on M, we consider the differential equation

$$\frac{\partial \eta}{\partial t} = \underline{v}(\eta(t)) \tag{2.2}$$

together with the initial condition:

$$\eta(0) = x, \quad x \in M. \tag{2.3}$$

Applying the theory of ordinary differential equation, we have the following.

Proposition 2.7. *A C^1-vector field $\underline{v}$ on M, which vanishes outside of a compact set contained in M, generates a unique 1-parameter group of diffeomorphisms $\varphi_t(x)$ of M, where $\varphi_t(x)$ is the solution of* (2.2), (2.3) *at time t.*

This allows to prove the following.

Theorem 2.8. *Let f be a smooth real valued function on a manifold M and $a, b \in \mathbb{R}$, $a < b$. Assume that the set $f^{-1}([a, b])$ is compact and does not contain critical points of f. Then the sublevel $f_a = \{ x \in M : f(x) \le a \}$ is a **deformation retract** of the sublevel $f_b = \{ x \in M : f(x) \le b \}$, i. e., there exists a map (**retraction**) $r: f_b \longrightarrow f_a$ such that $i \circ r$ is homotopic to the identity map of f_b, where $i: f_a \longrightarrow f_b$ is the inclusion map.*

Proof. It is well known (see, e. g., [208]) that using a partition of unity it is possible to construct on every smooth compact finite dimensional manifold a Riemannian metric, namely a smooth positive-definite symmetric bilinear form on tangent vectors. This clearly induces a scalar product on $T_x M$, for every $x \in M$ for which we will use the notation $\langle v, w \rangle$ for two tangent vectors v and w.

We consider the vector field $\operatorname{grad} f$ on M which is characterized by the identity $\langle v, \operatorname{grad} f \rangle$ = derivative of f in the direction of the vector v.

Obviously, this vector field vanishes only at the critical points of f. Then it is well-defined the map $\rho : M \longrightarrow \mathbb{R}$, $\rho = \frac{1}{\langle \operatorname{grad} f, \operatorname{grad} f \rangle}$ in $f^{-1}([a,b])$ and $\rho \equiv 0$ outside of a compact neighborhood of $f^{-1}([a,b])$.

The vector field $v = \rho \operatorname{grad} f$ satisfies the hypothesis of Proposition 2.7 and hence generates a 1-parameter group of diffeomorphisms $\varphi_t : M \longrightarrow M$.

We have

$$\frac{df(\varphi_t(x))}{dt} = \left\langle \frac{d\varphi_t(x)}{dt}, \operatorname{grad} f \right\rangle = \langle x, \operatorname{grad} f \rangle = 1$$

$\forall x \in M$ and $\forall t \in \mathbb{R}$ such that $\varphi_t(x) \in f^{-1}([a,b])$. Hence the map $t \longmapsto f(\varphi_t(x))$ is linear as long as $f(\varphi_t(x))$ belongs to $[a,b]$.

Moreover, the diffeomorphism $\varphi_{b-a} : M \longrightarrow M$ carries f_a onto f_b. Finally, the family of maps $r_t : f_b \longrightarrow f_b$ defined by

$$r_t(x) = \begin{cases} x & \text{if } f(x) \le a \\ \varphi_{t(a-f(x))}(x) & \text{if } f(x) \in [a,b] \end{cases}$$

gives the desired retraction of f_b onto f_a since r_0 is the identity, while $r_1 = f_a$. $\qquad\square$

Before passing to the next theorem, we need some definitions.

We denote by B_k a k-dimensional ball centered at the origin: $B_k = \{x \in \mathbb{R}^n : \sum_{i=1}^{k} x_i^2 < 1 \text{ and } x_{k+1} = \cdots = x_N = 0\}$, in the context of this section it will be called a k-**cell**. The operation of **attaching a k-cell to a topological space** Y can be defined as follows.

We consider a continuous function $g : \overline{\partial B_k} \longrightarrow X$ which is a homeomorphism onto its image. The topological space obtained by taking the union of Y and the closed k-cell $\overline{B_k}$ with the equivalence relation which identifies any $x \in \partial B_k$ with its image $g(x) \in Y$ will be denoted by $Y \cup_g B_k$ (Y with a k-cell attached).

The next theorem describes the change of topology of the sublevels of f while crossing a nondegenerate critical point.

Theorem 2.9. *Let $f : M \longrightarrow \mathbb{R}$ be a smooth function and $\bar{x}$ a nondegenerate critical point of f with Morse index k and corresponding critical value $c = f(\bar{x})$. If there exists $\epsilon > 0$ such that $f^{-1}([c - \epsilon, c + \epsilon])$ is compact and has no critical points other than $\bar{x}$, then $f_{c+\epsilon}$ has the homotopy type of $f_{c-\epsilon}$ with a k-cell attached.*

Proof. By the Morse lemma (Lemma 2.5), there exists a neighborhood U of $\bar{x}$ and a local system of coordinates $(y_1, \ldots, y_n)$ such that

$$f(x) = f(\bar{x}) - \sum_{i=1}^{k} y_i^2 + \sum_{i=k+1}^{N} y_i^2,$$

$\forall x \in U$ and $y_1(\bar{x}) = \cdots = y_N(\bar{x}) = 0$.

By assumption, we can choose $\epsilon > 0$ such that $f^{-1}([c - \epsilon, c + \epsilon])$ is compact and contains only one critical point, namely $\bar{x}$. Moreover, ϵ can be taken so small that U is diffeomorphic to a neighborhood of the origin in $\mathbb{R}^N$ and contains the closed ball $\overline{B(\epsilon)} = \{(y_1, \ldots, y_n) : \sum y_i^2 \leq 2\epsilon\}$. Let B_k be the cell $B_k = \{(y_1, \ldots, y_n) : \sum_{i=1}^{k} y_i^2 \leq \epsilon$ and $y_{k+1} = \cdots = y_N = 0\}$.

The resulting situation in the neighborhood of $\bar{x}$ can be illustrated as in Figure 2.5.

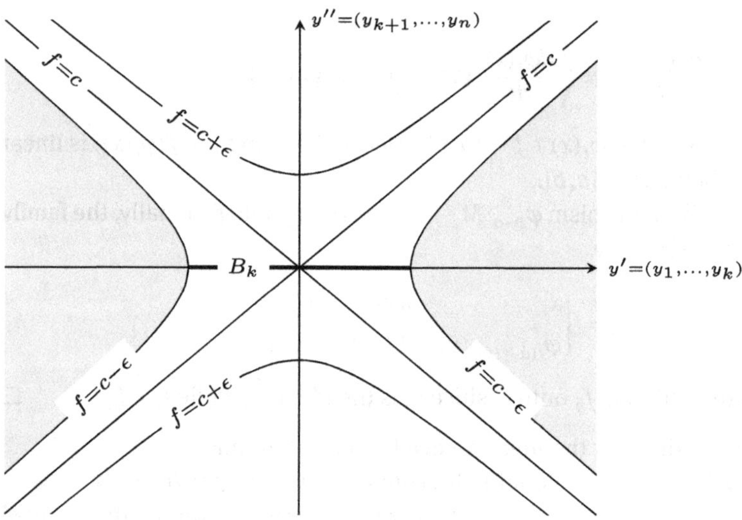

Figure 2.5: Attaching a k-cell to a sublevel.

The intersection $B_k \cap f_{c-\epsilon}$ is ∂B_k so that $f_{c-\epsilon}$ has a k-cell attached. We would like to prove that $f_{c-\epsilon} \cup_g B_k$ (g is the gluing map) is a deformation retract of $f_{c+\epsilon}$.

To this aim, we construct a new function $F \colon M \longrightarrow \mathbb{R}$ which coincides with f outside the neighborhood U and in U takes the form

$$F = f - \mu\left(\sum_{i=1}^{k} y_i^2 + 2 \sum_{i=k+1}^{N} y_i^2 \right)$$

where $\mu \colon \mathbb{R} \longrightarrow [0, +\infty)$ is a smooth function such that

$$\mu(0) > \epsilon, \quad \mu(r) = 0 \quad \text{for } r \geq 2\epsilon \quad \text{and} \quad -1 < \mu'(r) \leq 0 \quad \forall r \in \mathbb{R}.$$

Observe that the sublevel $f_{c+\epsilon}$ is the same as the sublevel $F_{c+\epsilon}$. Indeed outside the ellipsoid $E = \{\sum_{i=1}^{k} y_i^2 + 2 \sum_{i=k+1}^{n} y_i^2 \leq 2\epsilon\}$ the functions f and F coincide, since $\mu = 0$. In the interior of E, we have

$$F \leq f = c - \sum_{i=1}^{k} y_i^2 + 2 \sum_{i=k+1}^{N} y_i^2 \leq c + \frac{1}{2} \sum_{i=1}^{k} y_i^2 + 2 \sum_{i=k+1}^{N} y_i^2 \leq c + \epsilon.$$

Moreover, the functions F and f have the same critical points; indeed, setting $\xi = \sum_{i=1}^{k} y_i^2$ and $\eta = \sum_{i=k+1}^{n} y_i^2$ we have that $f = c - \xi + \eta$. Consequently,

$$\frac{\partial F}{\partial \xi} = -1 - \mu'(\xi + 2\eta) < 0, \qquad \frac{\partial F}{\partial \eta} = 1 - 2\mu'(\xi + 2\eta) \geq 1$$

so that

$$dF = \frac{\partial F}{\partial \xi}d\xi + \frac{\partial F}{\partial \eta}d\eta.$$

Since $d\xi$ and $d\eta$ are simultaneously zero only at the origin, it follows that F has no critical points in U other than the origin.

We know that $f_{c+\epsilon} = F_{c+\epsilon}$ and $F \leq f$; therefore $F^{-1}([c-\epsilon, c+\epsilon]) \subset f^{-1}([c-\epsilon, c+\epsilon])$. Moreover, $F(\bar{x}) = c - \mu(0) < c - \epsilon$. As a consequence, $F^{-1}([c-\epsilon, c+\epsilon])$ is compact and does not contain any critical point of F. Thus Theorem 2.9 implies that $F_{c-\epsilon}$ is a deformation retract of $F_{c+\epsilon} = f_{c+\epsilon}$, and hence the assertion will be proved if we show that $F_{c-\epsilon}$ has the same homotopy type of $f_{c-\epsilon} \cup_g B_k$.

Note that $F_{c-\epsilon} = f_{c-\epsilon} \cup H$, with $H = \overline{F_{c-\epsilon} \setminus f_{c-\epsilon}}$ as illustrated in Figure 2.6.

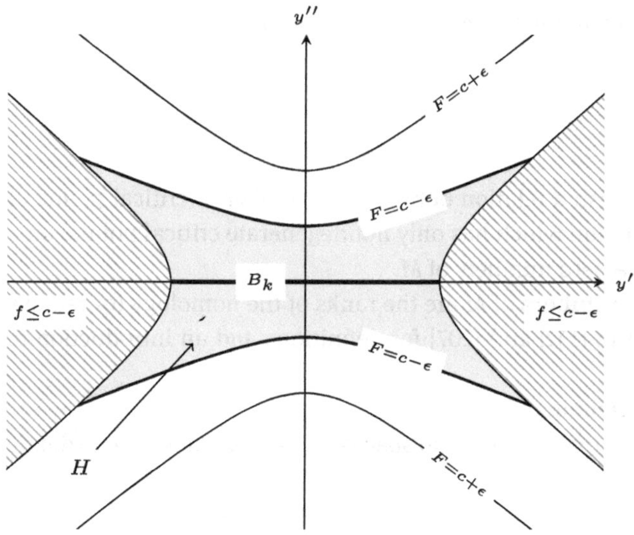

Figure 2.6: Attaching a k-cell to a sublevel.

We define a retraction r_t of $f_{c-\epsilon} \cup H$ into $f_{c-\epsilon} \cup_g B_k$ in the following way:
(i)

$$r_t(y_1, \ldots, y_N) = (y_1, \ldots, y_k, ty_{k+1}, \ldots, ty_N) \quad \text{if } \xi \leq \epsilon.$$

Hence r_1 is the identity map, r_0 maps the set $\{\xi \leq \epsilon\}$ into B_k and $r_t(F_{c-\epsilon}) \subset F_{c-\epsilon}$ since $\frac{\partial F}{\partial \eta} > 0$.

(ii) $r_t(y_1, \ldots, y_N) = (y_1, \ldots, y_k, s_t y_{k+1}, \ldots, s_t y_N)$ if $\epsilon \leq \xi \leq \eta + \epsilon$,

$$\text{where} \quad s_t = t + (1 - t)\left(\frac{\xi - \epsilon}{\eta}\right)^{\frac{1}{2}}.$$

Hence r_1 is the identity map, r_0 maps the set $\{\epsilon \leq \xi \leq \eta + \epsilon\}$ into the set $f^{-1}(c - \epsilon)$ and r_t is defined as in (i) for $\xi = \epsilon$.

(iii) $r_t(y_1, \ldots, y_N) = (y_1, \ldots, y_N)$ if $\xi \geq \eta + \epsilon$, i. e. in $f_{c-\epsilon}$.

It is easy to see that r_t is the same as in (ii) if $\xi = \eta + \epsilon$.

By definition, r_t maps $f_{c-\epsilon} \cup H$ into $f_{c-\epsilon} \cup_g B_k$ and the assertion is proved. $\qquad \square$

In the same way, it could be proved that if the function f has q nondegenerate critical points with c as corresponding critical value and with Morse indices $k_1, \ldots, k_q$ then the sublevel $f_{c+\epsilon}$ has the same homotopy type as $f_{c-\epsilon} \cup_{g_1} B_{k_1} \cup \cdots \cup_{g_q} B_{k_q}$. This indicates that, starting with the sublevel f_a, for $a < \min_M f$, it is possible somehow "*to construct*" the manifold M attaching cells as the value a growths crossing critical values of f, until when a reaches the maximum of f. This is the case of the torus and the height function considered in the Example 2.3, as shown in Figure 2.7.

2.1.3 Morse inequalities

In this section, we will establish a relation between the number of critical points of a Morse function (i. e., a function which has only nondegenerate critical points) on a compact manifold M and the Betti numbers of M.

We recall that the **Betti numbers** of M are the ranks of the homology (or relative homology) groups. We refer to [132] and [207] for definitions and an introduction to homology theory.

Let us start with some preliminaries.

We consider a triple of compact topological spaces $X \supset Y \supset Z$. Then we can define an exact homology sequence:

$$\longrightarrow H_{k+1}(X, Y) \xrightarrow{\partial_k} H_k(Y, Z) \longrightarrow H_k(X, Z) \longrightarrow$$

$$\longrightarrow H_k(X, Y) \xrightarrow{\partial_{k-1}} \cdots \longrightarrow H_0(X, Y) \longrightarrow 0 \qquad (2.4)$$

where $H_i(\cdot, \cdot)$ are the relative homology groups (see [207], Chapter IV) over a field G.

Assuming that the ranks of the homology groups in (2.4) are all finite we denote by $P_t(A, B)$ the Poincaré polynomial (or series) of the topological pair (A, B), i. e.,

$$P_t(A, B) = \sum_{i \geq 0} \beta_i(A, B) t^i$$

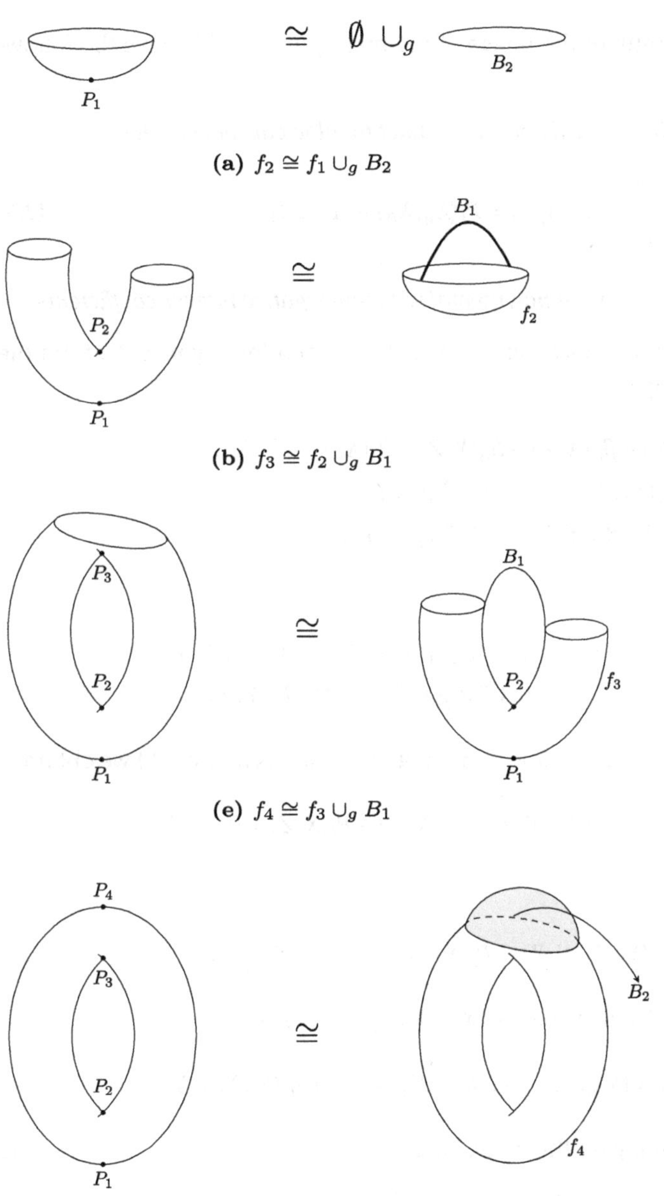

(a) $f_2 \cong f_1 \cup_g B_2$

(b) $f_3 \cong f_2 \cup_g B_1$

(e) $f_4 \cong f_3 \cup_g B_1$

(f) $f_5 \cong f_4 \cup_g B_2$

Figure 2.7: Attaching cells as crossing critical values.

where $\beta_i(A, B)$ are the relative Betti numbers of the pair. Next, we consider the formal polynomial (or series)

$$q_t(X, Y, Z) = \sum_{i \geq 0} d_i(X, Y, Z) t^i \tag{2.5}$$

where $d_i(X, Y, Z)$ is the rank of the image of the map ∂_i in (2.4). We have the following.

Lemma 2.10. *Let $X_0 \subset X_1 \subset \cdots \subset X_n$, $n + 1$ compact topological spaces. Then*

$$\sum_{j=1}^{n} P_t(X_j, X_{j-1}) = P_t(X_n, X_0) + (1 + t)Q_t \tag{2.6}$$

where $Q_t = \sum_{j=2}^{n} q_t(X_j, X_{j-1}, X_0)$ is a polynomial with nonnegative integer coefficients.

Proof. For any triple of topological spaces $X \supset Y \supset Z$ and for any $m \in \mathbb{N}$, from the exact sequence (2.4) we get:

$$\beta_0(X, Y) - \beta_0(X, Z) + \beta_0(Y, Z) - \beta_1(X, Y) + \beta_1(X, Z)$$
$$- \beta_1(Y, Z) + \cdots + (-1)^m \beta_m(X, Y) - (-1)^m \beta_m(X, Z)$$
$$+ (-1)^m \beta_m(Y, Z) - (-1)^m d_m(X, Y, Z) = 0.$$

Thus

$$(-1)^m d_m(X, Y, Z) = (-1)^{m-1} d_{m-1}(X, Y, Z) + (-1)^m \beta_m(X, Y)$$
$$- (-1)^m \beta_m(X, Z) + (-1)^m \beta_m(Y, Z).$$

Multiplying both sides by $(-1)^m t^m$ and summing on the index m, by (2.5) we obtain

$$q_t(X, Y, Z) = -t q_t(X, Y, Z) + P_t(X, Y) - P_t(X, Z) + P_t(Y, Z)$$

or, equivalently

$$P_t(X, Y) + P_t(Y, Z) = P_t(X, Z) + (1 + t)q_t(X, Y, Z).$$

If we take as triple (X, Y, Z) the triple (X_j, X_{j-1}, X_0), $j \geq 2$, we get

$$P_t(X_j, X_{j-1}) = P(X_j, X_0) - P_t(X_{j-1}, X_0) + (1 + t)q_t(X_j, X_{j-1}, X_0).$$

Then (2.6) follows, summing on the index $j \geq 2$. $\qquad\square$

For the next result, we need the property of the homology of a pair (X, Y) which relates it to the homology of the quotient space X/Y obtained identifying Y with only one selected point $y_0 \in Y$.

More precisely, it holds (see [132, 207]).

Proposition 2.11. *Assume that the topological pair (X, Y) satisfies:*
(i) *X is an Hausdorff space;*
(ii) *Y is closed in X;*
(iii) *$\forall x \in X/Y$ there exist two disjoints open sets U and V such that $x \in U$ and $Y \subset V$;*

(iv) *there exists an open neighborhood I of Y such that Y is a deformation retract of I and $I \neq Y$.*

Then $H_q(X, Y) \simeq H_q(X/Y)$, $\forall q > 0$ and $H_0(X, Y) = H_0^\#(X/Y)$ where $H_0^\#(X/Y) = \{0\}$ if X/Y is path-connected and $H_0^\#(X/Y)$ is a free group with $(r-1)$ generators if X/Y has r connected components.

An example of a topological pair satisfying the conditions (i)–(iv) is the pair $(B_k, \partial B_k)$, where B_k is the unit ball in $\mathbb{R}^k$.

Now we consider a compact manifold M as in the previous sections and a smooth Morse function $f: M \longrightarrow \mathbb{R}$. Let $a_0 < a_1 < \cdots < a_n$ be real numbers such that $f_{a_0} = \emptyset$, $f_{a_n} = M$ and each sublevel f_{a_i} contains exactly i critical points, each one being the only one in the set $f_{a_i} \setminus f_{a_{i-1}}$. By Theorem 2.9, we have

$$H_*(f_{a_i}, f_{a_{i-1}}) = H_*(f_{a_{i-1}} \cup_g B_{\lambda_i}, f_{a_{i-1}}) = H_*(B_{\lambda_i}, \partial B_{\lambda_i}) \tag{2.7}$$

where λ_i is the Morse index of the only critical point in $f_{a_i} \setminus f_{a_{i-1}}$.

Moreover, by Proposition 2.11

$$H_q(B_{\lambda_i}, \partial B_{\lambda_i}) = \begin{cases} \mathbb{Z} & \text{if } q = \lambda_i \\ \{0\} & \text{if } q \neq \lambda_i, \end{cases}$$

here $\mathbb{Z}$ is taken as the unitary commutative ring of the coefficients for the homology groups H_q.

Hence the Poincaré polynomial of the pair $(f_{a_i}, f_{a_{i-1}})$ is t^{λ_i}.

Applying (2.6) to the spaces $\emptyset = f_{a_0} \subset f_{a_1} \subset \cdots \subset f_{a_n} = M$, by (2.7) we get

$$\sum_{i=1}^{n} t^{\lambda_i} = \sum_{i=1}^{n} P_t(f_{a_i}, f_{a_{i-1}}) = P_t(M, \emptyset) + (1 + t)Q_t(f).$$

This procedure can be generalized to the case of more than one critical points corresponding to the same critical value obtaining:

$$P_t(f) = \sum_{i=0}^{m} \alpha_i t^i = \sum_{i=0}^{m} \beta_i(M)t^i + (1 + t)Q_t(f) \tag{2.8}$$

where m is the dimension of M and α_i is the number of critical points of f with Morse index i. The polynomial $P_t(f) = \sum_{i=0}^{m} \alpha_i t^i$ is also called the Morse polynomial of f.

In particular, from (2.8) we deduce that $\alpha_i \geq \beta_i(M)$, i. e., any Morse function f on a compact manifold M has at least $\beta_i(M)$ critical points with index i. This, for example, allows to claim that any Morse function on the torus has at least four critical points: a maximum point, a minimum point and two saddle points.

Now let us give a geometric explanation of (2.8), independently of Lemma 2.10. Let c be a critical value of f to which corresponds only one critical point with Morse index λ. Obviously, passing from the sublevel $f_{c-\epsilon}$ to the sublevel $f_{c+\epsilon}$ a power t^λ should be added to the Morse polynomial of f. Let us try to understand how the Poincaré polynomial of the manifold changes at crossing the critical level c. By Theorem 2.9, we know that the sublevel $f_{c+\epsilon}$ is obtained by $f_{c-\epsilon}$ adding a λ-cell. An important observation is that the operation of attaching a λ-cell to a space can either "increase" the homology in dimension λ or "decrease" the homology in dimension $(\lambda - 1)$. In Figure 2.8, there are some examples for the case $\lambda = 1, 2$.

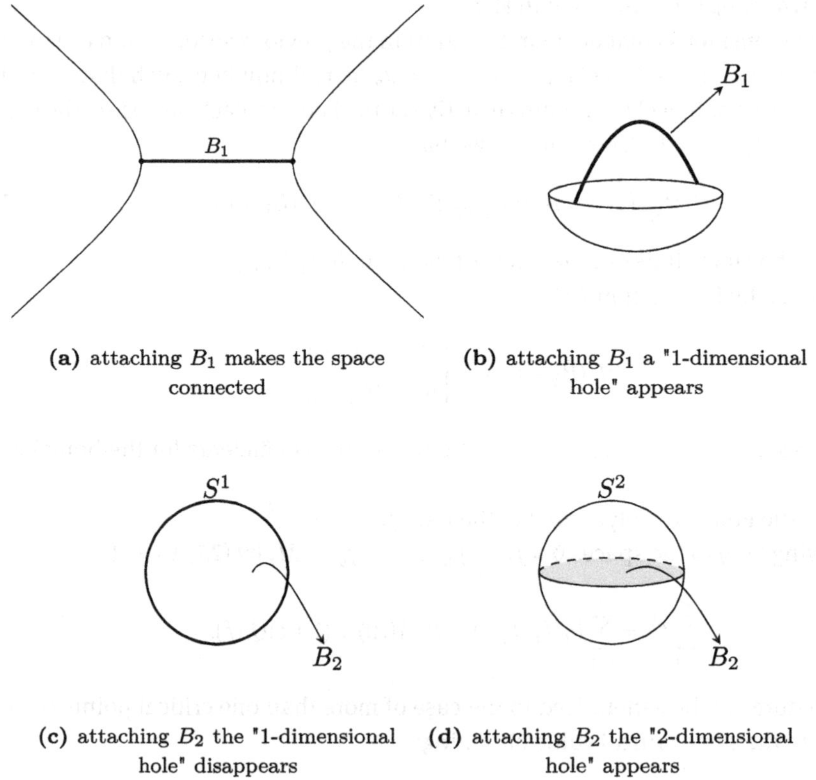

(a) attaching B_1 makes the space connected

(b) attaching B_1 a "1-dimensional hole" appears

(c) attaching B_2 the "1-dimensional hole" disappears

(d) attaching B_2 the "2-dimensional hole" appears

Figure 2.8: Changing the homology of sets.

As a consequence, we have

$$P_t(f_{c+\epsilon}) = \begin{cases} P_t(f_{c+\epsilon}) + t^{\lambda} \\ \text{or} \\ P_t(f_{c-\epsilon}) - t^{\lambda-1} \end{cases} \tag{2.9}$$

Then, denoting by $P_t(f|_{f_a})$ the Morse polynomial of the restriction of f to the sublevel f_a and assuming that $P_t(f|_{f_{c-\epsilon}}) = P_t(f_{c-\epsilon})$ (which holds, for example, if c is the minimum of f) from (2.9) we get

$$P_t(f|_{f_{c+\epsilon}}) = P_t(f|_{f_{c-\epsilon}}) + t^\lambda = P_t(f_{c-\epsilon}) + t^\lambda =$$

$$= \begin{cases} P_t(f_{c+\epsilon}) \\ \text{or} \\ P_t(f_{c-\epsilon}) + (1+t)t^{\lambda-1} \end{cases} \tag{2.10}$$

This explains why the term $(1+t)$ appears and how we get (2.8) as the value c increases. The presence of this term $(1+t)$ in (2.8) is important and allows to deduce more information on the critical points of f, other than the inequalities $\alpha_i \geq \beta_i(M)$. For example, if $M = S^2$ and f is a function which has two minimum points, then by (2.8) we have

$$2 + \alpha_1 t + \alpha_2 t^2 = 1 + t^2 + (1+t)(a_0 + a_1 t + a_2 t^2)$$

with $a_i \geq 0$. This implies that $a_0 \geq 1$, and hence $\alpha_1 \geq 1$ which means that the function f must have at least one saddle point.

Obviously, it could happen that $Q_t(f) = 0$, i.e., $P_t(f) = P_t(M)$ in which case f is called a **perfect Morse function**.

It is worth observing that all this depends on the choice of the coefficients in the homology group. A class of perfect Morse functions is given by those ones which do not have critical points with consecutive Morse index. Finally, to find a perfect Morse function on a manifold M allows to compute the homology of M.

2.2 Infinite dimensional Morse theory

2.2.1 Critical groups and Morse lemma

Let H be an Hilbert space and M a C^2-Hilbert manifold (see, e. g., [14], Section 6.1 for the definition).

The basic Morse theory for a C^2-function $f: M \longrightarrow \mathbb{R}$ is set up in two steps:

(i) a local study of the behavior of the function f near its critical points, through the definition of the *critical groups*;

(ii) a global study which focuses on the relationship between some numbers (Morse numbers) related to the number of critical points of f and the topological properties of the underlying manifold.

In analogy with the finite dimensional case, we define the Morse index of a critical point u of f (i. e., a point where $f'(u) = 0$) as the maximal dimension of a subspace of H on which the bilinear form

$$B(v, w) = (f \circ x^{-1})''(x(u))(v, w)$$

is negative definite, where x is a chart at u. If M is the whole Hilbert space, the **Morse index** of u is just the maximal dimension of a subspace of H on which $f''(u)$ is negative definite.

The *nullity* of u is the supremum of the dimensions of subspaces of H on which the bilinear form B is zero. Finally, u is said to be a **nondegenerate critical point** if the linear operator $L: H \longrightarrow H^* \simeq H$ defined by

$$(Lv, w) = B(v, w), \quad \forall v, w \in H$$

is invertible, where $(\cdot, \cdot)$ is the inner product on H. Note that, by the chain rule, the above definitions are independent of the choice of the chart x. Moreover, by the implicit function theorem, a nondegenerate critical point is isolated.

We now define the **critical groups of an isolated critical point** and show their relation with the Morse index (see also [22]).

Definition 2.12. Let u be an isolated critical point of f and let $c = f(u)$. We define

$$C_q(f, u) = H_q(f_c \cap U, f_c \cap U \setminus \{u\}), \quad q = 0, 1, \ldots$$

the qth critical group of f at u, where U is a neighborhood of u which does not contain another critical point and $H_q(\cdot, \cdot)$ denotes the qth relative homology group over a field G as in the previous section.

According to the excision property of the singular homology theory, the critical groups are independent of the choice of the neighborhood U.

We show that for a nondegenerate critical point, the critical groups depend only on the Morse index.

Theorem 2.13. *Let u be an isolated critical point of $f \in C^2(M)$. Then*
(i) *if u is a local minimum point of f*

$$C_q(f, u) = \begin{cases} G & \text{if } q = 0 \\ 0 & \text{if } q \neq 0 \end{cases}$$

(ii) *if u is a nondegenerate critical point with Morse index $m(u) = k \geq 1$, then*

$$C_q(f, u) = \begin{cases} G & \text{if } q = k \\ 0 & \text{if } q \neq k \end{cases}$$

Proof. (i) Let B be a closed neighborhood of u so small that $f(v) > c = f(u)$, for every $v \in B \setminus \{u\}$. Then, by definition,

$$C_q(f, u) = H_q(f_c \cap B, f_c \cap B \setminus \{u\}) = H_q(\{u\}, \emptyset) = \delta_{q,0} G$$

and the assertion is proved.

(ii) Let x be a chart at u and let B be a closed neighborhood of u contained in the domain of x. Then, for $c = f(u)$,

$$C_q(f, u) = H_q(f_c \cap B, f_c \cap B \setminus \{u\})$$

and it is enough to consider the case when M is an open subset of the space H. For simplicity, we assume $u = 0$ and $c = f(u) = 0$.

Let $L: H \longrightarrow H$ be the self-adjoint operator defined by

$$(Lv, w) = f''(0)(v, w), \quad \forall v, w \in H.$$

Since, by assumption, L is invertible then the space H is the orthogonal sum of H^+ and H^- which are the subspaces where L is positive definite and negative definite, respectively. Then we take a closed ball B with center at 0 so small that

$$B \cap H^- \subset f_0 \tag{2.11}$$
$$B \cap H^+ \cap f_0 = \{0\} \tag{2.12}$$
$$f''(v)(w, w) \geq 0 \quad \forall v \in B, \forall w \in H^+. \tag{2.13}$$

Let us define

$$\eta: [0, 1] \times B \longrightarrow B, \quad \eta(t, v) = (1 - t)v + tP(v)$$

where P is the orthogonal projector onto H^-.

For $v \in F_0 \cap B$, let us write $g(t) = f(\eta(t, v))$. It follows from (2.13) that, for all $t \in [0, 1]$,

$$g''(t) = f''((1 - t)v + tP(v))((I - P)v) \geq 0$$

so that g is convex on $[0, 1]$. But $g(0) = f(v) \leq 0$, since $v \in f_0$ and $g(1) = f(P(v)) \leq 0$, by (2.11). Thus $f(\eta(t, v)) = g(t) \leq 0$, for all $t \in [0, 1]$. Moreover, if $\eta(t, v) = 0$, for some $t \in [0, 1]$ and some $v \in f_0 \cap B$, then $v = 0$, because $\eta(t, v) = 0$ implies $P(v) = 0$ so that, by (2.12), $v = 0$.

Finally, $H^- \cap B \setminus \{0\}$ is a deformation retract of $f_0 \cap B \setminus \{0\}$ and $H^- \cap B$ is a deformation retract of $f_0 \cap B$. Since, by assumption, the Morse index of u is k, $\dim H^- = k$ and we get

$$H_q(f_0 \cap B, f_0 \cap B \setminus \{0\}) \simeq H_q(f_0 \cap B, H^- \cap B \setminus \{0\})$$
$$\simeq H_q(H^- \cap B, H^- \cap B \setminus \{0\})$$
$$\simeq H_q(B^k, S^{k-1})$$
$$\simeq \delta_{q,k} G. \qquad \square$$

Remark 2.14. The statement of Theorem 2.13 holds also in Banach spaces. We refer to [60, 154] for the proof (see also [6]).

To study the local behavior of a function $f \in C^2(M)$ at a nondegenerate critical point, it is crucial to prove a Morse lemma, analogous to Lemma 2.5 for the finite dimensional case. We state it here and refer to [60] for the proof.

Theorem 2.15 (Infinite dimensional Morse lemma). *Let u be a nondegenerate critical point of a function $f \in C^2(M)$. Then there exist a neighborhood U of u and a local diffeomorphism $\Phi \colon U \longrightarrow T_u(M)$ ($T_u(M)$ is the tangent space at u) with $\Phi(u) = 0$, such that*

$$f \circ \Phi^{-1}(\xi) = f(u) + \frac{1}{2}(d^2 f(u)\xi, \xi), \quad \forall \xi \in \Phi(U).$$

2.2.2 Morse inequalities

In order to prove some kind of Morse identity, like (2.10) in the infinite dimensional case and a result analogous to that of Theorem 2.9, we need to define the Morse numbers of a function. Though everything could be done for a function f defined on a C^2-Finsler manifold modelled on a Banach space, we will only consider the case of a C^2-Hilbert manifold M and refer to [60] for the general case.

Let f be a real valued function on M of class C^1 and assume that f has only isolated critical values and that each of them corresponds to a finite number of critical points. We denote by c_i, $i \in \mathbb{Z}$, the critical values of f and by $K_{c_i} = \{z_j^i\}_{j=1}^m$, $i \in \mathbb{Z}$, the corresponding finite sets of critical points. Choosing $\epsilon_i \in (0, \eta_i)$, $\eta_i = \min\{c_{i+1} - c_i, c_i - c_{i-1}\}$, $i \in \mathbb{Z}$, we define the following.

Definition 2.16. For a pair of regular values $a < b$, the number

$$M_q(a, b) = \sum_{a < c_i < b} \operatorname{rank} H_q(f_{c_i + \epsilon_i}, f_{c_i} - \epsilon_i)$$

is called the **qth Morse number** of the function f with respect to (a, b), $q \in \mathbb{Z}$.

We recall the well-known definition of the compactness Palais–Smale condition, for the function f, (PS) in short.

Definition 2.17. A function $f \colon M \longrightarrow \mathbb{R}$, $f \in C^1(M)$ is said to satisfy the (PS) condition if every sequence $\{u_n\} \subset M$, such that $\{f(u_n)\}$ is bounded and $f'(u_n) \longrightarrow 0$, admits a convergent subsequence. We say that f satisfies the (PS) condition at the level c, $(\text{PS})_c$ in short, if $\{u_n\}$ has a converging subsequence whenever $f(u_n) \longrightarrow c$ and $f'(u_n) \longrightarrow 0$.

By using the notion of pseudo-gradient vector fields, it is possible to prove several deformation theorems, more or less analogous to Theorem 2.8, for C^1-functions satisfying the (PS) or $(\text{PS})_c$ condition. We recall here two different versions, for whose proof we refer to [14, 60, 196].

Theorem 2.18. *If $f \in C^1(M)$ satisfies the $(PS)_c$, for every c in an interval $[a, b]$ which does not contain any critical point of f, then f_a is a strong deformation retract of f_b.*

Theorem 2.19. *Let $f \in C^1(M)$ satisfy the $(PS)_c$ condition for a number $c \in \mathbb{R}$ and assume that N is a closed neighborhood of the set K_c of critical points of f in $f^{-1}(c)$. Then there exist a continuous map $\eta: [0,1] \times M \longrightarrow M$ and constants $\bar{\epsilon} > \epsilon > 0$ such that:*
(i) $\eta(0, u) = u$, for all $u \in M$;
(ii) $\eta(t, u) = u$, for all $t \in [0, 1]$ and $f(u) \notin [c - \bar{\epsilon}, c + \bar{\epsilon}]$;
(iii) $\eta(t, \cdot)$ is a homeomorphism of M onto M for each $t \in [0, 1]$;
(iv) $\eta(1, f_{c+\epsilon} \setminus N) \subset f_{c-\epsilon}$;
(v) $f \circ \eta(t, u)$ is nonincreasing in t, for all $(t, u) \in [0, 1] \times M$.

By the deformation theorems, we have that for functions satisfying the (PS) condition, the Morse numbers are well-defined, i. e., they are independent of the choice of $\{\epsilon_i\}$.

As a consequence of the deformation theorem and the homotopy invariance of the homology groups, we have a connection between the critical groups and the Morse numbers.

Theorem 2.20. *Assume that c is an isolated critical value of $f \in C^1(M)$ and $K_c = \{z_j\}_{j=1,\dots,m}$. Then, for $\epsilon > 0$ sufficiently small:*

$$H_*(f_{c+\epsilon}, f_{c-\epsilon}) \cong H_*(f_c, f_c \setminus K_c) \cong \bigoplus_{j=1}^{m} C_*(f, z_j)$$

Hence, for regular values $a < b$ of f we have

$$M_q(a, b) = \sum_{a < c_i < b} \sum_{j=1}^{m_i} \operatorname{rank} C_q(f, z_j^i), \quad q \in \mathbb{N}$$

Proof. See [60]. $\qquad\qquad\qquad\qquad\qquad\qquad\qquad\qquad\qquad\qquad\qquad\qquad\qquad\quad$ $\square$

To formulate the Morse inequalities, let us define, for regular values $a < b$, of f:

$$\beta_q = \beta_q(a, b) = \operatorname{rank} H_q(f_b, f_a), \quad q \in \mathbb{N}$$

Theorem 2.21. *Let $f \in C^1(M)$ satisfy the $(PS)_c$ condition, for any $c \in [a, b]$, with a, b regular values of f. Assuming that $z_1, \dots, z_l$, $l \in \mathbb{N}$, are the only critical points in $f^{-1}([a, b])$, then*

$$\sum_{q=0}^{+\infty} M_q t^q = \sum_{q=0}^{+\infty} \beta_q t^q + (1 + t)Q(t)$$

where Q is a formal series with nonnegative coefficients and $M_q = M_q(a, b) = \sum_{j=1}^{l} \operatorname{rank} C_q(f, z_j)$, for $q \in \mathbb{N}$.

We again refer to [60] for the proof of the previous theorem.

We conclude pointing out that a clear picture of the changes in the topology of the level sets passing through a critical value to which correspond only nondegenerate critical points can also be obtained in infinite dimension. We refer to ([60], Theorem 4.4) for the precise statement and the details.

2.2.3 Morse index of mountain pass critical points

Here, we estimate the Morse index of a special type of critical point of a function f, obtained by a famous minimax theorem due to A. Ambrosetti and P. Rabinowitz (see [15]).

We state everything for functionals defined in a Banach space. For the case of functions on Finsler manifolds modelled on Banach spaces, we refer to [60]. Let E be a Banach space and $F\colon E \longrightarrow \mathbb{R}$ a C^1-functional. We say that F satisfies the Palais–Smale compactness condition in E or the Palais–Smale condition at the level $c \in \mathbb{R}$ if for F the conditions of Definition 2.17 hold with obvious changes.

The following theorem gives a way of proving the existence of a critical point of the functional F if it satisfies some geometrical assumption.

Theorem 2.22 (Mountain pass theorem). *Let $F \in C^1(E)$ a functional satisfying:*

$$F(0) = 0 \quad and \quad \exists r,\rho > 0 \ such \ that$$
$$F(u) \geq \rho \ \forall u \in S_r = \{ u \in E : \|u\| = r \}; \tag{2.14}$$

$$\exists e \in E \ with \ \|e\| > r \ such \ that \ F(e) \leq 0. \tag{2.15}$$

Let

$$c = \inf_{y \in \Gamma} \max_{t \in [0,1]} F(y(t)) \tag{2.16}$$

where Γ is the set of all paths joining $u = 0$ and $u = e$, i. e.,

$$\Gamma = \{ y \in C([0,1],E) : y(0) = 0, \ y(1) = e \}$$

Then, if F satisfies the (PS)$_c$ *condition, the number c is a positive critical level of F, namely there exists $u \in E$ such that $F(u) = c$ and $F'(u) = 0$, in particular $u \neq 0$ and $u \neq e$.*

The original proof of this theorem is contained in [15] and it is also included in every book about variational methods; we refer, for example, to [14, 196, 225].

The aim of this section is to show that at a mountain pass level c (i. e., c given by (2.16)) of a C^2-functional F there exists a critical point with Morse index at most one.

As defined in Section 2.2.1, the Morse index $m(u)$ of a critical point u of F is the maximal dimension of a subspace on which $F''(u)$ is negative definite. Thus we define the subspaces:

$$E^0(u) = \ker F''(u) \tag{2.17}$$

$$E^-(u) = \{\, v \in E : \langle F''(u)v, v \rangle < 0 \,\} \tag{2.18}$$

$$E^+(u) = \{\, v \in E : \langle F''(u)v, v \rangle > 0 \,\} \tag{2.19}$$

and u will be a nondegenerate critical point of F if $E^0(u) = \{0\}$.

Theorem 2.23. *Let F be a C^2-functional satisfying the hypothesis of Theorem 2.21 and let $K_c = \{z \in E : F'(z) = 0\} \cap F^{-1}(c)$, c given by (2.16). Then, if K_c is discrete, there exists $u \in K_c$ with Morse index $m(u) \leq 1$. Moreover, if $K_c = \{u\}$ and u is nondegenerate, then $m(u) = 1$.*

To prove this theorem, we need some preliminary result.

Lemma 2.24. *Let $F\colon E \longrightarrow \mathbb{R}$ be a C^2-functional satisfying the hypotheses of Theorem 2.22 and let c be the critical mountain pass level defined by (2.16). Then, if K_c is discrete, there exists a critical point $u \in K_c$, such that $C_1(F, u) \neq 0$, where $C_1(F, u)$ denotes the first critical group of F at u, as in Definition 2.12.*

Proof. We refer to Theorem 1.5 of [60]. □

Lemma 2.25. *Let $F\colon E \longrightarrow \mathbb{R}$ a C^2-functional and assume that u is a critical point of F with finite Morse index $m(u)$. Then*

$$C_q(f, u) = 0, \quad \text{for every } q \leq m(u) - 1.$$

Proof. See [154]. □

We can now prove 2.23.

Proof of Theorem 2.23. If u is the only critical point in K_c and is nondegenerate, then applying Theorem 2.13 and Remark 2.14 as well as Lemma 2.24, we get that $m(u) = 1$.

In the general case, by Lemma 2.24, we know that there exists a critical point $u \in K_c$ such that $C_1(F, u) \neq 0$.

On the other side, by Lemma 2.25 we have that $C_q(F, u) = 0$, for every $q \leq m(u) - 1$. Combining this information, we obtain $m(u) \leq 1$. □

Another way of getting Theorem 2.23 without using explicitly the critical groups is shown in ([14], Theorem 12.31) in the case when E is an Hilbert space and K_c reduces to only one nondegenerate critical point.

It combines different ingredients already used to prove Theorem 2.13, Lemma 2.24 and Lemma 2.25 and shows more clearly the connection between the geometric construction of the mountain pass critical level and the Morse index of a corresponding critical point. Therefore, we outline it below.

Alternative proof of Theorem 2.23. We assume that E is a Hilbert space and $K_c = \{u\}$ and u is nondegenerate. Moreover, for simplicity, we take $u = 0$ while the point 0 where $F(0) = 0$ in the statement of Theorem 2.22 will be denoted by u_0 and write E^+ and E^- instead of $E^{\pm}(u)$ as defined in (2.18) and (2.19). Hence $E = E^+ \oplus E^-$ and, arguing by contradiction, we assume that $\dim E^- \geq 2$. By the Morse lemma in infinite dimension (see [60], Lemma 4.1), up to a smooth change of coordinates, we have that

$$F(u) = c - \left\| u^- \right\|^2 + \left\| u^+ \right\|^2$$

where $u = u^+ + u^-$, $u^- \in E^-$ and $u^+ \in E^+$.

Given $\beta > \alpha > 0$, we consider a neighborhood $U_{\alpha,\beta}$ of $u = 0$, defined by

$$U_{\alpha,\beta} = \left\{ u = u^- + u^+ : \left\| u^- \right\| < \alpha, \left\| u^+ \right\| < \beta \right\}.$$

Then, for every $u \in \overline{U}_{\alpha,\beta}$ with $\| u^+ \| = \beta$ we have

$$F(u) \geq c - \alpha^2 + \beta^2.$$

Thus, given $d > c$, there exists $\epsilon > 0$ such that, if $\alpha < \beta \leq \epsilon$, then

$$\inf\{F(u) : u \in \overline{U}_{\alpha,\beta}, \ \| u^+ \| = \beta\} \geq d.$$

Taking α and β sufficiently small, we can also assume that

$$\inf\{F(u) : u \in \overline{U}_{\alpha,\beta}\} > 0.$$

Note that, since $F(u_0) = 0$ and $F(e) \leq 0$, neither u_0 nor e belong to $\overline{U}_{\alpha,\beta}$. By the definition of the mountain pass critical level c, taken $\delta \in {]0, d - c[}$, we can find $y \in \Gamma$ such that

$$\sup\{F(y(t)) : t \in [0,1]\} \leq c + \delta.$$

By the deformation lemma, Theorem 2.19, we can find a deformation η such that:
a) $\eta \circ y \in \Gamma$ (Γ defined in the statement of Theorem 2.22);
b) $\eta(F_{c+\delta} \setminus U_{\alpha,\beta}) \subset F_{c-\delta}$.

As a consequence of b), we have that the path y must necessarily intersect $U_{\alpha,\beta}$, otherwise we get a contradiction with the definition of the critical level c. Since y connects the point u_0 and e which do not belong to $U_{\alpha,\beta}$, we deduce that there exist $t_1 < t_2 \in (0,1)$ such that the points $v_i = y(t_i)$, $i = 1, 2$, belong to $\partial U_{\alpha,\beta}$, while $y(t) \notin \overline{U}_{\alpha,\beta}$ for all $t < t_1$ or $t > t_2$.

Since

$$F(v_i) \leq c + \delta < d \leq \inf\{F(u) : u \in \overline{U}_{\alpha,\beta}, \ \| u^+ \| = \beta\}$$

it follows that $\|v_i^-\| = \alpha$ and $\|v_i^+\| < \beta$, $i = 1, 2$, for $v_i = v_i^- + v_i^+$, i. e., $v_i^\mp$ are the projections of v_i on the spaces E^- and E^+. Let $\pi_i \subset \partial U_{\alpha,\beta}$ denote the segments connecting v_i with v_i^-. Since we are assuming that $\dim E^- \geq 2$, we can connect v_1^- and v_2^- with an arc σ contained in $\partial U_{\alpha,\beta} \cap E^-$. This allows to construct a new path $\tilde{\gamma}$ defined as follows:

$$\tilde{\gamma}(t) = \begin{cases} \gamma(t) & \text{if } t \in [0, t_1] \cup [t_2, 1] \\ \pi_1 \cup \sigma \cup \pi_2 & \text{if } t \in [t_1, t_2] \end{cases}$$

and $\tilde{\gamma}$ belongs to Γ.

Since for $v \in \partial U_{\alpha,\beta} \cap E^-$ one has that $\|v^-\| = \alpha$ and $\|v^+\| = 0$, then

$$F(v) = c - \alpha^2, \quad \forall v \in \partial U_{\alpha,\beta} \cap E^-.$$

This implies that

$$\sup_{\sigma} F(u) < c.$$

On the other side $\sup_{\pi_i} F(u) \leq c + \delta$, $i = 1, 2$, so that

$$\sup_{\tilde{\gamma}} F(u) \leq c + \delta,$$

and the path $\tilde{\gamma}$ does not intersect $U_{\alpha,\beta}$. Then, as before, applying the deformation lemma we get, as in a) and b), a new path $\eta \circ \tilde{\gamma} \in \Gamma$ such that $\sup\{F(\tilde{\gamma}(t)), t \in [0, 1]\} \leq c - \delta$, which is a contradiction with the definition of the mountain pass level c. Hence for the critical point $u \in K_c$ it must be $m(u) \leq 1$.

To rule out that $m(u) = 0$, we observe that in such case, since u is nondegenerate, then u is a local minimum for F and $E = E^+$. Taking $U_{\alpha,\beta}$ such that $\inf\{F(v) : v \in \partial U_{\alpha,\beta}\} \geq d > c$, the previous arguments can be repeated, reaching again a contradiction. $\qquad\square$

We conclude by mentioning that similar arguments can be used to compute the Morse index of critical points obtained by other mini-max procedure. We refer to [60, 121] and the references therein for the statements and details.

3 Morse theory for semilinear elliptic equations

In this chapter, we introduce the Morse index of solutions of semilinear elliptic equations and show some first properties and applications. In particular, we will consider the case of least energy solutions, both positive or sign changing, for which the Morse index can be explicitly computed.

As an application, we will show the uniqueness of the positive solution of Morse index one in planar convex bounded domains as obtained in [158].

Finally, we will prove estimates for the Morse index of symmetric sign changing solutions, both in the autonomous and nonautonomous case, following the papers [7] and [179].

3.1 Introduction

Let Ω be a Lipschitz domain in $\mathbb{R}^N$, $N \geq 2$, and Γ_0, Γ relatively open disjoint subsets of $\partial\Omega$ satisfying the conditions (1.28) and (1.29) of Chapter 1.

Though in this chapter we will focus mostly on Dirichlet boundary conditions, we give the main definitions and some general results in the case of mixed Dirichlet–Neumann problems since they will be studied in Chapter 6.

Let us consider the following semilinear elliptic boundary value problem:

$$\begin{cases} -\Delta u = f(x, u) & \text{in } \Omega \\ u = 0 & \text{on } \Gamma_0 \\ \dfrac{\partial u}{\partial v} = g(x, u) & \text{on } \Gamma \end{cases} \tag{3.1}$$

where v denotes the outer normal to $\partial\Omega$.

We assume that $f = f(x, s): \overline{\Omega} \times \mathbb{R} \longrightarrow \mathbb{R}$ and $g = g(x, s): \overline{\Gamma} \times \mathbb{R} \longrightarrow \mathbb{R}$ are locally Hölder continuous functions in $\overline{\Omega} \times \mathbb{R}$ and they are differentiable with respect to the second variable with

$$f, \frac{\partial f}{\partial s} \in C^0(\overline{\Omega} \times \mathbb{R}); \quad g, \frac{\partial g}{\partial s} \in C^0(\overline{\Gamma} \times \mathbb{R}) \tag{3.2}$$

We consider bounded weak solutions of (3.1), i. e., functions $u \in H_0^1(\Omega \cup \Gamma) \cap L^\infty(\Omega)$, where $H_0^1(\Omega \cup \Gamma)$ is the Sobolev space defined in Chapter 1, such that

$$\int_\Omega \nabla u \cdot \nabla \varphi \, dx = \int_\Omega f(x, u)\varphi \, dx + \int_\Gamma g(x', u)\varphi \, dx', \quad \forall \varphi \in H_0^1(\Omega \cup \Gamma). \tag{3.3}$$

Let us observe that since f is locally Hölder continuous, by standard elliptic regularity results $u \in C^2(\Omega)$. Corresponding to such a solution of (3.1), we consider the quadratic

https://doi.org/10.1515/9783110538243-003

form

$$Q_u(\psi) = \int_\Omega |\nabla\psi|^2 \, dx - \int_\Omega \frac{\partial f}{\partial s}(x,u)|\psi|^2 \, dx - \int_\Gamma \frac{\partial g}{\partial s}(x',u)|\psi|^2 \, dx' \tag{3.4}$$

for any $\psi \in C_c^1(\Omega \cup \Gamma)$.

The Morse index of a solution is defined as follows.

Definition 3.1. Let $u \in H_0^1(\Omega \cup \Gamma) \cap L^\infty(\Omega)$ be a weak solution of (3.1). We say that:
(i) u is stable (or has zero Morse index) if $Q_u(\psi) \geq 0 \ \forall \psi \in C_c^1(\Omega \cup \Gamma)$;
(ii) u has Morse index equal to the integer $m(u) \geq 1$ if $m(u)$ is the maximal dimension of a subspace of $C_c^1(\Omega \cup \Gamma)$ where the quadratic form $Q_u(\psi)$ is negative definite;
(iii) u has infinite Morse index if, for any integer $k \geq 1$ there exists a k-dimensional subspace of $C_c^1(\Omega \cup \Gamma)$ where Q_u is negative definite.

The above definition is the same as the one given in Chapter 2 if we observe that a weak solution of (3.1) is a critical point of the functional

$$J(v) = \frac{1}{2} \int_\Omega |\nabla v|^2 \, dx - \int_\Omega F(x,v) \, dx - \int_\Gamma G(x',v) \, dx' \tag{3.5}$$

in the space $H_0^1(\Omega \cup \Gamma) \cap L^\infty$, where $F(x,v) = \int_0^v f(x,s) \, ds$ and $G(x,v) = \int_0^v g(x,s) \, ds$.

Thus the quadratic form (3.4) is the one corresponding to the second derivative of the functional J, i. e., $Q_u(\psi) = J''(u)(\psi,\psi)$.

In general, to use the Morse index in applications it is convenient to relate it to the number of negative eigenvalues of a suitable linear operator. Therefore, we consider the following eigenvalue problem:

$$\begin{cases} -\Delta w - \dfrac{\partial f}{\partial s}(x,u)w = \lambda w & \text{in } \Omega \\[2mm] w = 0 & \text{on } \Gamma_0 \\[2mm] \dfrac{\partial w}{\partial v} - \dfrac{\partial g}{\partial s}(x,u)w = \lambda w & \text{on } \Gamma \end{cases} \tag{3.6}$$

where u is a given solution of (3.1). The linear problem (3.6) is the one corresponding to the linearization of (3.1) at a solution u. The linear operator $L_u : H_0^1(\Omega\cup\Gamma) \to (H_0^1(\Omega\cup\Gamma))^*$ defined, for any $v \in H_0^1(\Omega \cup \Gamma)$ by

$$L_u(v)(z) = J''(u)(v,z) = \int_\Omega \nabla v \cdot \nabla z \, dx - \int_\Omega \frac{\partial f}{\partial s} vz \, dx - \int_\Gamma \frac{\partial g}{\partial s} vz \, dx', \quad z \in H_0^1(\Omega \cup \Gamma)$$

is the so-called **linearized operator at the solution** u and will be denoted by

$$L_u = \left(-\Delta - \frac{\partial f}{\partial s}; \ \frac{\partial}{\partial v} - \frac{\partial g}{\partial s} \right) \tag{3.7}$$

according to the notations of Chapter 1 (see (1.74)). In the case of Dirichlet problems, i. e., when $\Gamma = \emptyset$, we will use the simple notation

$$L_u = -\Delta - \frac{\partial f}{\partial s}$$

Using the eigenvalue theory developed in Chapter 1, we obtain, in bounded domains, the following characterization of the Morse index.

Theorem 3.2. *Let Ω be as in* (3.1) *and assume further that it is bounded. Then the Morse index of a solution u to* (3.1) *equals the number of negative eigenvalues of problem* (3.6).

Proof. Let us denote by $\mu(u)$ the number of negative eigenvalues of (3.6).

If the quadratic form Q_u defined in (3.4) is negative definite on a k-dimensional supspace of $C_c^1(\Omega \cup \Gamma)$, then, by (iii) of Theorem 1.42 in Chapter 1 we have that the kth eigenvalue λ_k of (3.6) is negative. Hence $\mu(u) \geq m(u)$.

On the other hand, if there are k negative eigenvalues of (3.6), by (i) of Theorem 1.44 there is a k-dimensional supspace of $C_c^1(\Omega \cup \Gamma)$ where the quadratic form Q_u is negative definite, hence $m(u) \geq \mu(u)$. $\qquad\square$

We stress that Theorem 3.2 holds in bounded domains because of the spectral theory developed in Chapter 1.

Remark 3.3. If one of the derivative $\frac{\partial f}{\partial s}$, $\frac{\partial g}{\partial s}$ is nonpositive (or, more generally, if the linearized operator L_u is coercive in $H_0^1(\Omega \cup \Gamma)$), then other choices of eigenvalue problems could be possible to evaluate the Morse index of a solution u. For example, if $\frac{\partial f}{\partial s} \leq 0$, a modification of the construction done in Section 1.4 of Chapter 1 yields a compact operator in the space $L^2(\Gamma)$ and a corresponding sequence of eigenvalues of the problem

$$\begin{cases} -\Delta w + c(x)w = 0 & \text{in } \Omega \\ w = 0 & \text{on } \Gamma_0 \\ \dfrac{\partial w}{\partial \upsilon} + d(x)w = \lambda w & \text{on } \Gamma \end{cases} \tag{3.8}$$

where $c = -\frac{\partial f}{\partial s}$ and $d = -\frac{\partial g}{\partial s}$.

This case occurs, in particular, in the study of harmonic functions subjected to nonlinear boundary conditions (see, e. g., [1], [27] and the references therein). In this case $f(x, u) \equiv 0$, so that the previous eigenvalue problem becomes

$$\begin{cases} -\Delta w = 0 & \text{in } \Omega \\ w = 0 & \text{on } \Gamma_0 \\ \dfrac{\partial w}{\partial \upsilon} + d(x)w = \lambda w & \text{on } \Gamma \end{cases}$$

If instead $\frac{\partial g}{\partial s} \leq 0$, modifying the construction of Section 1.4 of Chapter 1, we get a compact linear operator in $L^2(\Omega)$ and a corresponding sequence of eigenvalues of the problem:

$$\begin{cases} -\Delta w + c(x)w = \lambda w & \text{in } \Omega \\ w = 0 & \text{on } \Gamma_0 \\ \dfrac{\partial w}{\partial v} + d(x)w = 0 & \text{on } \Gamma \end{cases} \tag{3.9}$$

In both cases, the eigenvalues share the same variational characterization as the one given in Chapter 1, so that the Morse index of a solution can be characterized by the number of negative eigenvalues of (3.8) or (3.9).

Some other eigenvalue problems with weights have been considered in the literature (see [18, 120, 173] and the references therein). In particular, when $\Gamma = \partial\Omega$ in [173] the eigenvalue problem

$$\begin{cases} -\Delta w_j + c(x)w_j = \lambda_j m(x)w_j & \text{in } \Omega \\ \dfrac{\partial w_j}{\partial v} + d(x)w_j = \lambda_j n(x)w_j & \text{on } \partial\Omega \end{cases}$$

with positive weights m, n is considered and the eigenvalues sequence constructed by constrained minimization.

In that paper, the weights can also vanish in part of the domain, but the nonnegativity of the coefficients c, d is assumed (or more generally coercivity of the corresponding linear operator).

The general linear eigenvalue problem (3.6) that we consider, whose corresponding spectral theory has been developed in Chapter 1, with the *same* eigenvalue parameter in the equation and in the nonlinear boundary condition, has the advantage of not requiring the nonnegativity of the coefficients c and d. This is important while studying the Morse index of solutions of nonlinear problems as (3.1), since many choices of the nonlinearities f and g lead to negative or sign changing coefficients $c = -\frac{\partial f}{\partial s}$, $d = -\frac{\partial g}{\partial s}$ in the linearization and to noncoercive linear operators.

3.2 Positive solutions of Dirichlet problems

Let us consider the Dirichlet semilinear elliptic problem:

$$\begin{cases} -\Delta u = f(x, u) & \text{in } \Omega \\ u = 0 & \text{on } \partial\Omega \end{cases} \tag{3.10}$$

where Ω is a smooth bounded domain in $\mathbb{R}^N$, $N \geq 2$. Obviously, (3.10) is a special case of (3.1) when $\Gamma = \emptyset$.

Under some hypotheses on the function $f(x, s)$, we will show the existence of positive solutions of (3.10) with Morse index one.

Finally, as an application of Morse theory, we will prove a uniqueness property of Morse index one solutions.

3.2.1 Existence of a positive solution by the mountain pass theorem

The standard regularity assumptions on f usually required are that f is a Carathéodory function (see [14], Chapter 1, or [196] for the definition) and that it is locally Hölder continuous. However, since our aim is to study the Morse index of solutions, we will always assume that $f : \overline{\Omega} \times \mathbb{R} \longrightarrow \mathbb{R}$ is differentiable in the second variable and (3.2) holds.

Let us start observing that if f satisfies the following superlinear condition:

$$\frac{\partial f}{\partial s}(x, s) > \frac{f(x, s)}{s} \quad \forall x \in \Omega, \ \forall s \in \mathbb{R} \setminus \{0\} \tag{3.11}$$

then every solution u of (3.10) has Morse index $m(u)$ greater than or equal to one. Indeed, multiplying the equation (3.10) by u and integrating on Ω we have

$$\int_\Omega |\nabla u|^2 \, dx = \int_\Omega f(x, u) u \, dx.$$

Hence, testing the quadratic form Q_u defined in (3.4) on the function u itself we get

$$Q_u(u) = \int_\Omega \left[f(x, u) - \frac{\partial f}{\partial s}(x, u) u \right] u \, dx = \int_{\Omega \cap [u \neq 0]} \left[\frac{f(x, u)}{u} - \frac{\partial f}{\partial s}(x, u) \right] u^2 \, dx.$$

Thus if (3.11) holds and u does not vanish in Ω we have

$$Q_u(u) < 0$$

so that $m(u) \geq 1$.

If u changes sign, we can apply the same procedure in each nodal region of u, i. e., in each connected component of the set $\mathcal{Z}(u) = \{ x \in \Omega : u(x) \neq 0 \}$ obtaining that the restriction of u to each nodal domain, extended to zero to the whole Ω, gives a direction in the space $H_0^1(\Omega)$ where the quadratic form $Q(u)$ is negative. This implies that, if (3.11) holds, then

$$m(u) \geq n(u) \tag{3.12}$$

where $n(u)$ denotes the number of nodal regions of the solution u.

Now we will apply the mountain pass theorem (Theorem 2.22 of Chapter 2) to find a positive solution of (3.10).

We assume that $f(x, s)$ satisfies the following additional hypotheses:

(i)

there exists $a_1 \in L^{\frac{2N}{N+2}}(\Omega)$ and $a_2 > 0$ such that

$$|f(x, s)| \leq a_1(x) + a_2|s|^p \quad \forall (x, s) \in \Omega \times \mathbb{R}$$

$$\text{for } 1 < p < \frac{N+2}{N-2} = 2^* - 1 \text{ if } N \geq 3, \; p > 1 \text{ if } N = 2 \tag{3.13}$$

$$\left(2^* = \frac{2N}{N-2} \text{ is the critical Sobolev exponent and } (2^*)' = \frac{2N}{N+2} \right);$$

(ii)

$$f(x, s) = o(|s|) \quad \text{as } s \to 0, \text{ uniformly in } x \in \Omega; \tag{3.14}$$

(iii)

there exist $\alpha > 2$ and $r \geq 0$ such that for $|s| \geq r$

$$0 < \alpha F(x, s) \leq s f(x, s) \quad \forall x \in \Omega \tag{3.15}$$

where F is the primitive of f which appears in (3.5).

The assumption (3.13) which states that f has subcritical growth is needed to prove the Palais–Smale compactness condition for the functional J in (3.5) (with $\Gamma = \emptyset$). The hypotheses (3.14) and (3.15) imply that the functional J satisfies the geometrical conditions of the mountain pass theorem.

We have the following.

Theorem 3.4. *If f satisfies* (3.2), (3.13)–(3.15) *then the Dirichlet problem* (3.10) *has a nontrivial classical positive solution u. Moreover, the Morse index $m(u)$ is less than or equal to one and if* (3.11) *holds $m(u) = 1$.*

Proof. The existence of a nontrivial weak solution u of (3.10) can be obtained by a standard application of the mountain pass theorem to the functional

$$J(u) = \frac{1}{2} \int_\Omega |\nabla u|^2 \, dx - \int_\Omega F(x, u) \, dx \tag{3.16}$$

in the space $H_0^1(\Omega)$. The proof can be found in many books on variational methods (e. g., in [196], Theorem 2.15 or in [212], Theorem 6.2). Then, by standard regularity theorems, it is easy to show that u belongs to $C^2(\Omega) \cap C^0(\overline{\Omega})$ and it is a classical solution of (3.10).

In order to get a positive solution, we consider the function

$$f^+(x, s) = \begin{cases} f(x, s) & \text{if } s \geq 0 \\ 0 & \text{if } s < 0 \end{cases}, \quad x \in \Omega$$

and observe that the functional

$$J^+(u) = \frac{1}{2} \int_\Omega |\nabla u|^2 \, dx - \int_\Omega F^+(x, u) \, dx,$$

where F^+ is the primitive of f^+ such that $F^+(0) = 0$, satisfies the same hypotheses as the functional J. Thus, applying the mountain pass theorem, we get a critical point $\tilde{u}$ of J^+ in $H_0^1(\Omega)$. It is easy to see that $\tilde{u} \geq 0$, because using $\tilde{u}^-$ as a test function in the equation: $(J^+)'(\tilde{u}) = 0$ we get

$$\int_\Omega |\nabla \tilde{u}^-|^2 \, dx = \int_\Omega f^+(x, \tilde{u}) \tilde{u}^- \, dx = 0.$$

Then, again by regularity theorems, $\tilde{u}$ is a classical solution of (3.10) and, by the strong maximum principle, $\tilde{u} > 0$ in Ω.

By Theorem 2.23 of Chapter 2, we know that at the mountain pass critical level there exists a critical point u whose Morse index is less than or equal to one.

Finally, if (3.11) holds we know the Morse index $m(\tilde{u})$ must be greater than or equal to one. Hence $m(\tilde{u}) = 1$. □

An important example of a nonlinearity $f(x, s)$ which satisfies all hypotheses of Theorem 3.4, as well as (3.11) is $f(x, s) = |s|^{p-1}s$, with $1 < p < \frac{N+2}{N-2}$ if $N \geq 3$, $1 < p$ if $N = 2$. In this case, (3.10) becomes the famous **Lane–Emden problem:**

$$\begin{cases} -\Delta u = |u|^{p-1}u & \text{in } \Omega \\ u = 0 & \text{on } \partial\Omega \end{cases} \tag{3.17}$$

3.2.2 Existence of a positive solution by constrained minimization

Another way of getting a positive solution of (3.10) is by looking for critical points of the functional J in (3.16) on a suitable Hilbert manifold in the space $H_0^1(\Omega)$. To this aim, let us observe that any critical point of J belongs to the **Nehari manifold** $\mathcal{N}$ defined by

$$\mathcal{N} = \left\{ v \in H_0^1(\Omega) \setminus \{0\} : \langle J'(v), v \rangle = \int_\Omega |\nabla v|^2 \, dx - \int_\Omega f(x, v)v \, dx = 0 \right\}.$$

We will show that, under some hypotheses on the function f, the set $\mathcal{N}$ is a C^1-Hilbert manifold of codimension one in $H_0^1(\Omega)$ and is a natural constraint for the functional J, in the sense that any critical point of the restriction of J on $\mathcal{N}$ is a critical point of J in the whole space $H_0^1(\Omega)$.

Proposition 3.5. *Let us assume that f satisfies (3.11), (3.13)–(3.15). Then $\mathcal{N}$ is a C^1-Hilbert manifold of codimension one and:*

(i) $\exists r > 0$ *such that* $B_r \cap \mathcal{N} = \emptyset$;
(ii) *any critical point of* $J|_{\mathcal{N}}$ *is a critical point of* J *on* $H_0^1(\Omega)$;
(iii) *for any* $u \in H_0^1(\Omega) \setminus \{0\}$ *there exists* $t(u) > 0$ *such that* $t(u)u \in \mathcal{N}$.

Proof. Let us set $G(v) = \langle J'(v), v \rangle$ so that $\mathcal{N}$ is the zero level set of G. Clearly, G is of class C^1 on $H_0^1(\Omega)$ and, for any $u \in \mathcal{N}$, we have

$$\langle G'(u), u \rangle = J''(u)(u, u) + \langle J'(u), u \rangle = J''(u)(u, u) = Q_u(u) < 0$$

by (3.11), as we have shown before.

Hence $G'(u) \neq 0$, for any $u \in \mathcal{N}$ which means that $\mathcal{N}$ is a C^1-Hilbert manifold of codimension one.

The manifold $\mathcal{N}$ separates $H_0^1(\Omega)$ into two components and, by (3.13), (3.14) the one which contains the origin also contains a small ball around the origin, so that (i) holds.

Let z be a critical point of J restricted to $\mathcal{N}$, then by the Lagrange multiplier theorem (see [150]) there exists $\lambda \in \mathbb{R}$ such that

$$J'(z) = \lambda G'(z).$$

Taking the scalar product with z we get $(J'(z), z) = \lambda \langle G'(z), z \rangle$ which is impossible unless $\lambda = 0$, since $z \in \mathcal{N}$. Thus $\mathcal{N}$ is a natural constraint for the functional J.

To prove (iii), let us observe that if we fix $u \in H_0^1(\Omega)$, with $\|u\| = 1$ then, by the hypotheses on f, $G(tu) > 0$ for $t > 0$ and t small, while $G(tu) < 0$ for t large. Therefore, there exists $t(u) > 0$ such that $t(u) \cdot u \in \mathcal{N}$. $\qquad\square$

Remark 3.6. It could be proved that, for any $u \in H_0^1(\Omega) \setminus \{0\}$, the number $t(u)$ in (iii) is unique and $t(u)u$ corresponds to the point where the functional $\Phi(t) = J(tu)$ achieves its maximum (see [225]).

The properties of the Nehari manifold allow to prove the following existence theorem.

Theorem 3.7. *Let* f *satisfy* (3.11), (3.13)–(3.15) *and*

$$\frac{1}{2}f(x, s)s - F(x, s) \geq c, \quad \forall x \in \Omega, s \in \mathbb{R} \tag{3.18}$$

for some constant $c \in \mathbb{R}$.

Then the Dirichlet problem (3.10) *has a nontrivial classical positive solution with Morse index equal to one.*

Proof. Let us start by proving that the functional J is bounded from below on the Nehari manifold $\mathcal{N}$. Indeed for any $u \in \mathcal{N}$, by (3.18), we have

$$J(u) = \int_\Omega \left[\frac{1}{2}uf(x, u) - F(x, u) \right] dx \geq c|\Omega|.$$

As in Theorem 3.4, we have that J satisfies the Palais–Smale compactness condition (see e. g. [196]), consequently the infimum of J on $\mathcal{N}$ is achieved at a point $u \in \mathcal{N}$ (see e. g. [14], Theorem 7.12). Then u is a critical point of $J|_{\mathcal{N}}$, and hence by (ii) of Proposition 3.5 it is also a critical point of J in $H_0^1(\Omega)$ and so a weak solution of (3.10).

Proceeding as in the proof of Theorem 3.4, i. e., using the modified functional J^+, we get a positive classical solution.

Since u is a minimizer of a C^2-functional on a Hilbert manifold of codimension one, we have that the Morse index $m(u)$ is less or equal than one. Indeed, denoting by $T(u)$ the tangent space of the manifold $\mathcal{N}$ at u we have that

$$J''(u)(v,v) \geq 0 \quad \forall v \in T(u). \tag{3.19}$$

This holds because for every $v \in T(u)$ we consider a C^1-curve $y \colon [-1,1] \longrightarrow \mathcal{N}$ such that $y(0) = u$ and $y'(0) = v$. Since $J'(u)(v) = 0$, we get that $J \circ y \colon [-1,1] \longrightarrow \mathbb{R}$ is twice differentiable at $t = 0$ with derivative

$$\left.\frac{\partial^2}{\partial t^2}(J \circ y)\right|_{t=0} = \langle J''(u)v, v \rangle.$$

Recalling that $J(u)$ is the minimum of J on $\mathcal{N}$, we infer that $\frac{\partial^2}{\partial t^2}(J \circ y)|_{t=0} \geq 0$. Then $m(u) \leq 1$, since $T(u)$ has codimension one and using (3.11) we get $m(u) = 1$, as in Theorem 3.4. $\qquad\square$

Definition 3.8. A solution u of (3.10) which minimizes the functional J on the Nehari manifold $\mathcal{N}$ is called *least-energy solution*. Since all solutions of (3.10) belong to $\mathcal{N}$, this is equivalent to say that u has the least energy among all solutions of (3.10).

Let us denote by α the minimum of J on the Nehari manifold $\mathcal{N}$. It could be proved that

$$\alpha = \inf_{y \in \Gamma} \max_{t \in [0,1]} J(y(t))$$

where $\Gamma = \{ y \in C^0([0,1], H_0^1(\Omega)) : y(0) = 0, J(y(1)) < 0 \}$ (see [225], Theorem 4.2).

This interesting min-max characterization of α also would allow to prove that a minimizer u of J on $\mathcal{N}$ has Morse index $m(u) \leq 1$ (and hence $m(u) = 1$ if (3.11) holds), with the same arguments used to prove that a mountain pass critical point has Morse index less than or equal to one (see Section 2.2.3 of Chapter 2).

3.2.3 Uniqueness of solutions of Morse index one of Lane-Emden problems

Let us consider the Lane–Emden Dirichlet problem (3.17) for positive solutions and its more general version obtained by adding a linear perturbation:

$$\begin{cases} -\Delta u = u^p + \lambda u & \text{in } \Omega \\ u > 0 & \text{in } \Omega \\ u = 0 & \text{on } \partial\Omega \end{cases} \tag{3.20}$$

where Ω is a bounded smooth domain in $\mathbb{R}^N$, $N \geq 2$, $\lambda \in \mathbb{R}$ and $p > 1$.

An important question related to the study of (3.20) is to know whether the solution is unique, whenever it exists. It is not difficult to provide cases when (3.20) admits more than one solution, as in the case of the annulus or more general annular domains (see, e. g., [58, 129, 157]) or in some nonconvex domains as the ones dumb-bell shaped (see [86]). In the case of convex domains, a conjecture has been formulated, with its roots in the papers [86] and [122].

Conjecture 1. *If Ω is a bounded and convex domain, then there exists only one solution of (3.20) for the full range of λ and $p > 1$ for which solutions exist.*

Note that a necessary condition to have a solution of (3.20) is $\lambda < \lambda_1(\Omega)$ where $\lambda_1(\Omega)$ is the first eigenvalue of the operator $-\Delta$ in $H_0^1(\Omega)$. This is easily proved by multiplying (3.20) by the first eigenfunction of $-\Delta$ and integrating.

Moreover, no solutions of (3.20) exist if $N \geq 3$, Ω is star-shaped, $p \geq \frac{N+2}{N-2}$ and $\lambda \leq 0$ as a consequence of the famous Pohozaev's identity [192].

On the other side, if $\lambda < \lambda_1(\Omega)$, if $N \geq 2$ or if $N \geq 3$ and $1 < p < \frac{N+2}{N-2}$, a solution of (3.20) always exists, using the mountain pass theorem or the constrained minimization method described in the previous section.

Let us also observe that, if $0 < p < 1$, then uniqueness holds in any smooth bounded domain [44, 46].

When Ω is a ball the conjecture has been proved for the full range of the values of λ and p for which existence holds, mainly exploiting ODE techniques. Indeed, the famous symmetry result of Gidas, Ni, Nirenberg [122] asserts that every positive solution of (3.20) is radial and radially decreasing (see Chapter 6) so that (3.20) can be rewritten as an ordinary differential equation. When $\lambda = 0$, the proof is quite easy and can be obtained by rescaling arguments and the uniqueness for O. D. E. initial value problem (see also [122]). Instead, when $\lambda \neq 0$ the uniqueness in the ball is much more difficult to obtain and the complete result is spread in several papers [4, 5, 182, 209, 226].

When Ω is not a ball, only few results are available. In the case $\lambda = 0$, some are of perturbative type like that of [228] for domains close to a ball or that of [133] where the exponent p is close to $\frac{N+2}{N-2}$ in dimension $N \geq 3$ and the domain Ω is assumed to be symmetric and convex in N orthogonal directions. We also quote [67] for a partial result in star-shaped domains.

As regards general results, the only ones to our knowledge are those contained in [74, 86, 158], again for the case $\lambda = 0$. In [74] and [86] the case of domains in $\mathbb{R}^2$, symmetric and convex in two orthogonal directions are considered and, by different methods, it is proved that only one positive solution exists for any exponent $p > 1$.

Instead in the paper [158] only the uniqueness of the least energy solution in convex domains in the plane is proved. However, the proof is very interesting and applies more generally to solutions of Morse index one. Since it is a nice application of Morse theory, we will detail it below.

Finally, let us quote two recent results for the case when Ω is a square in $\mathbb{R}^2$ obtained by a "computer-assisted" proof; see [175, 176]. They apply also when $\lambda \in [0, \lambda_1(\Omega)]$ and seem to be the only ones available for $\lambda \neq 0$.

Now, following [158], we will show that in a convex domain in $\mathbb{R}^2$ problem (3.20) with $\lambda = 0$ has only one positive solution with Morse index one.

We start by proving some preliminary results. The first one holds in any bounded domain in any dimension and for every $\lambda < \lambda_1(\Omega)$.

Lemma 3.9. *Let Ω be any smooth bounded convex domain in $\mathbb{R}^N$, $N \geq 2$, then there exists $p_0 > 1$ such that (3.20) has only one positive solution for any $p \in (1, p_0]$ and for any $\lambda < \lambda_1(\Omega)$.*

Proof. If u_1 and u_2 are two distinct solutions, then $w = u_1 - u_2$ must change sign, otherwise the identity

$$0 = \int_\Omega u_1(-\Delta u_2 - \lambda u_2) - u_2(\Delta u_1 - \lambda u_1)\, dx = \int_\Omega u_1 u_2(u_2^{p-1} - u_1^{p-1})\, dx$$

deduced from (3.20) would imply $u_1 \equiv u_2$.

Now let u_n be a solution of (3.20) with $p = p_n$, $p_n \searrow 1$ and set $M_n = \max u_n = u_n(Q_n)$, for some $Q_n \in \Omega$. By using the method of moving plane (see Chapter 6), we can show that, since Ω is convex, there exists a neighborhood V of $\partial\Omega$ such that $Q_n \notin V$, $\forall n$.

We claim that $M_n^{p_n-1} \longrightarrow \lambda_1(\Omega) - \lambda$ as $n \to +\infty$.

First of all, we show that $M_n^{p_n-1}$ is bounded. Suppose that $M_n^{p_n-1} \longrightarrow +\infty$ and consider the function

$$\tilde{u}_n(x) = \frac{1}{M_n} u_n\left(\frac{x + Q_n}{M_n^{\frac{p_n-1}{2}}}\right).$$

which satisfies the equation $-\Delta \tilde{u}_n = \tilde{u}_n^{p_n} + \frac{\lambda}{M_n^{p_n-1}}\tilde{u}_n$ in $M_n^{\frac{p_n-1}{2}}\Omega - Q_n$.

By standard elliptic estimates, $\tilde{u}_n$ converges uniformly to a function $\tilde{u} \in C^2(K)$, for any compact set K in $\mathbb{R}^N$ and $\tilde{u}$ satisfies

$$\begin{cases} -\Delta\tilde{u} = \tilde{u} & \text{in } \mathbb{R}^N \\ \tilde{u} > 0 & \text{in } \mathbb{R}^N \\ \tilde{u}(0) = 1 \end{cases}$$

Let λ_R and Φ_R be respectively the first eigenvalue and the corresponding positive eigenfunction of the operator $-\Delta$ in the ball $B_R(0)$ with zero Dirichlet boundary condition.

Using the Hopf's boundary lemma and the fact that $\lambda_R \longrightarrow 0$ as $R \to +\infty$, for R large we have

$$0 > \int_{\partial B_R(0)} \tilde{u}\frac{\partial \Phi_R}{\partial v}\, d\sigma = (1 - \lambda_R) \int_{B_R(0)} \tilde{u}\Phi_R\, dx > 0$$

which is a contradiction and shows that $M_n^{p_n-1}$ is bounded. Thus, up to a subsequence, $M_n^{p_n-1} \longrightarrow \mu$.

Let $\bar{u}_n = \frac{u_n}{M_n}$, which is a solution of the problem

$$\begin{cases} -\Delta \bar{u}_n = M_n^{p_n-1}\bar{u}_n^{p_n} + \lambda \bar{u}_n & \text{in } \Omega \\ \bar{u}_n = 0 & \text{on } \partial\Omega \end{cases}$$

By elliptic estimates, $\bar{u}_n$ converges in $C^2(\Omega) \cap C^0(\overline{\Omega})$ to a function $\bar{u}$ which satisfies

$$\begin{cases} -\Delta \bar{u} = \mu\bar{u} + \lambda\bar{u} & \text{in } \Omega \\ \bar{u} > 0 & \text{in } \Omega \\ \bar{u} = 0 & \text{on } \partial\Omega \end{cases}$$

Hence $\mu + \lambda = \lambda_1(\Omega)$ and $\bar{u} = \varphi_1$, the first eigenfunction of $-\Delta$ in Ω, so that $\mu = \lambda_1(\Omega) - \lambda$ and the claim is proved.

Now suppose that the assertion of the theorem is false, i. e., let us assume that u_n and v_n are two distinct solutions of (3.20) with $p = p_n \searrow 1$. Since $\bar{u}_n \longrightarrow \varphi_1$ uniformly, then $\bar{u}^{p_n-1} \longrightarrow 1$, and hence the convergence of $M_n^{p_n-1}$ to $\lambda_1 - \lambda$ implies that

$$u_n^{p_n-1} \longrightarrow \lambda_1 - \lambda \text{ uniformly in any compact set of } \Omega.$$

Obviously, the same happens to the sequence $\bar{v}_n^{p_n-1}$ where $\bar{v}_n = \frac{v_n}{\|v_n\|_\infty}$. The functions $w_n = \frac{u_n - v_n}{\|u_n - v_n\|_\infty}$ satisfy

$$\begin{cases} -\Delta w_n = g_n w_n + \lambda w_n & \text{in } \Omega \\ w_n = 0 & \text{on } \partial\Omega \end{cases}$$

where $g_n = \frac{u_n^{p_n} - v_n^{p_n}}{u_n - v_n} \longrightarrow \lambda_1 - \lambda$ for what we have proved before. Again by standard elliptic estimates we deduce that $w_n \longrightarrow \varphi_1$ uniformly. This is not possible since φ_1 does not change sign while we have shown, at the beginning of the proof, that w_n must change sign. $\qquad\square$

The next result applies to Morse index one solutions of (3.20) with $\lambda = 0$, in convex domains in the plane. It shows that this type of solutions are nondegenerate, i. e., zero is not an eigenvalue of the linearized operator (3.7) which, in the case of the Lane–Emden problem becomes

$$L_u = -\Delta - p|u|^{p-1} \tag{3.21}$$

for a solution u of (3.20).

Theorem 3.10. *Assume that Ω is a smooth bounded convex domain in $\mathbb{R}^2$ and u is a classical solution of (3.20), for $\lambda = 0$, with Morse index $m(u) = 1$. Then u is nondegenerate.*

Proof. Since $m(u) = 1$, only the first eigenvalue of L_u is negative. Therefore, arguing by contradiction, we assume that the second eigenvalue $\lambda_2 = \lambda_2(L_u)$ is zero. Then a corresponding eigenfunction φ_2 solves

$$\begin{cases} -\Delta\varphi_2 - pu^{p-1}\varphi_2 = 0 & \text{in } \Omega \\ \varphi_2 = 0 & \text{on } \partial\Omega \end{cases} \tag{3.22}$$

Consider a point $Q = (x_0, y_0) \in \mathbb{R}^2$ which will be chosen later and the function:

$$w(x,y) = (x - x_0)\frac{\partial u}{\partial x} + (y - y_0)\frac{\partial u}{\partial y}$$

which satisfies

$$-\Delta w - pu^{p-1}w = 2u^p \quad \text{in } \Omega \tag{3.23}$$

Multiplying (3.20) by φ_2 and (3.22) by u, integrating and subtracting, we get

$$(p-1)\int_\Omega u^p \varphi_2\, dx = 0. \tag{3.24}$$

Instead, multiplying (3.22) by w and (3.23) by φ_2, integrating and subtracting we have

$$-\int_{\partial\Omega} w\frac{\partial\varphi_2}{\partial v}\, ds = -2\int_\Omega u^p\varphi_2\, dx = 0 \tag{3.25}$$

by (3.24).

Now we observe that by Courant's theorem, the nodal line

$$l(\varphi_2) = \{\, x \in \Omega : \varphi_2(x) = 0 \,\},$$

of the eigenfunction φ_2 divides Ω into two regions. Then either the closure of $l(\varphi_2)$ does not touch $\partial\Omega$ or it does at two points. If the first case happens, then we choose the point Q as any point inside Ω and we deduce that

$$w(x,y) < 0 \quad \text{for any point } (x,y) \in \partial\Omega.$$

Then by (3.25), we get a contradiction since $\frac{\partial\varphi_2}{\partial v}$ has only one sign on $\partial\Omega$, and by Hopf's boundary lemma cannot be identically zero.

In the case when the closure of $l(\varphi_2)$ touches $\partial\Omega$, then we denote by $P_i \in \partial\Omega$, $i = 1, 2$, the two points in $\overline{l(\varphi_2)} \cap \partial\Omega$. Note that if they coincide we are back in the previous case. Again we distinguish two possibilities: either the tangent lines to $\partial\Omega$

at the points P_i are parallel or not. In the second case, we choose the point Q in the definition of the function w as the intersection of the two tangent lines and observe that w and $\frac{\partial \varphi_2}{\partial v}$ both simultaneously change sign at P_i, $i = 1, 2$. This implies that the product $w \frac{\partial \varphi_2}{\partial v}$ has only one sign on $\partial\Omega$, and hence (3.25) leads again to a contradiction.

If instead the two tangent lines are parallel, we assume that the x-axis is the common direction and set $w = \frac{\partial u}{\partial x}$. Then, repeating the same argument as before, we obtain

$$\int_{\partial\Omega} w \frac{\partial \varphi_2}{\partial v} \, ds = 0$$

which is again a contradiction to the Hopf's boundary lemma. $\qquad\square$

We can now prove the uniqueness result.

Theorem 3.11. *Let Ω be a smooth bounded convex domain in $\mathbb{R}^2$. Then for any $p > 1$ and $\lambda = 0$ there exists only one solution of (3.20) with Morse index one.*

Proof. By Lemma 3.9, we know that there exists $p_0 > 1$ such that (3.20) has a unique solution for $p \in (1, p_0)$. Obviously, this solution is the least energy solution which has Morse index one, and hence is nondegenerate by Theorem 3.10. Let $(1, \bar{p})$ be the maximal interval with this uniqueness property. If $\bar{p} = +\infty$, the assertion is proved; otherwise, since all solutions are nondegenerate, we have that there is only one solution of Morse index one also for $p = \bar{p}$. This is a consequence of the implicit function theorem and of the fact that the Morse index of a solution does not change, as p varies, as long as it stays nondegenerate (since by Theorem 1.44 the eigenvalues depend continuously on the coefficients).

Arguing by contradiction, let us assume that there exists a sequence $p_n \searrow \bar{p}$ and two distinct solutions u_n, v_n of (3.20) with Morse index one for $p = p_n$. By standard elliptic estimates, we have that u_n, v_n both converge in $C^2(\Omega)$ to the unique solution $\bar{u}$ corresponding to $p = \bar{p}$. Set

$$w_n = u_n - v_n \quad \text{and} \quad \bar{w}_n = \frac{w_n}{\|w_n\|_{H_0^1(\Omega)}}. \tag{3.26}$$

Then $\bar{w}_n$ satisfies

$$\begin{cases} -\Delta\bar{w}_n = \alpha_n(x)\bar{w}_n & \text{in } \Omega \\ \bar{w}_n = 0 & \text{on } \partial\Omega \end{cases} \tag{3.27}$$

where $\alpha_n(x) = \int_0^1 p_n(tu_n(x) + (1-t)v_n(x))^{p_n-1} \, dt$.

Moreover, $\bar{w}_n \longrightarrow \bar{w}$ weakly in $H_0^1(\Omega)$ and $\bar{w} \neq 0$. In fact by (3.27), we have

$$1 - \int_\Omega |\nabla\bar{w}_n|^2 \, dx = \int_\Omega \alpha_n\bar{w}_n^2 \, dx = \bar{p} \int_\Omega \bar{u}^{\bar{p}-1}\bar{w}^2 \, dx + o(1) \tag{3.28}$$

which implies $\bar{w} \neq 0$. Passing to the limit in (3.27), we get

$$\begin{cases} -\Delta\bar{w} = \bar{p}\bar{u}^{\bar{p}-1}\bar{w} & \text{in } \Omega \\ \bar{w} \neq 0 & \text{in } \Omega \\ \bar{w} = 0 & \text{on } \partial\Omega \end{cases} \qquad (3.29)$$

which is a contradiction since we assumed $\bar{u}$ to be nondegenerate. $\qquad\square$

Remark 3.12. In the recent paper [99], using some uniform estimates obtained in [146], it is proved that when the exponent p is large all solutions of (3.20) in convex planar domain have Morse index one. Thus combining this result with Theorem 3.11, we get that the uniqueness conjecture is true for $p > p_1$, for some exponent $p_1 > 1$. The proof of [99] relies on a careful analysis of the asymptotic behavior of solutions of (3.20) performed in [103, 104] and [98].

Remark 3.13. The proof of Theorem 3.11 relies on the nondegeneracy of the solutions and on the fact that the uniqueness holds when the exponent p is close to one. This last fact is true regardless of the Morse index and also in higher dimension, while the proof of the nondegeneracy strongly uses the fact that the Morse index of the solution is one and that the domain is planar.

3.3 Sign changing solutions of Dirichlet problems

In this section, we consider again the general semilinear elliptic problem (3.10) and study the existence and the Morse index of sign changing solutions.

In the case when the nonlinear term is superlinear, but not necessarily odd, the existence of sign changing solutions in general bounded domains is not obvious. The first result is due to Castro, Cassio and Neuberger in [57], then other results have been obtained in [20] and in [23]. In the next section, we will use the approach of [23] to prove the existence of a least-energy sign changing solution and show that its Morse index is two. In the other sections, we will study the case of symmetric solutions to give an estimate of their Morse index.

Often we will refer to a sign changing solution as a **nodal solution** since it must have at least two nodal domains.

3.3.1 Existence of a solution with Morse index two by constrained minimization

We study the Dirichlet problem (3.10) under some hypotheses on the function $f(x,s)$. Some of them are the same as for the case of positive solutions but for the reader's convenience we prefer to state precisely what is needed for the existence of sign changing solutions.

So let us consider the problem (3.10) where Ω is a smooth bounded domain in $\mathbb{R}^N$, $N \geq 2$. We assume that the function $f(x,s)$ satisfies the following assumptions:

(i)

$$f \in C^1(\Omega \times \mathbb{R}, \mathbb{R}), \quad f(x,0) = 0 \quad \forall x \in \Omega; \tag{3.30}$$

(ii)

$$\text{there exists } p \in \left(1, \frac{N+2}{N-2}\right) \text{ if } N \geq 2 \text{ or } p > 1 \text{ if } N = 2,$$

$$\text{such that } \left|\frac{\partial f}{\partial s}(x,s)\right| \leq c(1 + |s|^{p-1}) \quad \forall x \in \Omega, \ s \in \mathbb{R}; \tag{3.31}$$

(iii)

$$\frac{\partial f}{\partial s}(x,s) > \frac{f(x,s)}{s} \quad \forall \, x \in \Omega, \ s \neq 0; \tag{3.32}$$

(iv)

$$\text{there exist } R > 0 \text{ and } \alpha > 2 \text{ such that}$$

$$0 < \alpha F(x,s) \leq s f(x,s), \quad \forall x \in \Omega, \ |s| \geq R$$

$$\text{where } F(x,s) = \int_0^s f(x,s)\,ds \text{ is a primitive of f;} \tag{3.33}$$

(v)

$$\text{the second Dirichlet eigenvalue } \mu_2 \text{ of the operator}$$

$$-\Delta - \frac{\partial f}{\partial s}(x,0) \quad \text{on } \Omega \text{ is positive.} \tag{3.34}$$

Note that (3.32) is the same as (3.11) which is important to estimate the Morse index of a solution of (3.10).

We consider again the functional J associated to (3.10)

$$J(v) = \frac{1}{2}\int_\Omega |\nabla v|^2\,dx - \int_\Omega F(x,v)\,dx, \quad v \in H_0^1(\Omega). \tag{3.35}$$

Then we define the set:

$$M = \left\{ u \in H_0^1(\Omega) : u^+ \neq 0, \ u^- \neq 0, \ \langle J'(u), u^+ \rangle = \langle J'(u), u^- \rangle = 0 \right\}$$

where u^+ and u^- are the positive and negative part of u, i.e., $u^+ = \max\{u, 0\}$, $u^- = \max\{-u, 0\}$.

The set M is not a manifold in $H_0^1(\Omega)$ but we will show that $M \cap H$ is a manifold where $H = H_0^1(\Omega) \cap H^2(\Omega)$, endowed with the scalar product from $H^2(\Omega)$. Obviously, M contains all sign changing solutions of (3.10) and is contained in the Nehari manifold $\mathcal{N}$ previously defined. It is often called the **nodal Nehari set**.

Proposition 3.14. *The set $M \cap H$ is a C^1-manifold of codimension two in H.*

For the proof of the above result, we need the following lemma whose proof is contained in [23].[1]

Lemma 3.15. *Let us define the functionals:*

$$Q_\pm(u) = \int_\Omega \nabla u \cdot \nabla u^\pm \, dx = \pm \int_\Omega |\nabla u^\pm|^2 \, dx$$

$$\Psi_\pm(u) = \int_\Omega f(x,u)u^\pm \, dx,$$

$u \in H_0^1(\Omega)$. *Then, setting $f'(x,s) = \frac{\partial f}{\partial s}(x,s)$, we have:*

a) $Q_\pm$ *is differentiable at $u \in H$, with derivative $Q'_\pm \in H^{-1}(\Omega)$ given by*

$$Q'_\pm(u)v = \pm \int_{[\pm u > 0]} ((-\Delta u)v + \nabla u \cdot \nabla v) \, dx.$$

b) $Q_\pm|_H \in C^1(H)$.

c) $\Psi_\pm \in C^1(H_0^1(\Omega))$ *with derivative given by*

$$\Psi'_\pm(u)v = \pm \int_{[\pm u > 0]} f'(x,u)uv \, dx \pm \int_{[\pm u > 0]} f(x,u)v \, dx$$

In particular,

$$Q'_+(u)u^- = Q'_-(u)u^+ = \Psi'_+(u)u^- = \Psi'_-(u)u^+ = 0$$

$$Q'_\pm(u)u^\pm = \pm 2\int_\Omega \nabla u \nabla u^\pm \, dx = 2\int_\Omega |\nabla u^\pm|^2 \, dx$$

$$\Psi'_\pm(u)u^\pm = \int_\Omega f'(x,u)(u^\pm)^2 \, dx \pm \int_\Omega f(x,u)u^\pm \, dx.$$

Proof of Proposition 3.14. Let us observe that $M \cap H$ is the intersection of the zero level sets of the functionals $g_\pm$ defined by

$$g_\pm(u) = \langle J'(u), u^\pm \rangle = Q_\pm(u) - \Psi_\pm(u), \quad u \in H, \tag{3.36}$$

i. e. $M \cap H = \{ u \in H : u^+ \neq 0, u^- \neq 0, g_+(u) = 0 = g_-(u) \}$.

---

[1] Note that in the paper [23] the authors make the convention that $u^- = \min\{u, 0\}$ and, therefore, the lemma appears slightly different.

The previous lemma implies that $g_{\pm}|_H \in C^1(H)$ and, since $\int_\Omega \nabla u \nabla u^{\pm}\, dx = \int_\Omega f(x,u)u^{\pm}\, dx$ for $u \in M \cap H$, we have

$$g'_+(u)u^+ = \int_\Omega \left(|\nabla u^+|^2 - f'(x,u)(u^+)^2\right) dx, \quad g'_+(u)u^- = 0$$

$$g'_-(u)u^- = \int_\Omega \left(|\nabla u^-|^2 - f'(x,u)(u^-)^2\right) dx, \quad g'_-(u)u^+ = 0$$

for $u \in M \cap H$.

Hence the assumption (3.32) yields

$$g'_+(u)u^+ = \int_\Omega \left[\nabla u \nabla u^+ - f'(x,u)(u^+)^2\right] dx = \int_{[u>0]} \left[\frac{f(x,u)}{u} - f'(x,u)\right](u^+)^2\, dx < 0$$

and

$$g'_-(u)u^- = \int_\Omega \left[-\nabla u \nabla u^- - f'(x,u)(u^-)^2\right] dx = \int_\Omega \left[-f(x,u)u^- - f'(x,u)(u^-)^2\right] dx$$

$$= \int_{[u<0]} \left[\frac{f(x,u)}{u} - f'(x,u)\right](u^-)^2\, dx < 0$$

for any $u \in M \cap H$.

Approximating u^+ and u^- by functions in H, we deduce that the linear map $(g'_+(u), g'_-(u))$ from H into $\mathbb{R}^2$ is surjective, for every $u \in M \cap H$. Thus the assertion is proved. $\qquad\square$

Let us set

$$\beta = \inf_{u \in M} J(u). \tag{3.37}$$

Our aim is to show that β is achieved by a function $\bar{u}$ which is a sign changing solution of (3.10) and, therefore, will be called **least energy nodal solution**. For the proof of the existence of a minimizer for (3.37), we follow the approach of [23], which combines variational and topological methods.

We recall some notation from [20] and [23]. We set

$$X = \left\{ u \in C^1(\overline{\Omega}) : u|_{\partial\Omega} = 0 \right\}$$

and

$$J^c = \left\{ u \in H^1_0(\Omega) : J(u) \le c \right\}, \quad J^c_X = X \cap J^c$$

and

$$S_c = \left\{ u \in H^1_0(\Omega) : J(u) = c,\ J'(u) = 0 \right\}.$$

Let P denote the closed cone of nonnegative functions in X. As in [20], we consider the sets:

$$SC^- = \{\, u \in X : u \text{ is a sign changing subsolution of (3.10)} \,\},$$
$$SC^+ = \{\, u \in X : u \text{ is a sign changing supersolution of (3.10)} \,\},$$
$$\mathcal{I} = \{(0,0)\} \cup \{\, (u,v) \in SC^- \times SC^+ : u < v \,\} \subset X \times X,$$
$$A = \bigcup_{(u,v)\in\mathcal{I}} ((u+P) \cup (v-P)) \subset X.$$

Using the gradient vector field $-\nabla J$ we obtain a flow $\eta \colon \mathcal{O} \longrightarrow H_0^1(\Omega)$ defined on an open subset $\mathcal{O} \subset \mathbb{R} \times H_0^1(\Omega)$ which satisfies

$$\begin{cases} \dfrac{\partial}{\partial t}\varphi(t,u) = -\nabla J(\eta(t,u)) \\[2mm] \eta(0,u) = u \end{cases}$$

for all $(t,u) \in \mathcal{O}$. By means of this flow the following deformation lemma can be proved (see [20]).

Lemma 3.16. *Suppose that $S_c \subset A$, for some $c > 0$. Then there exists $\epsilon > 0$ and a homotopy $h \colon (J_X^{c+\epsilon} \cup A) \times [0,1] \longrightarrow J_X^{c+\epsilon} \cup A$ such that:*
(i) $h_t(J_X^d \cup A) \subset J_X^d \cup A, \forall d \le c + \epsilon, t \in [0,1];$
(ii) $h_1(J_X^{c+\epsilon} \cup A) \subset J_X^{c-\epsilon} \cup A.$

The existence of a least-energy nodal solution is given by the following theorem.

Theorem 3.17. *Under the assumptions (3.30)–(3.34), the infimum β (see (3.37)) is achieved by a function $\bar{u}$ which is a sign changing solution of (3.10) with Morse index two and exactly two nodal domains.*

Proof. We have assumed (3.34) but, in order to simplify the proof, we suppose the stronger hypothesis that the first eigenvalue μ_1 of the operator $-\Delta - \frac{\partial f}{\partial s}(x,0)$ is positive. This is obviously the case when $\frac{\partial f}{\partial s}(x,0) = 0$, so that μ_1 is just the first eigenvalue of $-\Delta$. For the more general case, we refer to [23].

Let e_1 be the first normalized Dirichlet eigenfunction of the operator $-\Delta - \frac{\partial f}{\partial s}(x,0)$ on Ω and consider the set

$$S_r = \left\{\, w \in E_1^\perp, \ \|w\|_{H_0^1(\Omega)} = r \,\right\} \subset H_0^1(\Omega)$$

where $E_1 = \mathbb{R}e_1$ is the space spanned by e_1.

Observe that $\alpha = \inf J(S_r) > 0$ and $S_r \cap A = \emptyset$ (see [20]). Setting $y = \frac{\alpha}{2}$, we consider the inclusion

$$j_c \colon (J_X^c \cup A, J_X^y \cup A) \longrightarrow (H_0^1(\Omega), H_0^1(\Omega) \setminus S_r)$$

for any $c \geq y$, which is well-defined since $S_r \cap A = \emptyset$. It induces a homomorphism:

$$j_c^* : H^2(H_0^1(\Omega), H_0^1(\Omega) \setminus S_r) \longrightarrow H^2(J_X^c \cup A, J_X^y \cup A)$$

where $H^*(C, D)$ stands for the Alexander–Spanier cohomology of the pair with $D \subset C$ with integer coefficients.

Next, we prove that

$$H^2(H_0^1(\Omega), H_0^1(\Omega) \setminus S_r) \cong \mathbb{Z}. \tag{3.38}$$

Indeed, the pair $(H_0^1(\Omega), H_0^1(\Omega) \setminus S_r)$ is the same as the product pair $(E_1, E_1 \setminus \{0\}) \times (E_1^\perp, E_1^\perp \setminus S_r)$.

The Künneth theorem (see [207]) shows that

$$H^2(H_0^1(\Omega), H_0^1(\Omega) \setminus S^r) \cong H^1(E_1^\perp, E_1^\perp \setminus S_r) \cong \tilde{H}^0(E_1^\perp \setminus S_r) \cong \mathbb{Z}$$

which proves (3.38). Now we can define $\bar{c} = \inf\{c \geq y : j_c^*$ is injective$\}$. Then $\bar{c} \geq \alpha$, since $J_X^c \cup A \subset H_0^1(\Omega) \setminus S_r$, hence $j_c^* = 0$ for $c < \alpha$.

Next, we show

$$\bar{c} \leq \beta \tag{3.39}$$

with β given by (3.37). To prove this, let $\epsilon > 0$ and choose $u \in M$ such that $J(u) < \beta + \frac{\epsilon}{2}$. Since M is contained in the Nehari manifold $\mathcal{N}$, by Remark 3.6 we have

$$J(\lambda u^+ - \mu u^-) \leq J(u) \quad \text{for every } \lambda, \mu \geq 0.$$

Now we recall that, by the assumption (3.33) on $f(x, s)$,

$$\lim_{t \to 0} J(tu) = -\infty.$$

Therefore, there exists some number $R > 0$ such that

$$J(\lambda u^+ - \mu u^-) \leq 0 \quad \text{whenever } \max\{\lambda, \mu\} \geq R.$$

Approximating u^+ and u^- with suitable functions $v_1, v_2 \in P$, we get

$$J(\lambda v_1 - \mu v_2) \leq \beta + \epsilon \quad \text{for } 0 \leq \lambda, \ \mu \leq R,$$
$$J(\lambda v_1 - \mu v_2) \leq y \quad \text{if } \max\{\lambda, \mu\} \geq R.$$

Now we consider the sets

$$C = \{\lambda v_1 - \mu v_2 : 0 \leq \lambda, \mu \leq R\} \subset [v_1, v_2] \subset X,$$
$$\partial C = \{\lambda v_1 - \mu v_2 \in C : \min\{\lambda, \mu\} = 0 \text{ or } \max\{\lambda, \mu\} = R\}$$

where $[v_1, v_2]$ is the span of $\{v_1, v_2\}$.

We have the following inclusions:

$$(C, \partial C) \xhookrightarrow{i} (J_X^{\beta+\epsilon} \cup A, J_X^y \cup A) \xrightarrow{j_{\beta+\epsilon}} (H_0^1(\Omega), H_0^1(\Omega) \setminus S_r).$$

We claim that the induced map

$$i^* \circ j_{\beta+\epsilon}^* \colon H^2(H_0^1(\Omega), H_0^1(\Omega) \setminus S_r) \longrightarrow H^2(C, \partial C)$$

is an isomorphism.

We choose $e_2 \in E_1^\perp$ with $\|e_2\|_{H_0^1(\Omega)} = 1$ and consider the sets

$$C_1 = \{ \lambda e_1 + \mu e_2 : |\lambda| \le R, \ 0 \le \mu \le R \} = (B_R \cap E_1) \times [0, R] \cdot e_2,$$
$$\partial C_1 = \{ \lambda e_1 + \mu e_2 : |\lambda| = R \text{ or } \mu \in \{0, R\} \}.$$

Clearly, $(C, \partial C)$ can be deformed to $(C_1, \partial C_1)$ within $(H_0^1(\Omega), H_0^1(\Omega) \setminus S_r)$. This shows that $i^* \circ j_{\beta+\epsilon}^*$ is an isomorphism if, and only if, the inclusion

$$i_1 \colon (C_1, \partial C_1) \hookrightarrow (H_0^1(\Omega), H_0^1(\Omega) \setminus S_r)$$

induces an isomorphism. Now

$$(C_1, \partial C_1) \cong (E_1 \cap B_R, E_1 \cap S_R) \times ([0, R]e_2, \{0, R\}e_2)$$
$$(H_0^1(\Omega), H_0^1(\Omega) \setminus S_r) \cong (E_1, E_1 \setminus \{0\}) \times (E_1^\perp, E_1^\perp \setminus S_r)$$

Since the inclusions

$$(E_1 \cap B_R, E_1 \cap S_R) \hookrightarrow (E_1, E_1 \setminus \{0\}),$$
$$([0, R] \cdot e_2, \{0, R \cdot e_2\}) \hookrightarrow (E_1^\perp, E_1^\perp \setminus S_r)$$

induce isomorphisms on cohomology levels, the chain follows by the naturality of the Künneth maps.

Now, since $i^* \circ j_{\beta+\epsilon}^*$ is an isomorphism, $j_{\beta+\epsilon}^*$ is injective, and thus $\bar{c} \le \beta + \epsilon$, for $\epsilon > 0$ arbitrary. Hence (3.39) holds.

Next, we show that

$$S_{\bar{c}} \not\subset A \tag{3.40}$$

where $S_{\bar{c}}$ was defined as the set of the critical points of J at $\bar{c}$ level. Indeed, if this was not true then by Lemma 3.16 we can take $\epsilon > 0$ and a homotopy $h \colon (J_X^{\bar{c}+\epsilon} \cup A) \times [0, 1] \longrightarrow J_X^{\bar{c}+\epsilon} \cup A$ such that $h_1^* \circ j_{\bar{c}-\epsilon}^* = j_{\bar{c}+\epsilon}^*$, where

$$h_1^* \colon H^2(J_X^{\bar{c}-\epsilon} \cup A, J_X^y \cup A) \longrightarrow H^2(J_X^{\bar{c}+\epsilon} \cup A, J_X^y \cup A)$$

is induced by $h_1: (J_X^{\bar c+\epsilon} \cup A, J_X^y \cup A) \longrightarrow (J_X^{\bar c-\epsilon} \cup A, J_X^y \cup A)$. Hence, since $j_{\bar c+\epsilon}^*$ is injective, $j_{\bar c-\epsilon}$ has to be injective as well. This however contradicts the definition of $\bar c$, so that (3.40) holds.

Now let $\bar u \in S_{\bar c} \setminus A$, then $\bar u$ is a sign changing solution of (3.10). In particular, $\bar u$ belong to the nodal Nehari set M and, therefore, $\bar c = J(\bar u) \geq \beta$, which together with (3.39) implies that $\bar u$ is a minimum for (3.37).

Let us show that the Morse index $m(\bar u)$ is exactly 2. By (3.32), since $\bar u$ changes sign, $m(\bar u) \geq 2$ (see (3.12)). To show that $m(\bar u) \leq 2$, let us observe that $\bar u \in H$, by elliptic regularity. Then, since by Proposition 3.14, we have that $M \cap H$ is a C^1-manifold of codimension two, arguing exactly as in the proof of Theorem 3.7, we get that $m(\bar u) \leq 2$.

Finally, since $\bar u$ changes sign and has Morse index two, by (3.12) $\bar u$ must have exactly two nodal domains. $\qquad\square$

3.3.2 Estimates of Morse index for symmetric sign changing solutions: the autonomous case

In this section, we consider smooth bounded domains $\Omega \subset \mathbb{R}^N$, $N \geq 2$, symmetric with respect to an hyperplane which, without loss of generality, can be assumed to pass through the origin. More precisely, we denote by

$$T_i = \left\{ x = (x_1, \dots, x_N) \in \mathbb{R}^N, \, x_i = 0 \right\}, \quad i \in \{1, \dots, N\} \tag{3.41}$$

the hyperplane orthogonal to the unit vector, $e_i = \{0, \dots, 0, 1, 0, \dots, 0\}$ and denote by

$$\Omega_i^- = \{ x \in \Omega, \, x_i < 0 \} \quad \text{and} \quad \Omega_i^+ = \{ x \in \Omega, \, x_i > 0 \} \tag{3.42}$$

the two subdomains in which Ω is divided by T_i.

We assume that Ω is symmetric with respect to T_i, for some $i = 1, \dots, N$, so that Ω_i^- is just the reflection of Ω_i^+ with respect to T_i.

Then we consider the autonomous version of problem (3.10) which we rewrite as

$$\begin{cases} -\Delta u = f(u) & \text{in } \Omega \\ u = 0 & \text{on } \partial\Omega \end{cases} \tag{3.43}$$

where $f = f(x, s) = f(s)$ satisfies the hypothesis (3.30)–(3.34).

As it will be clear from the proof, all results of this section strongly depend on the fact that the nonlinear term f does not depend on the x-variable.

We consider a sign changing solution u which has the same symmetry as Ω, i. e., u is even in the x_i-variable, $i \in \{1, \dots, N\}$.

Note that such a symmetric solution of (3.43) always exists under the above assumptions on f, applying the variational methods of Section 3.3.1 in the subspace of $H_0^1(\Omega)$ given by the functions even in the x_i-variable, $i \in \{1, \dots, N\}$.

In particular, we will consider radial solutions of (3.43) when Ω is either a ball or an annulus centered at the origin.

The aim of this section is to prove estimates from below on the Morse index of symmetric sign changing solutions whose nodal sets does not touch $\partial\Omega$, following the ideas of [7]. This will be based on the general fact that each symmetry direction for a nodal solution u and for the domain, as defined above, produces a negative eigenvalue for the linearized operator at u which we recall to be defined, in the case of the Dirichlet problem, as

$$L_u = -\Delta - f'(u). \tag{3.44}$$

Moreover, from the estimate on the Morse index we will deduce information on the nodal set of the solution.

Let us denote by λ_k the eigenvalues of the operator L_u in Ω with homogeneous Dirichlet boundary conditions and by μ_i the first eigenvalue of L_u in the domain Ω_i^-, $i \in \{1,\ldots,N\}$. Thus, there exists a function ψ_i solution of

$$\begin{cases} -\Delta\psi_i - f'(u)\psi_i = \mu_i\psi_i & \text{in } \Omega_i^- \\ \psi_i = 0 & \text{on } \partial\Omega_i^- \\ \psi_i > 0 & \text{in } \Omega_i^- \end{cases} \tag{3.45}$$

Our proofs will be based on the study of the sign of μ_i when u is symmetric with respect to the x_i-direction. Let us start pointing out the following properties (earlier considered in [70] for the case of the eigenvalues of the Laplacian).

Proposition 3.18. *If u is a solution of (3.43) even in the x_i-variable, for some $i \in \{1,\ldots,N\}$ then the odd extension of ψ_i to Ω, defined by*

$$\tilde{\psi}_i(x) = \begin{cases} \psi_i(x) & \text{if } x \in \overline{\Omega_i^-} \\ -\psi_i(x_1,\ldots,x_{i-1},-x_i,x_{i+1},\ldots,x_N) & \text{if } x \in \Omega_i^+ \end{cases}$$

is an eigenfunction for the linearized operator L_u with corresponding eigenvalue μ_i. Hence $\mu_i = \lambda_{\beta(i)}$ for some $\beta(i) \geq 2$.

If u is even in k variables, say $x_1,\ldots,x_k$, $1 \leq k \leq N$, then the eigenfunctions $\tilde{\psi}_1,\ldots,\tilde{\psi}_k$ are linearly independent so they produce k eigenvalues $\lambda_{\beta(1)},\ldots,\lambda_{\beta(k)}$ of L_u in Ω.

Proof. Since ψ_i vanishes on the hyperplane T_i, we have

$$\frac{\partial\psi_i}{\partial x_j} = 0, \quad \frac{\partial^2\psi_i}{\partial x_j^2} = 0 \quad \text{on } T_i \cap \Omega \text{ for } j \neq i$$

and, by the equation in (3.45),

$$\frac{\partial^2\psi_i}{\partial x_i^2} = 0 \quad \text{on } T_i \cap \Omega.$$

Hence, by the symmetry of u in the x_i-variable, it follows easily that the functions $\tilde{\psi}_i$ are Dirichlet eigenfunctions of L_u in Ω with corresponding eigenvalue $\mu_i = \lambda_{\beta(i)}$. Since $\tilde{\psi}_i$ changes sign (it has precisely two nodal domains, namely Ω_i^- and Ω_i^+), then $\beta(i) \geq 2$. The second part of the assertion is a consequence of the oddness of $\tilde{\psi}_i$ in the x_i variable. $\qquad\square$

The general statement on the Morse index of a symmetric solution is the following (see [7]).

Theorem 3.19. *Let Ω be symmetric with respect to the hyperplane T_i and convex in the x_i-direction for some $i \in \{1,\dots,N\}$ (i.e. if x and y belong to Ω and lay on the same line parallel to the x_i-axis then the segment joining x and y is contained in Ω).*

If u is a sign changing solution of (3.43) even in the x_i-variable and such that the closure of its nodal set $N(u) = \{ x \in \Omega : u(x) = 0 \}$ does not intersect $\partial\Omega$ then the eigenvalue μ_i (see (3.45)) is negative. Moreover, for the Morse index $m(u)$, it holds

$$m(u) \geq k + 2 \tag{3.46}$$

where $k \in \{1,\dots,N\}$ is the number of variables x_i such that u is even in x_i.

Proof. Let us consider the function $\frac{\partial u}{\partial x_i}$ in Ω_i^-. By the symmetry of u in the x_i variable, we have that $\frac{\partial u}{\partial x_i} = 0$ on $\partial\Omega_i^- \cap T_i$. On the other hand, since $\overline{N(u)}$ does not touch $\partial\Omega$ we have that u does not change sign near $\partial\Omega$, so we can assume $u > 0$ near the boundary.

Then the smoothness of $\partial\Omega$ and the assumption that Ω is convex in the x_i direction imply that

$$\frac{\partial u}{\partial x_i} \geq 0 \quad \text{on } \partial\Omega_i^- .$$

In addition, $\frac{\partial u}{\partial x_i}$ satisfies

$$L_u\left(\frac{\partial u}{\partial x_i}\right) - \Delta\left(\frac{\partial u}{\partial x_i}\right) - f'(u)\frac{\partial u}{\partial x_i} = 0. \tag{3.47}$$

Since u changes sign in Ω_i^-, necessarily $\frac{\partial u}{\partial x_i}$ must change sign and be negative somewhere in Ω_i^-. Hence, there exists a domain D strictly contained in Ω_i^- such that

$$\frac{\partial u}{\partial x_i} < 0 \text{ in } D \quad \text{and} \quad \frac{\partial u}{\partial x_i} = 0 \text{ on } \partial D. \tag{3.48}$$

We deduce from (3.47)–(3.48) that the first eigenvalue of L_u in D is zero and as a consequence, the first eigenvalue μ_i of L_u in Ω_i^- is negative as we wanted to prove.

Then (see Proposition 3.18) the corresponding eigenvalue $\lambda_{\beta(i)} = \mu_i$ of L_u in Ω is negative, and to it there corresponds an eigenfunction $\tilde{\psi}_i$ odd with respect to x_i.

On the other side, under the assumption (3.32) the first eigenvalue λ_1 of L_u in Ω is negative. Moreover, since u is even in x_i and changes sign also the second eigenvalue of L_u in the space of even functions must be negative, again by (3.32). Summing up all the symmetric directions, we get that the Morse index $m(u)$ satisfies (3.46). □

The assumption that Ω is convex in the x_i-direction is not essential, but simplifies the proof otherwise other parts of $\partial\Omega_i^-$, neither on $\partial\Omega$ nor on T_i, should be considered. Below we will consider the case of an annulus and it will be clear how to deal with domains which have holes. In particular, Theorem 3.19 applies to nodal radial solutions in the ball which are even in all variables $x_1, \ldots, x_N$ and whose nodal sets do not touch $\partial\Omega$ so that, using also Proposition 3.18, we deduce that their Morse index is at least $N + 2$. However, in the radial case, we can have a better estimate which also takes into account the number of nodal regions of u and the number of the radial negative eigenvalue of L_u. This has been derived in [106] in the case when Ω is a ball and also works when the domain is an annulus.

Let Ω be a ball or an annulus centered at the origin and u a nodal radial solution of (3.43). We denote by $m_{\mathrm{rad}}(u)$ the *radial Morse index of u*, i. e., the number of negative *radial eigenvalues* of the linearized operator L_u.

Theorem 3.20. *If Ω is a ball or an annulus in $\mathbb{R}^N$, $N \geq 2$ and u is a radial solution of (3.43) with $n = n(u) \geq 2$ nodal domains then*

$$m(u) \geq m_{\mathrm{rad}} + N(n - 1). \tag{3.49}$$

Since the assumption (3.32) implies that

$$m_{\mathrm{rad}}(u) \geq n$$

(as for (3.11), (3.12)), then from (3.49) we deduce

$$m(u) \geq n + N(n - 1). \tag{3.50}$$

As for Theorem 3.19, the proof is based on using the partial derivatives of u to produce negative eigenvalues whose corresponding eigenfunctions are odd with respect to a hyperplane passing through the origin. However, since we now consider also the case of an annulus and want to get a more precise estimate we need some extra work.

Proof of Theorem 3.20. For any $i = 1, \ldots, N$, we consider the hyperplanes T_i and the domains Ω_i^- as defined in (3.41), (3.42). Note that now Ω_i^- is a *half-ball* or a *half-annulus* determinated by T_i, for each $i = 1, \ldots, N$.

Let us fix $n \in \mathbb{N}$, $n \geq 2$ and let us denote by u_n a radial solution of (3.43) having n nodal domains. We denote by $A_1, \ldots, A_n$ the nodal regions of u_n counting them starting from the outer boundary in such a way that ∂A_1 contains $\partial\Omega$, if Ω is a ball, or the outer boundary of Ω, if Ω is an annulus. Since u_n is radial we have that A_j are annuli for $j \in \{1, \ldots, n - 1\}$ while A_n is a ball if Ω is a ball or another annulus if so is Ω.

Let us first consider the case of the ball so that

$$A_j = \{ x \in \Omega : R_{j+1} < |x| < R_j \}, \quad j = 1, \ldots, n-1$$

$$A_n = \{ x \in \Omega : |x| < R_n \}$$

where $R_j, j = 2, \ldots, n$ are the nodal radii and R_1 is the radius of the ball Ω.

We consider the derivatives $\frac{\partial u_n}{\partial x_i}, i = 1, \ldots, N$, which, as in (3.47), satisfy the equation

$$L_{u_n}\left(\frac{\partial u_n}{\partial x_i} \right) = 0 \quad \text{in } \Omega. \tag{3.51}$$

Using the symmetry of u_n, we have

$$\frac{\partial u_n}{\partial x_i} = 0 \quad \text{on } \overline{\Omega} \cap T_i. \tag{3.52}$$

Then we consider the *half-nodal regions*

$$A_{i,j}^- = A_j \cap \Omega_i^-, \quad j = \cdots, n \text{ and } i = 1, \ldots, N. \tag{3.53}$$

To simplify the notation, let us fix $i = 1$ and focus on the function $\frac{\partial u_n}{\partial x_1}$ in the sets $A_{1,j}^-$, that we simply denote by A_j^-, whatever we prove for $\frac{\partial u_n}{\partial x_1}$ will hold, with obvious changes, for the other derivatives $\frac{\partial u_n}{\partial x_i}, i = 2, \ldots, N$.

Let us observe that for each nodal region A_j, writing $u_n(r) = u_n(|x|)$ there exists at least one value $r_j \in (R_j, R_{j+1}), j = 1, \ldots, n-1$, such that

$$\frac{du_n}{dr}(r_j) = 0. \tag{3.54}$$

Notice that if the nonlinearity $f = f(s)$ satisfies the condition $sf(s) \geq 0$ then r_j is the unique radius in (R_j, R_{j+1}) such that (3.54) holds in $A_j, j = 1, \ldots, n-1$.

Then, since u_n is radial we have that $\frac{\partial u_n}{\partial x_1} \equiv 0$ on the spheres

$$S_j = \left\{ x \in \mathbb{R}^N : |x| = r_j \right\}, \quad j = 1, \ldots, n-1. \tag{3.55}$$

Let us fix one $r_j \in (R_j, R_{j+1})$ for each $j = 1, \ldots, n-1$ (i. e., just one value of the radius in the interval (R_j, R_{j+1}) such that (3.54) holds) and consider the sets

$$N_j^- = \left\{ x \in \mathbb{R}^N : r_j > |x| > r_{j-1} \right\} \cap \Omega_1^-, \quad j = 1, \ldots, n-2$$

We observe that for $j = 1, \ldots, n-2$, by (3.51) and (3.54) we have

$$\begin{cases} L_{u_n}\left(\dfrac{\partial u_n}{\partial x_1} \right) = 0 & \text{in } N_j^- \\[2mm] \dfrac{\partial u_n}{\partial x_1} = 0 & \text{on } \partial N_j^- \end{cases} \tag{3.56}$$

Thus $\frac{\partial u_n}{\partial x_1}$ is an eigenfunction of the linearized operator L_{u_n} in N_j^- corresponding to the zero eigenvalue which is the first one or a higher one according to the fact that $\frac{\partial u_n}{\partial x_1}$ changes sign or not in N_j^-.

Moreover, also in the set

$$N_{n-1}^- = \left\{ x \in \mathbb{R}^N : r_{n-1} > |x| \geq 0 \right\} \cap \Omega_1^-$$

the function $\frac{\partial u_n}{\partial x_1}$ satisfies (3.56) (for $j = n - 1$). Hence also in the set N_{n-1}^- zero is an eigenvalue for L_{u_n} with corresponding eigenfunction $\frac{\partial u_n}{\partial x_1}$.

In conclusion, we have obtained $(n - 1)$ adjacent regions where an eigenvalue of L_{u_n} is zero. This implies that in the domain $N^- = \bigcup_{j=1}^{n-1} N_j^-$ the hth eigenvalue λ_h of L_{u_n} is zero for some $h \geq n - 1$.

Since N^- is strictly contained in Ω_1^-, by construction we have that the hth eigenvalue λ_h of L_{u_n} in Ω_1^- is negative for some $h \geq n - 1$. In particular, $\lambda_{n-1} = \lambda_{n-1}(L_{u_n}) < 0$ in Ω_1^- and so are all $\lambda_k = \lambda_k(L_{u_n})$ in Ω_1^- for $k \leq n - 1$. Reflecting by oddness with respect to T_1, the corresponding eigenfunctions we get eigenfunctions of L_{u_n} in the whole Ω corresponding to the same $(n - 1)$ negative eigenvalues λ_k, $k = 1, \ldots, n - 1$.

Repeating the same arguments for all $i = 1, \ldots, N$, we get at least $(n - 1)$ negative eigenvalues $\lambda_k(u_n)$ in the domains Ω_i^-, for each $i = 1, \ldots, N$, which give eigenvalues of L_{u_n} in the whole Ω whose corresponding eigenfunctions are odd with respect to T_i, $i = 1, \ldots, N$.

Note that, by symmetry,

$$\lambda_k(L_{u_n}, \Omega_i^-) = \lambda_k(L_{u_n}, \Omega_s^-) \quad \text{for } i \neq s, \ i, s = 1, \ldots, N, \ k = 1, \ldots, n - 1$$

but the corresponding eigenfunctions are linearly independent, because they are odd with respect to orthogonal axes.

Hence the multiplicity of each eigenvalue λ_k of L_{u_n} in Ω is at least N and so we have got at least $N(n - 1)$ negative eigenvalues. Since the eigenfunctions we have found are not radial, adding $m_{\mathrm{rad}}(u_n)$, we get the estimate (3.49) and then (3.50), by (3.32).

The case when Ω is an annulus follows in a similar, slightly easier, way, since the only difference is that the last nodal domain A_n is an annulus, so that it does not need to be treated in a different way with respect to the other regions A_j, $j = 1 \ldots, n - 1$. $\square$

Remark 3.21. In the case when the nonlinearity in (3.43) is $f(u) = |u|^{p-2}u$, $p \in (1, \frac{N+2}{N-2})$ if $N \geq 3$ and $p \in (1, +\infty)$ if $N = 2$, and the domain Ω is a ball, a result of [143] (see also [23] for $N = 2$) shows that for a radial solution u of (3.43) with n nodal domains the radial Morse index $m_{\mathrm{rad}}(u)$ is exactly n. Hence, in this case, (3.49) and (3.50) are equivalent.

An interesting "symmetry breaking" result can be immediately deduced from Theorem 3.20.

Corollary 3.22. *If Ω is a ball or an annulus in $\mathbb{R}^N$, $N \geq 2$, then a least energy nodal solution of (3.43) cannot be radial.*

Proof. By Theorem 3.17, we know that a least energy nodal solution has Morse index two, and hence, by (3.49), cannot be radial. □

More generally, the results of Theorem 3.19 allow to give information on the nodal set $N(u)$ of a solution u of (3.43) which is symmetric in some directions and has a low Morse index.

Corollary 3.23. *Let Ω be symmetric and convex in k directions, say $x_1, \ldots, x_k$, $k \in \{1, \ldots, N\}$ and u be a solution of (3.43) with the nonlinearity f which is either convex or has its first derivative f' convex. Assume further that u is even in the k variables $x_1, \ldots, x_k$ and*

$$m(u) < k + 2. \tag{3.57}$$

Then the closure of the nodal set $N(u)$ intersects $\partial\Omega$. In particular, if Ω is a ball or an annulus, a least energy solution of (3.43) has this property.

Proof. By Theorem 3.19, we have the estimate (3.46) which contradicts (3.57) unless $\overline{N(u)}$ intersects $\partial\Omega$.

If u is a least energy solution then, by Theorem 3.17 its Morse index $m(u)$ is equal to 2. On the other side, by the symmetry results of Chapter 6 (see Definition 6.5, Theorem 6.20 and Theorem 6.22) under the convexity assumptions on the nonlinearity $f(s)$, we know that any solution with Morse index less or equal to N is foliated Schwarz symmetric, which in particular, means that it is axially symmetric. Therefore, u is symmetric in $(N-1)$ directions, and hence $\overline{N(u)} \cap \partial\Omega \neq \emptyset$; otherwise, by Theorem 3.19 we would have $m(u) \geq (N-1) + 2 > 2$. □

3.3.3 Estimates of Morse index for symmetric sign changing solutions: the nonautonomous case

The results of the previous section strongly rely on the fact that the equation in (3.43) is autonomous. Indeed, in the proofs of Theorem 3.19 and Theorem 3.20 it is used that the derivatives $\frac{\partial u}{\partial x_i}$ are solutions of the equation (3.47), which is obtained by (3.43) differentiating with respect to x_i. Of course, this is not true if the nonlinearity f depends on the x-variable. For this reason, the proofs of the above theorems do not extend to the general problem (3.10).

In the special case when $f(x, s) = |x|^\alpha f(s)$, $\alpha \geq 0$ and Ω is a bounded radially symmetric domain, some Morse index estimates have been obtained.

First, in [179] the 2-dimensional case is considered and with a simple change of coordinates some bounds from below have been derived, showing also that the Morse index goes to infinity as $\alpha \to +\infty$. Later some partial results in all dimensions, including the case of some Schrödinger–Hénon systems have been obtained in [168]. Finally, in [12] more complete results are proved by studying a related singular eigenvalue problem.

However, it is still an open question to understand whether Morse index bounds can be obtained for nodal solutions of (3.10) in the presence of general nonlinearities $f(x,s)$.

Here, we describe the results of [179] which are obtained in a simple way using the estimates for the autonomous case of Section 3.3.2.

Let us consider the following problem:

$$\begin{cases} -\Delta u = |x|^\alpha f(u) & \text{in } \Omega \\ u = 0 & \text{on } \partial\Omega \end{cases} \tag{3.58}$$

where $\alpha > 0$, $\Omega \subset \mathbb{R}^2$ is either a ball or an annulus centered at the origin and $f \colon \mathbb{R} \longrightarrow \mathbb{R}$ is $C^{1,\beta}$ on bounded sets of $\mathbb{R}$.

We also assume the following condition on f:

$$f'(s) > \frac{f(s)}{s} \quad \forall s \in \mathbb{R} \setminus \{0\} \tag{3.59}$$

which is equivalent to (3.32) for our type of nonlinearities. In the case when $f(u) = |u|^{p-1}u$, with $p > 1$, (3.58) is the well-known Hénon equation studied in [144]

$$\begin{cases} -\Delta u = |x|^\alpha |u|^{p-1}u & x \in \Omega \\ u = 0 & \text{on } \partial\Omega \end{cases} \tag{3.60}$$

which has been extensively analyzed since the work of Ni [181]. A part from its mathematical interest, it appears in several applications, in particular in astrophysics [144, 178].

The existence of a positive radial solution of (3.60) is obtained in [181] when Ω is a ball centered at zero in $\mathbb{R}^N$, $N \geq 3$ and $1 < p < \frac{N+2+2\alpha}{N-2}$ or $N = 2$ and $p = 1$, by using the mountain pass theorem. For the same range of exponents, using the arguments of [23], it is possible to prove the existence of a nodal radial solution, which has the least energy among all nodal radial solutions.

In the case when Ω is an annulus, the same existence results hold for any $p > 1$, since no lack of compactness occurs in the setting of radial functions.

For the more general problem (3.58) with $\Omega \subset \mathbb{R}^2$, applying the results of [23] we have that a least energy nodal solution exists, also in general bounded domains, and has Morse index 2, if f satisfies (3.59) and the conditions:

$$f(0) = 0 \text{ and } \exists p > 1 : |f'(s)| \leq c(1 + |s|^{p-1}), \ \forall s \in \mathbb{R}, \tag{3.61}$$

$$\exists R > 0, \ \theta > 2 : 0 < \theta \int_0^s f(t)\,dt \leq sf(s), \ \forall |s| \geq R. \tag{3.62}$$

Therefore, in radial domains the question whether the least energy nodal solution is radial or not arises. As for the autonomous case, the answer will be deduced by an estimate of the Morse index for nodal radial solutions of (3.58). Indeed, we have the following result.

Theorem 3.24. *Let u be a radial sign changing solution of (3.58). Then the Morse index $m(u)$ is greater than or equal to 3. Moreover, if (3.59) holds, then the Morse index of u is at least $n(u) + 2$, where $n(u)$ denotes the number of nodal regions of u.*

As a consequence of this theorem, we get the following symmetry breaking result.

Corollary 3.25. *Assume (3.59), (3.61) and (3.62). Then any least energy nodal solution of (3.58) is not radially symmetric.*

Let us point out that Corollary 3.25 was already shown for the Hénon problem (3.60) for every dimension $N \geq 2$ but only for particular values of α: for α large in [24] by a comparison of energy argument and for α small in [37, 38] by an asymptotic analysis as $\alpha \to 0$, of the least energy nodal solutions.

In contrast with the symmetry breaking result of Corollary 3.25, we observe that least energy nodal solutions of (3.58) are foliated Schwartz symmetric by the results described in Chapter 6.

The proof of Theorem 3.24 is obviously different from that of Theorem 3.19 and Theorem 3.20. It relies on a suitable change of variable which works well in $\mathbb{R}^2$ and has already been used in [64] (see also [65, 127]).

Thus, before proving Theorem 3.24, let us introduce the change of variable and describe its properties.

For a point $x = (x_1, x_2) \in \mathbb{R}^2$, let us denote by (r, θ) its polar coordinates, namely:

$$x_1 = r \cos \theta, \quad x_2 = r \sin \theta, \quad r = \sqrt{x_1^2 + x_2^2}.$$

So, for a function u defined in a domain of $\mathbb{R}^2$ we can write

$$u(x_1, x_2) = u(r \cos \theta, r \sin \theta) = u(r, \theta).$$

We recall the following formulae:

$$\nabla_x = \left(\frac{\partial}{\partial x_1}, \frac{\partial}{\partial x_2} \right) = \left(\cos \theta \frac{\partial}{\partial r} - \frac{1}{r} \sin \theta \frac{\partial}{\partial \theta}, \sin \theta \frac{\partial}{\partial r} + \frac{1}{r} \cos \theta \frac{\partial}{\partial \theta} \right),$$

$$|\nabla_x|^2 = \left(\frac{\partial}{\partial x_1} \right)^2 + \left(\frac{\partial}{\partial x_2} \right)^2 = \left(\frac{\partial}{\partial r} \right)^2 + \frac{1}{r^2} \left(\frac{\partial}{\partial \theta} \right)^2 \tag{3.63}$$

and

$$\Delta_x = \frac{\partial^2}{\partial x_1^2} + \frac{\partial^2}{\partial x_2^2} = \frac{\partial^2}{\partial r^2} + \frac{1}{r} \frac{\partial}{\partial r} + \frac{1}{r^2} \frac{\partial^2}{\partial \theta^2}. \tag{3.64}$$

In order to define a change of variable $x \mapsto y$ in $\mathbb{R}^2$, we set $y = (y_1, y_2) \in \mathbb{R}^2$ with associate polar coordinates (s, σ), i. e.,

$$y_1 = s \cos \sigma, \quad y_2 = s \sin \sigma, \quad s = \sqrt{y_1^2 + y_2^2}$$

so that for a function v we write

$$v(y_1, y_2) = v(s \cos \sigma, s \sin \sigma) = v(s, \sigma).$$

Then for a number $k > 0$, we consider the following transformation:

$$T_k : \mathbb{R}^2 \longrightarrow \mathbb{R}^2, \quad T_k(y) = y|y|^{k-1}, \tag{3.65}$$

setting $T_k(0,0) = (0,0)$ and $x = T_k(y)$.

In polar coordinates, the transformation T_k reads

$$T_k(s, \sigma) = (s^k, \sigma), \quad \text{i. e. } r = s^k, \ \theta = \sigma. \tag{3.66}$$

Some properties of this map are summarized in the following lemma.

Lemma 3.26. *We have:*
(i) *T_k is a homeomorphism whose inverse is*

$$T_k^{-1} x = x|x|^{\frac{1}{k}-1}, \quad \text{i. e. } T_k^{-1} = T_{\frac{1}{k}}. \tag{3.67}$$

(ii) *In Cartesian coordinate, the Jacobian matrix of T_k is, $\forall y \neq 0$,*

$$J_{T_k}(y) = \frac{\partial(x_1, x_2)}{\partial(y_1, y_2)}(y) = |y|^{k-3} \begin{bmatrix} |y|^2 + (k-1)y_1^2 & (k-1)y_1 y_2 \\ (k-1)y_1 y_2 & |y|^2 + (k-1)y_2^2 \end{bmatrix},$$

and

$$|\det J_{T_n}(y)| = k|y|^{2k-2}. \tag{3.68}$$

(iii) *Given a function ψ defined on a subset of $\mathbb{R}^2$, we set $\varphi = \psi \circ T_k^{-1}$ and $x = T_k(y)$ for $y \neq 0$. Then ψ is differentiable at y if and only if φ is differentiable at x.*
(iv) *Let ψ, φ, x as before and r, s, σ and θ as in (3.66). Then*

$$\left(\psi_s^2 + \frac{1}{s^2} \psi_\sigma^2 \right) s^{2-2k} = k^2 \varphi_r^2 + \frac{1}{r^2} \varphi_\theta^2, \quad \forall s \neq 0 \tag{3.69}$$

which implies that

$$\min\{1, k^2\}|\nabla\varphi(x)|^2 \le |\nabla\psi(y)|^2 |y|^{2-2k} \le \max\{1, k^2\}|\nabla\varphi(x)|^2 \tag{3.70}$$

$\forall y \neq 0$.
Moreover, if ψ is radially symmetric, then

$$k^2 |\nabla\varphi(x)|^2 = |\nabla\varphi(y)|^2 |y|^{2-2k}, \quad \forall y \neq 0. \tag{3.71}$$

Proof. The statements (i), (ii) and (iii) are just matter of computation. Regarding (iv), the identity (3.69) follows from (3.65). From (3.69), we infer that

$$\min\{1, k^2\}\left(\varphi_r^2 + \frac{1}{r^2}\varphi_\theta^2\right) \le \left(\varphi_s^2 + \frac{1}{s^2}\varphi_\sigma^2\right) \le \max\{1, k^2\}\left(\varphi_r^2 + \frac{1}{r^2}\varphi_\theta^2\right)$$

which combined with (3.63) implies (3.70). If ψ is radially symmetric, it is also clear that (3.71) follows from (3.69) since $\psi_\sigma \equiv 0$ and $\varphi_\theta \equiv 0$. $\qquad\square$

Recalling that $\Omega \subset \mathbb{R}^2$ is either a ball or an annulus centered at the origin we set $\Omega_k = T_k^{-1}(\Omega)$.

Lemma 3.27. *Let $1 \le r < \infty$. Then*

$$S_k : L^r(\Omega_k) \longrightarrow L^r(\Omega, |x|^{\frac{2-2k}{k}}), \quad \textit{defined by } S_k\psi := \psi \circ T_k^{-1},$$

is a continuous linear isomorphism such that

$$\int_{\Omega_k} |\psi(y)|^r \, dy = k^{-1} \int_\Omega |\varphi(x)|^r |x|^{\frac{2-2k}{k}} \, dx, \quad \textit{with } \varphi = \psi \circ T_k^{-1}. \tag{3.72}$$

Proof. In the case when Ω is an annulus centered at the origin, (3.72) derives by the standard change of variable theorem, using (3.67) and (3.68).

In the case when $\Omega = B(0, R)$ is a ball centered at the origin and radius $R > 0$, the singularity at zero of T_k or T_k^{-1} causes no problem, since we can reduce the arguments to the previous case by approximation with annuli. Indeed,

$$\int_{B(0,R)} |h(z)| \, dz = \lim_{\delta \to 0^+} \int_{B(0,R)\backslash B(0,\delta)} |h(z)| \, dz, \quad \forall h \in L^1(B(0, R)).$$

Then the monotone convergence theorem, passing to the limit, gives the result for the ball. $\qquad\square$

In the same way, we can prove the following lemma.

Lemma 3.28. *Let $F : \mathbb{R} \longrightarrow \mathbb{R}$ be a continuous function. Then $F \circ \psi \in L^1(\Omega_k)$ if, and only if, $F \circ \varphi \in L^1(\Omega, |x|^{\frac{2-2k}{k}})$ with $\varphi = \psi \circ T_k^{-1}$. Moreover,*

$$\int_{\Omega_k} F(\psi(y)) \, dy = k^{-1} \int_\Omega F(\varphi(x)) |x|^{\frac{2-2k}{k}} \, dx. \tag{3.73}$$

We point out that if $k = \frac{2}{a+2}$, then $\frac{2-2k}{k} = \alpha$ and so the weights $|x|^{\frac{2-2k}{k}}$ at (3.73) and $|x|^\alpha$ at (3.58) coincide.

Lemma 3.29. *The map*

$$S_k \colon H_0^1(\Omega_k) \longrightarrow H_0^1(\Omega), \quad \text{defined by } S_k \psi := \psi \circ T_k^{-1},$$

is a continuous linear isomorphism. Moreover, setting $\varphi = \psi \circ T_k^{-1}$, *we have*

$$\min\left\{k, \frac{1}{k}\right\} \int_\Omega |\nabla\varphi(x)|^2 \, dx \le \int_{\Omega_k} |\nabla\psi(y)|^2 \, dy \le \max\left\{k, \frac{1}{k}\right\} \int_\Omega |\nabla\varphi(x)|^2 \, dx$$

for all $\psi \in H_0^1(\Omega_k)$ *and*

$$k \int_\Omega |\nabla\varphi(x)|^2 \, dx = \int_{\Omega_k} |\nabla\psi(y)|^2 \, dy, \quad \forall \psi \in H_{0,rad}^1(\Omega_k).$$

Proof. Here, we use (3.67)–(3.71) and proceed as in the proof of Lemma 3.27. $\square$

Now we consider the change of variable (3.65) restricted to radial functions. For a radial function $u\colon \Omega \subset \mathbb{R}^2 \longrightarrow \mathbb{R}$, we define the radial function $v\colon \Omega_k \longrightarrow \mathbb{R}$ by setting $v(y) = u(T_k(y))$, i. e.,

$$v(s) = u(s^k) = u(r), \quad r = s^k, \ r = |x|, \ s = |y|. \tag{3.74}$$

Then an easy computation yields

$$v_{ss}(s) + \frac{1}{s} v_s(s) = k^2 s^{2k-2}\left(u_{rr}(s^k) + \frac{1}{s^k} u_r(s^k)\right), \quad s > 0.$$

So, using the previous notation in polar coordinates, since $r = |x|$, $s = |y|$, $r = s^k$, we infer that

$$\Delta v(y) = k^2 |y|^{2k-2} \Delta u(T_k(y)) = k^2 |x|^{2-\frac{2}{k}} \Delta u(x). \tag{3.75}$$

Hence, if u is a radial solution of the Hénon type equation (3.58), then $v\colon \Omega_k \longrightarrow \mathbb{R}$ is a radial function that satisfies

$$\begin{cases} -\Delta v(y) = k^2 |y|^{2k-2+k\alpha} f(v(y)) & y \in \Omega_k \\ v = 0 & \text{on } \partial\Omega_k \end{cases}$$

Thus if we choose k such that

$$2k - 2 + k\alpha = 0, \quad \text{i. e., } k = \frac{2}{\alpha + 2}, \tag{3.76}$$

then we infer that

$$-\Delta v(y) = \left(\frac{2}{\alpha+2}\right)^2 f(v(y)), \quad y \in \Omega_k, \ v = 0 \text{ on } \partial\Omega_k. \tag{3.77}$$

This means that the map T_k, for $k = \frac{2}{\alpha+2}$ transforms radial solutions of the nonautonomous problem (3.64) into radial solutions of the autonomous problem (3.77). This will be very useful to estimate the Morse index of nodal solutions of (3.58), so to prove Theorem 3.24.

To this aim, we consider the quadratic form associated to a solution u of (3.58), i. e.,

$$Q_u(\varphi) = \int_\Omega |\nabla\varphi(x)|^2 - \int_\Omega |x|^\alpha f'(u)\varphi(x)^2 \, dx, \quad \varphi \in H_0^1(\Omega)$$

and the one relative to $v = u \circ T_k$, for $k = \frac{2}{\alpha+2}$, i. e.,

$$Q_v(\psi) = \int_{\Omega_k} |\nabla\psi(y)|^2 \, dy - \left(\frac{2}{\alpha+2}\right)^2 \int_{\Omega_k} f'(v)\psi(y)^2 \, dy, \quad \psi \in H_0^1(\Omega_k).$$

The crucial point for the proof of Theorem 3.24 is the following result.

Proposition 3.30. *Let $v, \psi \in H_0^1(\Omega_k)$ and set $u = v \circ T_k^{-1}$, $\varphi = \psi \circ T_k^{-1}$. Then*

$$Q_v(\psi) \geq \frac{2}{\alpha+2} Q_u(\varphi), \quad \forall\psi \in H_0^1(\Omega_k) \tag{3.78}$$

and

$$Q_v(\psi) = \frac{2}{\alpha+2} Q_u(\varphi), \quad \forall\psi \text{ radial in } H_0^1(\Omega_k).$$

Proof. It is a direct consequence of Lemma 3.28 and Lemma 3.29. $\qquad\square$

Proof of Theorem 3.24. Let u be a radial nodal solution of (3.58) and $v(y)$ the transformed function defined in (3.74) for $k = \frac{\alpha}{\alpha+2}$ which is a nodal radial solution of (3.77). Observe that the eigenvalue problem for the linearized operator associated to (3.77) is

$$\begin{cases} -\Delta\psi - \left(\dfrac{2}{\alpha+2}\right)^2 f'(v)\psi = \lambda\psi & \text{in } \Omega_k \\ \psi = 0 & \text{on } \partial\Omega_k \end{cases} \tag{3.79}$$

Hence, if ψ is a radial eigenfunction for (3.79), the function φ defined by $\varphi(s^{\frac{2}{\alpha+2}}) = \psi(s)$ is a radial eigenfunction for the eigenvalue problem

$$\begin{cases} -\Delta\varphi - |x|^\alpha f'(u)\varphi = \lambda\left(\dfrac{\alpha+2}{2}\right)^2 |x|^\alpha\varphi & \text{in } \Omega \\ \varphi = 0 & \text{on } \partial\Omega \end{cases} \tag{3.80}$$

by (3.75), (3.76).

We know, from [7] (see also Theorem 3.20) that the Morse index of v is at least 3 and greater than or equal to $n(u) + 2$, if (3.59) holds. More precisely, the problem (3.79) has a negative first eigenvalue $\lambda_{1,\mathrm{rad}}$ to which there corresponds a radial eigenfunction $\psi_{1,\mathrm{rad}}$ and two other negatives eigenvalues $\lambda_2 = \lambda_3$ with corresponding eigenfunction ψ_2 and ψ_3. By the proof of Theorem 3.20 (see also Theorem 3.19), we have that

$$\psi_2(y_1, y_2) \text{ is even in } y_2 \text{ and odd in } y_1,$$
$$\psi_3(y_1, y_2) \text{ is even in } t_1 \text{ and odd in } y_2.$$

Hence, in particular,

$$Q_v(\psi_{1,\mathrm{rad}}) < 0 \quad \text{and} \quad Q_v(\psi_i) < 0, \ i = 2, 3.$$

Moreover, if (3.59) holds then the radial eigenvalues for (3.79) $\lambda_{i,\mathrm{rad}}$ are also negative, for $i = 2, \ldots, n(u)$. Let us denote by $\psi_{i,\mathrm{rad}}$, $i = 2, \ldots, n(u)$, the associated radial eigenfunctions. As we have observed, the change of variable T_k guarantees that $\varphi_{i,\mathrm{rad}}$ defined by $\psi_{i,\mathrm{rad}} = \varphi_i(T_k(y))$ with $i = 1, 2, \ldots, n(u)$ are radial eigenfunctions of (3.80) for $\lambda = \lambda_{i,\mathrm{rad}}$. Even though φ_2 and φ_3 defined by $\varphi_i(x) = \psi(T_k^{-1}(x))$, $i = 2, 3$, are not eigenfunctions of (3.80), they correspond to directions in which the quadratic form induced by Q_u is negative definite, which follows from (3.78). Using the symmetries of $\varphi_{1,\mathrm{rad}}, \ldots, \varphi_{n(u),\mathrm{rad}}, \varphi_2, \varphi_3$, it is easy to see that they are all mutually orthogonal with respect to both the bilinear forms

$$(u, w) \longrightarrow \int_\Omega |x|^\alpha uw \, dx \quad \text{and}$$
$$(u, w) \longrightarrow \int_\Omega (\nabla u \nabla w - |x|^\alpha f'(u)uw) \, dx.$$

Therefore we deduce that $Q_u(w) < 0$ for every nonzero w in the span $[\varphi_{1,\mathrm{rad}}, \varphi_2, \varphi_3]$ or for every nonzero w in the span $[\varphi_{1,\mathrm{rad}}, \ldots, \varphi_{n(u),\mathrm{rad}}, \varphi_2, \varphi_3]$ if (3.59) holds. This proves the assertion. $\qquad \square$

Theorem 3.24 just proved gives an estimate on the Morse index of a nodal radial solution of (3.58), which is independent of the exponent α of the nonlinearity. It is an interesting question to see how the weight $|x|^\alpha$, and hence its exponent α influences the Morse index of a solution. In this direction, we describe the following result, also proved in [179].

Theorem 3.31. *Let $\alpha > 0$ be even and let u be a radial nodal solution of* (3.58). *Then u has Morse index greater than or equal to $\alpha + 3$. If in addition* (3.59) *holds, then the Morse index of u is at least $n(u) + \alpha + 2$.*

The proof of this theorem relies on a modification of the previous change of variable that works fine for the case when α is even. This change of variable is the key argument to prove the existence of many negative eigenvalues of the problem (3.80). A variant of it was used in [189] in higher dimensions to pass from doubly symmetric solutions of a supercritical problem in dimension $2m$, $m \geq 2$, to axially symmetric solutions of a subcritical problem in dimension $m+1$. To prove Theorem 3.31, there is not a change of dimension but a somehow similar idea is applied to create a correspondence between eigenfunctions of linearized operator of two different problems.

To the aim of proving Theorem 3.31, let us define this other change of variable in $\mathbb{R}^2$ which involves changing both polar coordinates r and θ.

Given $k > 0$ and $m \in \mathbb{N}$, we set

$$T_{k,m} \colon [0, \infty) \times [0, 2\pi] \longrightarrow [0, \infty) \times \left[0, \frac{2\pi}{m}\right],$$

$$T_{k,m}(s, \sigma) := \left(s^k, \frac{\sigma}{m}\right), \quad r = s^k, \quad \theta = \frac{\sigma}{m}. \tag{3.81}$$

Obviously, $T_{k,1}$ is just T_k of (3.66).

Consider any continuous function ψ defined on a radially symmetric domain Ω in $\mathbb{R}^2$ in the Cartesian coordinates (y_1, y_2). Then using the polar coordinates we can write

$$\psi(y_1, y_2) = \psi(s \cos \sigma, s \sin \sigma) = \psi(s, \sigma)$$

and we set

$$\varphi(x_1, x_2) = \varphi(r, \theta) = \psi(T_{k,m}^{-1}(r, \theta)).$$

Hence φ is a function defined for $\theta \in [0, \frac{2\pi}{m}]$ which, since $\psi(s, 0) = \psi(s, 2\pi)$, can be extended $\frac{2\pi}{m}$-periodically and continuously for all $\theta \in [0, 2\pi]$. We still denote this extension by φ and we observe that, if it is smooth, by direct computation, we have

$$k^2 r^{2 - \frac{2}{k}} \left[\varphi_{rr} + \frac{1}{r} \varphi_r + \frac{1}{r^2} \varphi_{\theta\theta} \right] = \psi_{ss} + \frac{1}{s} \psi_s + \frac{m^2 k^2}{s^2} \psi_{\sigma\sigma}.$$

Hence if we choose $k = \frac{1}{m}$, for the Laplacian in cartesian coordinates we have

$$m^{-2} |x|^{2(1-m)} \Delta\varphi(x) = \Delta\psi(y). \tag{3.82}$$

In view of the relation (3.82) involving the Laplacian of φ and ψ, we will apply the above procedure to work with the Hénon type equations (3.58) in the case when $\alpha = 2(m - 1)$, with $m \geq 2$, that is for every α even. Indeed

$$\alpha = 2(m - 1) \iff k = \frac{1}{m} = \frac{2}{\alpha + 2}$$

which coincides with the relation (3.76) between k and α.

Note that, in the complex plane, the above transformation $T_{\frac{1}{k}, m}$ is just the one which sends z into $z^{\frac{1}{m}}$, $z \in \mathbb{C}$.

With the above choice of α, we consider a radial nodal solution u of (3.58). By Theorem 3.24, we know that u has Morse index greater than or equal to 3 and at least $n(u) + 2$ if (3.59) is also satisfied. We will use the change of variable (3.81) with $k = \frac{1}{m}$ to construct $\alpha + 2 = 2m$ convenient nonradial directions on which the quadratic form $Q_u(w)$ is negative.

Proof of Theorem 3.31. Let $\alpha = 2(m - 1)$, with $m \geq 2$, $k = \frac{1}{m}$, and let u be a radial nodal solution of (3.58). Then, by (3.77) the radial function $v = u \circ T_k$ solves

$$\begin{cases} -\Delta v = \dfrac{1}{m^2} f(v) & \text{in } \Omega_k \\ v = 0 & \text{on } \partial\Omega_k \end{cases}$$

Therefore, by results of [7], see Theorem 3.20, there exists two eigenfunctions ψ_2 and ψ_3 for the eigenvalue problem

$$\begin{cases} -\Delta\psi - \dfrac{1}{m^2} f'(v)\psi = \lambda\psi & \text{in } \Omega_k \\ \psi = 0 & \text{on } \partial\Omega_k \end{cases} \tag{3.83}$$

with the following properties:

(i)   the corresponding eigenvalues $\lambda_2 = \lambda_3$ are negative;
(ii) ψ_2 is even with respect to y_2 and odd with respect to y_1, while ψ_3 is even with respect to y_1 and odd with respect to y_2;
(iii) $\psi_2(y_1, y_2) > 0$ if $y_1 > 0$, while $\psi_3(y_1, y_2) > 0$ if $y_2 > 0$.

Next, applying the change of variables (3.81), we consider the functions $\varphi_{m,i}(r, \theta) = \psi_i \circ T^{-1}_{\frac{1}{m},m}(r, \theta)$, $i = 2, 3$, extended by periodicity as before for all $\theta \in [0, 2\pi]$, so to have them defined on the whole Ω. Then, by the conditions $\frac{\partial\psi_2}{\partial\sigma} = 0$ and $\psi_3 = 0$ at $\sigma = 0$, we have that $\varphi_{m,i}$, $i = 2, 3$, are $C^2(\overline{\Omega})$-functions and by (3.82) they satisfy

$$\begin{cases} -\Delta\varphi - |x|^\alpha f'(u)\varphi = \lambda|x|^\alpha\varphi & \text{in } \Omega \\ \varphi = 0 & \text{on } \partial\Omega \end{cases} \tag{3.84}$$

with $\lambda = \lambda_i m^2$. Moreover, it is easy to see that both $\varphi_{m,i}$, $i = 2, 3$, have $2m$ nodal sets, each one being an angular sector of amplitude $\frac{\pi}{m}$. This means that each one is a first eigenfunction of (3.84) in that sector with corresponding eigenvalue $\lambda_i m^2 < 0$. In particular, $\varphi_{m,2}$ is the first eigenfunction in the sector

$$\Omega_{m,2} = \left\{ (x_1, x_2) = (r\cos\theta, r\sin\theta) \in \Omega, \ \theta \in \left[-\frac{\pi}{2m}, \frac{\pi}{2m} \right] \right\}$$

while $\varphi_{m,3}$ is the first eigenfunction in the sector

$$\Omega_{m,3} = \left\{ (x_1, x_2) = (r\cos\theta, r\sin\theta) \in \Omega, \ \theta \in \left[0, \frac{\pi}{m} \right] \right\}.$$

Then, by the monotonicity of the first eigenvalue with respect to the domain, we have that the first eigenvalue in $\Omega_{n,2}$ or $\Omega_{n,3}$ are also negative for every integer $1 \le n < m$, $\Omega_{n,i}$ defined as before, replacing m by n, for $i = 2, 3$.

The corresponding eigenfunctions, say $\varphi_{n,i}$, extended by oddness with respect to the anticlockwise part of the boundary of $\Omega_{n,i}$ and periodically, with angular period $\frac{2\pi}{n}$, give rise to other two eigenfunction for (3.84), for every $n \in \{1, \ldots, m\}$. By construction, their symmetry or antisymmetry, all these pairs of eigenfunctions are mutually orthogonal with respect to both the bilinear forms

$$(u, w) \longmapsto \int_\Omega |x|^\alpha uw \, dx \quad \text{and}$$

$$(u, w) \longmapsto \int_\Omega \left[\nabla u \nabla w - |x|^\alpha f'(u)uw \right] dx$$

so that we get $2m$ negative eigenvalues for (3.84) corresponding to nonradial directions. Counting also the first radial eigenvalue, which is negative, and the second, up to the $n(u)$-th radial eigenvalue which are also negative if (3.59) holds, we get the assertion, since $\alpha = 2(m - 1)$. $\qquad\square$

In the particular case of the Hénon equation (3.60), the same change of variable can be used to prove the uniqueness of radial solutions, up to multiplication by -1, having n nodal sets. Moreover, it can be shown that the least energy nodal radial solution of (3.60) is nondegenerate in the space of radial function (see [179], Theorem 1.5). This is achieved by passing again to the autonomous problem and arguing as in Proposition 4 of [188].

This nondegeneracy result, together with the property, by Theorem 3.31, that the Morse index of nodal radial solutions tends to $+\infty$ along sequences of even exponents $\alpha \to +\infty$, indicates that there should be infinitely many branches of nonradial nodal solutions of (3.60) bifurcating from the nodal radial solutions. This should derive by arguments similar to those described in Chapter 5, where some applications of Morse theory to bifurcation are outlined.

4 Morse index of radial solutions of Lane–Emden problems

In this chapter, we will compute exactly the Morse index of radial solutions of the classical Lane–Emden problem

$$\begin{cases} -\Delta u = |u|^{p-1}u & \text{in } B \\ u = 0 & \text{on } \partial B \end{cases} \tag{4.1}$$

where B is the unit ball of $\mathbb{R}^N$, $N \geq 2$, centered at the origin $0 \in \mathbb{R}^N$ and $1 < p < p_s$ with $p_s = \frac{N+2}{N-2}$ if $N \geq 3$ and $p_s = +\infty$ if $N = 2$.

The results we present in the next sections hold either for p sufficiently close to $p_s = \frac{N+2}{N-2}$, if $N \geq 3$ or for p sufficiently large if $N = 2$.

As we will see, there is a substantial difference between the case $N \geq 3$ and $N = 2$ which derives from the different asymptotic behavior of the solutions of (4.1) when p tends to the limit value p_s.

In order to compute the Morse index, we will use a spectral decomposition, for the linearized operator at a radial solution, which will be described in Section 4.1.

The same approach can be used to study the Morse index of radial solutions in annuli, in particular to the aim of proving existence of nonradial solutions bifurcating from the radial ones. This will be shown in Chapter 5.

We point out that, in the case of the annuli, Morse index estimates can be obtained by some nonexistence Liouville-type results for finite Morse index solutions (see [25]).

4.1 Spectral decomposition for the linearized operator at a radial solution

In this section, we describe a way of decomposing the spectrum of the linearized operator at a radial solution which has been used in several papers to study similar problems in an annulus [21, 129, 160, 185].

Since in our case we are dealing with problem (4.1) in a ball, we use an approximation procedure by annuli with a small hole in order to avoid singularities at the origin [105, 106]. Most of the proofs of this section are taken from [21, 105, 106] and [129].

From now on, we denote by u a radial solution of (4.1). The interesting case will be when u changes sign. Indeed problem (4.1) admits only one positive solution, as recalled in Section 3.2.3, which is radial by the symmetry result of [122]. Since it is unique it must be the least energy solution (see Section 3.2.2) and, therefore, its Morse index is equal to one. This means that the linearized operator $L_u \colon H^2(B) \cap H_0^1(B) \to L^2(\Omega)$

$$L_u(v) = -\Delta v - p|u(x)|^{p-1}v \tag{4.2}$$

https://doi.org/10.1515/9783110538243-004

has only one negative eigenvalue to which there corresponds a radial eigenfunction, since u is also the least energy solution of (4.1) in the space of radial functions.

Thus we assume that u is a radial nodal solution of (4.1).

Since the nonlinearity which defines the equation in (4.1) is $f(s) = |s|^{p-1}s$, $p > 1$, it satisfies all conditions stated in Section 3.3.1 and Section 3.3.2, so we can apply all results there described. In particular, if $n = n(u)$ denotes the number of nodal domains of u, by Theorem 3.20 we have

$$m(u) \geq m_{\mathrm{rad}}(u) + N(n-1) \tag{4.3}$$

where, as usual, $m(u)$ is the Morse index of u and $m_{\mathrm{rad}}(u)$ is its *radial Morse index*, i. e., the number of negative *radial* eigenvalues of L_u (i. e., eigenvalues which are associated to a radial eigenfunction).

In the special case of the Lane–Emden problem, the radial Morse index can be explicitly computed.

Theorem 4.1. *Let u be a radial solution of (4.1) with $n = n(u)$ nodal regions. Then*

$$m_{rad}(u) = n. \tag{4.4}$$

This has been proved in [23] when u has only two nodal regions (see also Section 3.3.1) and in ([143], Proposition 2.9) for any number of nodal domains.

Note also that for any fixed number $n \in \mathbb{N}^+$, there exists only one, up to the sign at the origin, radial solution of (4.1) with n nodal regions. This is a well-known result which is contained, for instance, in [145].

Let us denote by

$$\mu_1 < \mu_2 \leq \cdots \leq \mu_i \leq \ldots, \quad \mu_i \longrightarrow +\infty \quad \text{as } i \to +\infty$$

the eigenvalues of the linearized operator (4.2) counted according to their multiplicity.

It is useful to recall their min-max characterization (see Chapter 1)

$$\mu_i = \inf_{\substack{W \subset H_0^1(B) \\ \dim W = i}} \max_{\substack{v \in W \\ v \neq 0}} R(v), \quad i \in \mathbb{N}^+, \tag{4.5}$$

where $R(v)$ is the Rayleigh quotient

$$R(v) = \frac{Q_u(v)}{\int_B v(x)^2 \, dx} \tag{4.6}$$

and Q_u is the quadratic form associated to L_u, namely

$$Q_u(v) = \int_B \left[|\nabla v(x)|^2 - p|u(x)|^{p-1}v(x)^2 \right] dx. \tag{4.7}$$

Since u is a radial solution of (4.1), we can also consider the sequence of the radial eigenvalues of L_u that we denote by

$$\beta_i, \quad i \in \mathbb{N}^+$$

again counted according to their multiplicity.

For the β_i's, an analogous min-max characterization holds:

$$\beta_i = \inf_{\substack{W \subset H^1_{0,\mathrm{rad}}(B) \\ \dim W = i}} \max_{\substack{v \in W \\ v \neq 0}} R(v), \quad i \in \mathbb{N}^+ \tag{4.8}$$

where $R(v)$ is as in (4.6) and $H^1_{0,\mathrm{rad}}(B)$ denotes the subspace of $H^1_0(B)$ made by radial functions.

To study the spectrum of the linearized operator L_u, a suitable procedure is to decompose it as a sum of the spectrum of a radial weighted operator and the spectrum of the Laplace–Beltrami operator on the unit sphere. This leads to a weighted eigenvalue problem with a singularity at the origin. To bypass this difficulty, we first approximate the ball B by annuli with a small hole, showing that the number of the negative eigenvalues of the operator L_u stabilizes when the size of the hole is sufficiently small.

Therefore, we consider the annuli:

$$A_h = \left\{ x \in \mathbb{R}^N : \frac{1}{h} < |x| < 1 \right\}, \quad h \in \mathbb{N}^+ \tag{4.9}$$

and denote by

$$\mu_i^h, \quad i \in \mathbb{N}^+$$

the Dirichlet eigenvalues of L_u in A_h counted according to their multiplicity. Again they can be characterized as

$$\mu_i^h = \inf_{\substack{V \subset H^1_0(A_h) \\ \dim V = i}} \max_{\substack{v \in V \\ v \neq 0}} R^h(v) \tag{4.10}$$

where R^h is the corresponding Rayleigh quotient

$$R^h(v) = \frac{Q^h(v)}{\int_{A_h} v(x)^2 \, dx} \tag{4.11}$$

and $Q^h : H^1_0(A_h) \to \mathbb{R}$ is the associated quadratic form

$$Q^h(v) = \int_{A_h} \left[|\nabla v(x)|^2 - p|u_p(x)|^{p-1} v(x)^2 \right] dx.$$

Let us denote by k^h the number of negative eigenvalues μ_i^h of L_u in A_h. We also use the notation

$$\beta_i^h, \quad i \in \mathbb{N}^+$$

for the radial Dirichlet eigenvalues of L_u in A_h counted with their multiplicity. Again, we have

$$\beta_i^h = \inf_{\substack{V \subset H^1_{0,\mathrm{rad}}(A_h) \\ \dim V = i}} \max_{\substack{v \in V \\ v \neq 0}} R^h(v), \quad i \in \mathbb{N}^+ \tag{4.12}$$

where R^h is as in (4.11).

Finally, let k^h_{rad} be the number of radial eigenvalues of L_u in A_h.

It is easy to see, using the canonical embedding $H^1_0(A_h) \subset H^1_0(B)$ and the min-max characterizations (4.5), (4.10) and (4.8), (4.12), that the following inequalities hold:

$$\mu_i^h \geq \mu_i \quad \text{and} \quad \beta_i^h \geq \beta_i, \quad \forall i, h \in \mathbb{N}^+. \tag{4.13}$$

Similarly, we have

$$\mu_i^h \geq \mu_i^{h+1} \quad \text{and} \quad \beta_i^h \geq \beta_i^{h+1}, \quad \forall i, h \in \mathbb{N}^+. \tag{4.14}$$

Lemma 4.2. *Let $p \in (1, p_s)$ be fixed. Then*

$$\mu_i^h \searrow \mu_i \quad \text{and} \quad \beta_i^h \searrow \beta_i \quad \text{as } h \to +\infty, \quad \forall i \in \mathbb{N}^+$$

Proof. It is an immediate consequence of Corollary 1.47. $\qquad\square$

By Lemma 4.2 and (4.13), it follows that the number of negative eigenvalues (resp., negative radial eigenvalues) of the linearized operator L_u in B coincides with the number k^h (resp., k^h_{rad}) of the negative eigenvalues (resp., negative radial eigenvalues) of L_u in A_h, for h large. So we have the following.

Lemma 4.3. *Let $p \in (1, p_s)$ and let u be a solution to (4.1). Then there exists $h' \in \mathbb{N}^+$ such that:*
a) $m(u) = k^h$ *and* $m_{\mathrm{rad}}(u) = k^h_{\mathrm{rad}} \; \forall h \geq h'$.
b)   *In particular, if u is the least energy nodal radial solution of (4.1) then, by Theorem 3.17 applied in the space of radial functions, it follows that*

$$k^h_{\mathrm{rad}} = 2 \quad \forall h \geq h'.$$

In order to make a decomposition of the spectrum of the linearized operator L_u, we consider the auxiliary weighted linear operator $\tilde{L}_u^h \colon H^2(A_h) \cap H^1_0(A_h) \longrightarrow L^2(A_h)$ defined by

$$\tilde{L}_u^h(v) := |x|^2 (-\Delta v - p|u(x)|^{p-1} v), \quad x \in A_h \tag{4.15}$$

and denote by $\tilde{\mu}_i^h$, $i \in \mathbb{N}^+$, its eigenvalues counted with their multiplicity. Observe that the corresponding eigenfunctions ψ satisfy

$$\begin{cases} -\Delta\psi(x) - p|u(x)|^{p-1}\psi(x) = \tilde{\mu}_i^h \dfrac{\psi(x)}{|x|^2}, & x \in A_h \\ \psi = 0 & \text{on } \partial A_h \end{cases}$$

Since u is radial, we also consider the following linear operator: $\tilde{L}_{u,\mathrm{rad}}^h \colon H^2((\frac{1}{h},1)) \cap H_0^1((\frac{1}{h},1)) \longrightarrow L^2((\frac{1}{h},1))$ written in polar coordinates:

$$\tilde{L}_{u,\mathrm{rad}}^h(v) := r^2\left(-v'' - \frac{N-1}{r}v' - p|u(x)|^{p-1}v\right), \quad r \in \left(\frac{1}{h},1\right) \tag{4.16}$$

and denote by $\tilde{\beta}_i^h$, $i \in \mathbb{N}^+$, its eigenvalues counted with their multiplicity. Obviously, $\tilde{\beta}_i^h$ are nothing else than the radial eigenvalues of $\tilde{L}_u^h$. We also set

$$\tilde{k}^h := \#\left\{ i \in \mathbb{N}^+ \text{ such that } \tilde{\mu}_i^h < 0 \right\}, \tag{4.17}$$

$$\tilde{k}_{\mathrm{rad}}^h := \#\left\{ i \in \mathbb{N}^+ \text{ such that } \tilde{\beta}_i^h < 0 \right\}. \tag{4.18}$$

Denoting by $\sigma(\cdot)$ the spectrum of a linear operator we have the following decomposition.

Lemma 4.4.

$$\sigma(\tilde{L}_u^h) = \sigma(\tilde{L}_{u,\mathrm{rad}}^h) + \sigma(-\Delta_{S^{N-1}}) \quad \forall h \in \mathbb{N}^+ \tag{4.19}$$

where $\Delta_{S^{N-1}}$ denotes the Laplace–Beltrami operator on the unit sphere S^{N-1}, $N \geq 2$.

Proof. Let us denote by λ_k, $k = 0, 1, \ldots$, the eigenvalues of $-\Delta_{S^{N-1}}$. It is well known (see [34]) that

$$\lambda_k = k(k + N - 2) \quad \forall k \in \mathbb{N}. \tag{4.20}$$

Given $\mu \in \sigma(\tilde{L}_u^h)$, we consider an associated eigenfunction ψ satisfying:

$$\begin{cases} -\Delta\psi - p|u|^{p-1}\psi = \mu\dfrac{\psi}{|x|^2} & \text{in } A_h \\ \psi = 0 & \text{on } \partial A_h \end{cases}$$

Then we choose $k \in \mathbb{N}$ and an eigenfunction φ of $-\Delta_{S^{N-1}}$ associated to λ_k. The function

$$w(r) = \int_{S^{N-1}} \psi(r,\theta)\varphi(\theta)\,d\theta$$

satisfies

$$-w'' - \frac{N-1}{r}w' = \int_{S^{N-1}} \left(-\psi_{rr} - \frac{N-1}{r}\psi_r \right)\varphi \, d\theta$$

$$= \int_{S^{N-1}} \left(-\Delta\psi + \frac{1}{r^2}\Delta_{S^{N-1}}\psi \right)\varphi \, d\theta$$

$$= p|u|^{p-1}w + \frac{\mu}{r^2}w + \frac{1}{r^2}\int_{S^{N-1}} (\Delta_{S^{N-1}}\psi)\varphi \, d\theta.$$

Integrating the last term by parts, we get

$$-w'' - \frac{N-1}{r}w' - p|u|^{p-1}w = \frac{\mu - \lambda_k}{r^2}w,$$

which implies that the numbers $(\mu - \lambda_k)$ are eigenvalues of the operator $\tilde{L}^h_{u,\mathrm{rad}}$. Hence

$$\mu = (\mu - \lambda_k) + \lambda_k \in \sigma(\tilde{L}^h_{u,\mathrm{rad}}) + \sigma(-\Delta_{S^{N-1}}).$$

Vice versa, we consider $\beta \in \sigma(\tilde{L}^h_{u,\mathrm{rad}})$ and $\lambda_k \in \sigma(-\Delta_{S^{N-1}})$ and choose corresponding eigenfunctions w and φ.

Defining

$$\psi(x) = w(|x|)\varphi\left(\frac{x}{|x|}\right),$$

we have

$$-\Delta\psi = \left(-w'' - \frac{N-1}{r}w' \right)\varphi - \frac{w}{r^2}\Delta_{S^{N-1}}\varphi$$

$$= \left[p|u|^{p-1}w + \frac{\beta}{r^2}w \right]\varphi + \frac{\lambda_k}{r^2}w\varphi = p|u|^{p-1}\psi + \frac{\beta + \lambda_k}{r^2}\psi$$

which implies that $\beta + \lambda_k \in \sigma(\tilde{L}^h_u)$. $\qquad\square$

Thus, by (4.19), we can write

$$\tilde{\mu}^h_j = \tilde{\beta}^h_i + \lambda_k, \quad \text{for } i,j \in \mathbb{N}^+ \text{ and } k \in \mathbb{N}. \tag{4.21}$$

Note that in (4.21) only $\tilde{\beta}^h_i$ depend on the exponent p in (4.1) while, by (4.20), the eigenvalues λ_k depend only on the dimension N. Moreover, the multiplicity of λ_k is (see [34]):

$$N_k - N_{k-2} \tag{4.22}$$

where for $s \in \mathbb{Z}$

$$\begin{cases} N_s = \binom{N-1+s}{N-1} = \dfrac{(N-1+s)!}{(N-1)!s!}, & \text{if } s \geq 0 \\[2mm] N_s = 0, & \text{if } s < 0 \end{cases} \tag{4.23}$$

The next result shows that the number of the negative eigenvalues of the linearized operator L_u in the annulus A_h is the same as the one of the corresponding weighted operator.

Lemma 4.5. *We have*

$$k^h = \tilde{k}^h \quad and \quad k^h_{\mathrm{rad}} = \tilde{k}^h_{\mathrm{rad}}$$

where k^h, k^h_{rad}, $\tilde{k}^h$ and $\tilde{k}^h_{\mathrm{rad}}$ are as in Lemma 4.3 and in (4.17), (4.18).

Proof. We first show that $k^h \geq \tilde{k}^h$.

Let ψ be an eigenfunction for the operator $\tilde{L}^h_u$ corresponding to a negative eigenvalue $\tilde{\mu}^h < 0$. Hence

$$\begin{cases} -\Delta\psi - p|u|^{p-1}\psi = \tilde{\mu}^h \dfrac{\psi}{|x|^2} & \text{in } A_h \\ \psi = 0 & \text{on } \partial A_h \end{cases} \tag{4.24}$$

Multiplying (4.24) by ψ and integrating on A_h, we get

$$Q^h_u(\psi) = \int_{A_h} \left[|\nabla\psi(x)|^2 - p|u(x)|^{p-1}\psi(x)^2 \right] dx = \tilde{\mu}^h \int_{A_h} \frac{\psi(x)^2}{|x|^2}\, dx < 0$$

namely ψ makes the quadratic form Q^h_u negative. Then $k^h \geq \tilde{k}^h$ since the set of such eigenfunctions ψ is a space of dimension $\tilde{k}^h$.

To prove that $k^h \leq \tilde{k}^h$, let us assume, by contradiction, that $k^h > \tilde{k}^h$ and let W be the k^h-dimensional subspace spanned by the orthogonal eigenfunctions φ_i associated to the negative Dirichlet eigenvalues of L_u in A_h:

$$W = \mathrm{span}\{\varphi_1, \varphi_2, \ldots, \varphi_{k^h}\} \subset H^1_0(A_h).$$

By the variational characterization of the eigenvalues of $\tilde{L}^h_u$, we have

$$\tilde{\mu}^h_{k^h} \leq \max_{\substack{v \in W \\ v \neq 0}} \frac{\int_{A_h} (|\nabla v(x)|^2 - p|u(x)|^{p-1}v(x)^2)\, dx}{\int_{A_h} \frac{v(x)^2}{|x|^2}\, dx} < 0 \tag{4.25}$$

which gives a contradiction.

The statement for the number of negative radial eigenvalues can be proved in the same way. $\qquad\square$

Combining Lemma 4.3, Lemma 4.5 and (4.3), (4.4) we get the following.

Proposition 4.6. *Let u be a radial sign changing solution of (4.1) for $p \in (1, p_s)$. Then there exists $h' = h'(p) \in \mathbb{N}^+$ such that, for $h \geq h'$,*

$$m(u) = \tilde{k}^h \quad and \quad m_{\mathrm{rad}}(u) = \tilde{k}^h_{\mathrm{rad}}, \tag{4.26}$$

$$\tilde{k}^h \geq n + N(n-1) \quad \text{and} \quad \tilde{k}^h_{\text{rad}} = n \tag{4.27}$$

where $n = n(u)$ is the number of nodal domains of u.

Because of Proposition 4.6 and the decomposition (4.21) in order to evaluate the Morse index $m(u)$ of the solution u, we have to estimate the negative eigenvalues $\tilde{\beta}^h_i$ of the weighted operator $\tilde{L}^h_{u,\text{rad}}$ defined in (4.16) which, by (4.27), are only the first n ones. More precisely, by (4.21), we have to check how many $\tilde{\beta}^h_i$ summed up with the eigenvalues λ_k of the Laplace–Beltrami operator on S^{N-1} give negative numbers. In doing so, we also have to take into account the multiplicity of each eigenvalues λ_k, as in (4.22), (4.23).

4.2 Asymptotic analysis of radial solutions

To the aim of evaluating the negative eigenvalues $\tilde{\beta}^h_i$ of the operator $\tilde{L}^h_{u,\text{rad}}$ in (4.16) a good knowledge of the radial solution u is needed. This is possible when the exponent p is close to the limit exponent p_s $(= \frac{N+2}{N-2}$ if $N \geq 3$ and $= +\infty$ if $N = 2)$ by an accurate analysis of the asymptotic behavior of u as $p \to p_s$. This has been done in several papers and we summarize the results in the sequel, distinguishing between the two-dimensional and the higher dimensional case.

4.2.1 The case $N \geq 3$

Here, we analyze the asymptotic behavior of the radial solutions of (4.1) when the exponent p converges, from below, to the exponent $p_s = \frac{N+2}{N-2} = 2^* - 1$, where $2^* = \frac{2N}{N-2}$ is the well-known critical exponent for the embedding $H^1_0(\Omega) \hookrightarrow L^q(\Omega)$, for a bounded domain Ω.

The lack of compactness of this embedding when $q = 2^*$ is responsible of the blow-up and concentration phenomena which appear for the solution of (4.1), in general bounded domains, when $p \nearrow p_s$. This has been largely studied in the last decades starting with the papers [42, 43, 162, 163, 211] where concentration-compactness results were obtained. However, most of the literature deals with the case of positive solutions, while little is known for sign-changing ones. The reason is that there is a lack of a complete understanding of the finite energy nodal solutions of the "limit" problem:

$$-\Delta Z = |Z|^{2^*-2}Z \quad \text{in } \mathbb{R}^N, \quad N \geq 3 \tag{4.28}$$

which naturally arises in the study of the asymptotic behavior of solutions of (4.1). Instead, it is well known that (4.28) admits only one (up to rescaling) positive solution which is radial and is explicitly given by

$$U(x) = \left(\frac{N(N-2)}{N(N-2) + |x|^2} \right)^{\frac{N-2}{2}} \tag{4.29}$$

satisfying

$$U(0) = 1 \quad \text{and} \quad \int_{\mathbb{R}^N} |\nabla U|^2 \, dx = \int_{\mathbb{R}^N} U^{2^*} \, dx = S_N^{N/2} \tag{4.30}$$

where S is the best Sobolev constant for the embedding $H_0^1(\Omega) \hookrightarrow L^{2^*}(\Omega)$ for any bounded domain $\Omega \subset \mathbb{R}^N$, and also for the embedding $D^{1,2}(\mathbb{R}^N) \hookrightarrow L^{2^*}(\mathbb{R}^N)$ where $D^{1,2}(\mathbb{R}^N)$ is the space of function in $L^{2^*}(\mathbb{R}^N)$ whose gradient belongs to $L^2(\mathbb{R}^N)$.

On the other side, the nodal solutions of (4.28) are not all known, but many of them exist as proved in [63, 108, 110].

Because of this, the asymptotic behavior of sign-changing solutions in general bounded domains as $p \to p_s$ has been studied only for low energy solutions in [26], i. e., for solutions u_p satisfying

$$\int_{\Omega} |\nabla u_p|^2 \, dx \longrightarrow 2S^{N/2} \quad \text{as } p \to p_s. \tag{4.31}$$

In the case of radial solutions in the ball B, the asymptotic analysis can be made more accurately both for positive or nodal solutions because one can study the associate O. D. E. problem. For the positive solution, which is unique as recalled in Section 3.2.3, the asymptotic behavior has been studied in [16] (see also [138] for least energy solutions in general bounded domains). For sign changing solutions, a first analysis has been done in [143]. However, as explained in the previous section, in order to compute the Morse index of the radial solutions, we need many accurate estimates which have been obtained in [106] to which we refer for the proofs of the results we state below, in the case $N \geq 3$.

Since to study the asymptotic behavior of the radial solutions, the number of nodal regions will play a role, we will denote by u_p^n the unique radial solution of (4.1) having $n \in \mathbb{N}^+$ nodal regions and satisfying

$$u_p^n(0) > 0 \tag{4.32}$$

indicating also the dependence on the exponent p in (4.1).

We start by summarizing the basic properties of the solutions u_p^n. Note that also the case $n = 1$, i. e., when the solution is positive, is included.

Proposition 4.7. *Let $p \in (1, p_s)$ then:*
(i) $u_p^n(0) = \|u_p^n\|_\infty$;

(ii) *in each nodal region the map $r \mapsto u_p^n(r), r = |x|$ has exactly one critical point (which is either a local maximum or a local minimum point, and they alternate);*

(iii) $\int_B |\nabla u_p^n(y)|^2 \, dy = \int_B |u_p^n(y)|^{p+1} \, dy \longrightarrow n S^{N/2}$, *as* $p \to p_s$.

Now we denote by $r_{i,p}^n$ the ordered nodal radii of u_p^n, i. e., $u_p^n(r_{i,p}^n) = 0, i = 1, \ldots, n$:

$$0 < r_{1,p}^n < r_{2,p}^n < \cdots < r_{n-1,p}^n < r_{n,p}^n = 1. \tag{4.33}$$

Moreover, we denote by $s_{i,p}^n$ the unique maximum point of $|u_p^n|$ in each nodal region $B_{i,p}^n, i = 0, \ldots, n - 1$:

$$\begin{aligned}
B_{0,p}^n &= \left\{ x \in \mathbb{R}^N : |x| < r_{1,p}^n \right\}, \\
B_{i,p}^n &= \left\{ x \in \mathbb{R}^N : r_{i,p}^n < |x| < r_{i+1,p}^n \right\}, \quad i = 1, \ldots, n - 1 \text{ if } n \geq 2.
\end{aligned} \tag{4.34}$$

Finally, we consider the restriction of $|u_p^n|$ to the ith nodal region:

$$u_{i,p}^n = |u_p^n| \chi_{B_{i,p}^n}, \quad i = 0, \ldots, n - 1 \tag{4.35}$$

and define

$$M_{i,p}^n = \|u_{i,p}^n\|_\infty = |u_p^n(s_{i,p}^n)|, \quad i = 0, \ldots, n - 1 \tag{4.36}$$

The asymptotic behavior of these quantities is described in the following.

Proposition 4.8. *For any $i = 0, \ldots, n - 1$ and $p \to p_s$, we have*

$$\int_B |\nabla u_{i,p}^n(y)|^2 \, dy = \int_B |u_{i,p}^n(y)|^{p+1} \, dy \longrightarrow S^{N/2}; \tag{4.37}$$

$$\int_B |u_{i,p}^n(y)|^{2^*} \, dy \longrightarrow S^{N/2}; \tag{4.38}$$

$$\int_B |u_{i,p}^n(y)|^{\frac{N}{2}(p-1)} \longrightarrow S^{N/2}; \tag{4.39}$$

$$u_p^n \rightharpoonup 0 \quad in \ H_0^1(B); \tag{4.40}$$

$$M_{i,p}^n \longrightarrow +\infty; \tag{4.41}$$

$$s_{i,p}^n \longrightarrow 0 \quad (so \ that \ r_{i,p}^n \longrightarrow 0); \tag{4.42}$$

$$\frac{s_{i,p}^n}{r_{i+1,p}^n} \longrightarrow 0, \quad for \ i \neq n - 1, \ if \ n \geq 2; \tag{4.43}$$

$$r_{i,p}^n (M_{i-1,p}^n)^{\frac{p-1}{2}} \longrightarrow +\infty \quad if \ n \geq 2; \tag{4.44}$$

$$s_{i,p}^n (M_{i,p}^n)^{\frac{p-1}{2}} \longrightarrow 0 \quad if \ n \geq 2. \tag{4.45}$$

Some crucial estimates for $|u_p^n|$ in each nodal region are contained in the following.

Proposition 4.9. *We have*

$$|u_p^n(x)| \le \frac{M_{0,p}^n}{\left[1 + \frac{(M_{0,p}^n)^{p-1}}{N(N-2)}|x|^2\right]^{\frac{N-2}{2}}} \qquad \forall x \in B_{0,p}^n. \tag{4.46}$$

Moreover, for any $\alpha \in (0, \frac{N-2}{2})$ and $n \ge 2$ there exists $y = y(\alpha, n) \in (0,1)$, $y(\alpha, n) \to 1$ as $\alpha \to 0$ and $\delta_i = \delta_i(\alpha, n) \in (0, \frac{4}{N-2})$, $i = 1, \ldots, n-1$ such that for $p \ge p_s - \delta_i$ it holds

$$|u_p^n(x)| \le \frac{M_{i,p}^n}{\left[1 + \frac{2\alpha}{N(N-2)^2}(M_{i,p}^n)^{p-1}|x|^2\right]^{\frac{N-2}{2}}} \qquad \forall x \in C_{i,p}^n \tag{4.47}$$

where

$$C_{i,p}^n = \left\{ x \in \mathbb{R}^N : y^{-\frac{1}{N}} s_{i,p}^n < |x| < r_{i+1,p}^n \right\} (\subset B_{i,p}^n)$$

with $B_{i,p}^n$ and $M_{i,p}^n$ defined as in (4.34) and (4.36).

We consider now, for $n \in \mathbb{N}^+$, the *tail sets*

$$T_{i,p}^n = \bigcup_{j=1}^{n-1} B_{j,p}^n, \qquad i = 0, \ldots, n-1. \tag{4.48}$$

Hence $T_{0,p}^n = B$, $T_{1,p}^n = B \setminus B_{0,p}^n$, $\ldots$, $T_{n-1,p}^n = B_{n-1,p}^n$.
Then we consider the rescaled functions

$$z_{i,p}^n(x) = \frac{1}{M_{i,p}^n} u_p^n\left(\frac{|x|}{(M_{i,p}^n)^{\frac{p-1}{2}}}\right), \qquad x \in \tilde{T}_{i,p}^n = (M_{i,p}^n)^{\frac{p-1}{2}} T_{i,p}^n \tag{4.49}$$

for $i = 0, \ldots, n-1$, which are radial and solve

$$\begin{cases} -\Delta z_{i,p}^n = |z_{i,p}^n|^{p-1} z_{i,p}^n & \text{in } \tilde{T}_{i,p}^n \\ z_{i,p}^n = 0 & \text{on } \partial(\tilde{T}_{i,p}^n) \\ z_{i,p}^n(s_{i,p}^n) = 1 \quad \text{and} \quad (z_{i,p}^n)'(s_{i,p}^n) = 0 \end{cases} \tag{4.50}$$

Moreover, by (4.32), it holds

$$(-1)^i z_{i,p}^n > 0 \quad \text{in } \tilde{B}_{i,p}^n = (M_{i,p}^n)^{\frac{p-1}{2}} B_{i,p}^n. \tag{4.51}$$

The asymptotic behavior of these functions, as $p \to p_s$ is given by the following.

Proposition 4.10. *As $p \to p_s$,*

$$z_{0,p}^n \longrightarrow U \quad \text{in } C_{\text{loc}}^2(\mathbb{R}^N), \tag{4.52}$$

$$(-1)^i z_{i,p}^n \longrightarrow U \quad \text{in } C_{\text{loc}}^2(\mathbb{R}^N \setminus \{0\}) \text{ for } i = 1, \ldots, n-1 \ (n \ge 2) \tag{4.53}$$

where U is defined as in (4.29).

A part from the technical statements, the asymptotic behavior of the radial solutions of (4.1) as $p \to p_s$, which is deduced by the previous propositions, can be described as follows. All nodal regions $B_{i,p}^n$, $i = 0,\dots,n-1$, of u_p^n shrink to the origin, as $r_{i,p}^n \to 0$. Moreover, also the maximum or minimum points $s_{i,p}^n$ converge to 0 but faster than the nodal radius $r_{i+1,p}^n$, so that they do not see each other in the limit. The estimate (4.41) allows to use the local maxima $M_{i,p}^n$ as rescaling parameters in defining the rescaled functions $z_{i,p}^n$, while (4.44) implies that the limit domains for these functions, and the problem (4.50) that they solve, are $\mathbb{R}^N \setminus \{0\}$ or $\mathbb{R}^N$ for the rescaling in the first region. Therefore, the limit of the problems (4.50) is the problem (4.28) which explains why (4.52) and (4.53) hold. The unique positive solution U of (4.28) satisfying (4.30) is explicitly given in (4.29) and usually is referred as the *standard bubble*. Thus, the limit profile of the solutions u_p^n, as $p \to p_s$, looks like a superposition of n *bubbles* or, in other words, like a tower of n standard bubbles. Moreover, the restrictions of the solution u_p^n to each nodal domain carry the same energy as stated in (4.37) and (4.38).

This peculiar behavior of the nodal radial solutions also induces an interesting blow-up (in time) phenomenon in the associated parabolic problem with initial data close to the stationary radial solutions ([59], see also [171]).

4.2.2 The case $N = 2$

Now we consider the 2-dimensional case. We will only analyze the asymptotic behavior of the least energy nodal radial solution of (4.1) as the exponent $p \to p_s = +\infty$. This solution has only two nodal domains and *radial* Morse index two as it can be seen by repeating the minimizing procedure of Theorem 3.17 on the nodal Nehari set in $H_{0,\mathrm{rad}}^1(B)$.

The reason for analyzing only the least energy nodal solution relies on the fact that an accurate study of the asymptotic behavior of nodal solutions, as $p \to +\infty$, in dimension two is very difficult and has been done only for solutions with two nodal regions. Indeed, unlike the higher dimensional case, the sign-changing solutions can behave in a different way in each nodal domain, making so their analysis quite involved. We believe that a complete study of nodal radial solutions with any number of nodal regions should be possible though technically very complicated; however, it has not been done so far.

Since the least energy nodal radial solution has exactly two nodal domains, we denote it simply by u_p, neglecting the dependence of $n(u_p)$ as compared with the case $N \geq 3$.

The asymptotic behavior of u_p, as $p \to +\infty$, has been accurately studied in [137]. For general bounded domains, the asymptotic analysis of both positive and sign changing solutions to (4.1) has been done in [103] and [102], but the estimates are not as precise as in the radial setting.

We summarize below the main results on the asymptotic behavior of the solutions u_p and we refer to [137] for the proofs.

First, we recall two well-known properties of u_p, similar to (i) and (ii) of Proposition 4.7, for the case $N \geq 3$:

(i)

$$u_p(0) = \|u\|_\infty; \tag{4.54}$$

(ii)

$$\textit{in each nodal region the solution } u_p \textit{ has only one critical point}$$
$$\textit{(namely the maximum or the minimum point of } u_p). \tag{4.55}$$

From now on, we will assume that $u_p(0) > 0$ and denote by r_p the unique nodal radius of u_p and by s_p the unique minimum radius of u_p, i. e.,

$$r_p \in (0,1) \text{ is such that } u_p(r_p) = 0 \tag{4.56}$$

and

$$s_p \in (0,1) \text{ is such that } \|u_p^-\|_\infty = u_p^-(s_p) = -u_p(s_p) \tag{4.57}$$

where, as usual, u_p^- is the negative part of u_p.

Proposition 4.11. *Let (u_p) be a family of least energy radial nodal solutions to (4.1) with $u_p(0) > 0$. Let us define*

$$
\begin{aligned}
\left(\epsilon_p^+\right)^{-2} &:= p u_p(0)^{p-1}, \\
\left(\epsilon_p^-\right)^{-2} &:= p u_p(s_p)^{p-1}
\end{aligned}
\tag{4.58}
$$

and the rescaled functions

$$z_p^+(x) := p \frac{u_p(\epsilon_p^+ x) - u_p(0)}{u_p(0)}, \qquad x \in \frac{B}{\epsilon_p^+}, \tag{4.59}$$

$$z_p^-(x) := p \frac{u_p(\epsilon_p^- x) - u_p(s_p)}{u_p(s_p)}, \qquad x \in \frac{B}{\epsilon_p^-}. \tag{4.60}$$

Then

$$\epsilon_p^\pm \to 0, \tag{4.61}$$

$$z_p^+ \to U \quad \textit{in } C^1_{\mathrm{loc}}(\mathbb{R}^2), \tag{4.62}$$

$$z_p^- \to Z_l \quad \textit{in } C^1_{\mathrm{loc}}(\mathbb{R}^1 \setminus \{0\}) \tag{4.63}$$

as $p \to \infty$, where

$$U(x) := \log\left(\frac{1}{1 + \frac{1}{8}|x|^2}\right)^2 \tag{4.64}$$

is the regular solution of

$$\begin{cases} -\Delta u = e^U & \text{in } \mathbb{R}^2 \\ \displaystyle\int_{\mathbb{R}^2} e^U \, dx = 8\pi \\ U(0) = 0 \end{cases} \tag{4.65}$$

and

$$Z_l(x) := \log\left(\frac{2(\gamma + 2)^2 \delta^{\gamma+2}|x|^\gamma}{(\delta^{\gamma+2} + |x|^{\gamma+2})^2}\right), \tag{4.66}$$

with

$$l = \lim_{p \to +\infty} \frac{s_p}{\epsilon_p^-} \approx 7.1979, \quad \gamma = \sqrt{2l^2 + 4} - 2, \quad \delta = \left(\frac{\gamma + 4}{\gamma}\right)^{\frac{1}{\gamma+2}} l \tag{4.67}$$

is a singular radial solution of

$$\begin{cases} -\Delta Z = e^Z + H\delta_0 & \mathbb{R}^2 \\ \displaystyle\int_{\mathbb{R}^2} e^Z \, dx < +\infty \end{cases} \tag{4.68}$$

*where $H = -\int_0^l e^{Z_l(s)} s \, ds$ and δ_0 is the Dirac measure centered at 0.
Moreover,*

$$\frac{r_p}{\epsilon_p^+} \longrightarrow +\infty \quad \text{and} \quad \frac{\epsilon_p^-}{r_p} \longrightarrow +\infty. \tag{4.69}$$

Finally, there exists $c > 0$ such that

$$p|y|^2|u_p(y)|^{p-1} \leq c \quad \forall y \in B. \tag{4.70}$$

All previous statements have been proved in [137] with the exception of (4.70) which corresponds to the property P_3^k in ([103], Proposition 2.2).

Let us explain and comment on the results of Proposition 4.11.

As in the case of higher dimension, one would like to detect the limit profile of the solutions u_p, as $p \to +\infty$, by some limit problem in the whole $\mathbb{R}^2$. However, in dimension two there is not an obvious limit equation for two reasons:

(i) the nonlinearity $f(u) = |u|^{p-1}u$ does not converge to the exponential function, as $p \to +\infty$,

(ii) the L^∞ norm of the solution does not blow-up, as $p \to +\infty$, so that it cannot be used as rescaling parameter for the solution.

This motivates the choice of $\epsilon_p^\pm$ in (4.58) as a parameter to define the rescaled functions $z_p^\pm$ in (4.59), (4.60).

Let us recall that the asymptotic analysis of solutions of (4.1) in bounded domains $\Omega \subset \mathbb{R}^2$ started in [198] and [197] where the authors considered the case of families of least energy (hence positive) solutions and, in some domains, proved concentration results as well as some asymptotic estimates. However, they did not identify a "limit problem." This was done later in [3] (see also [114]) by showing that suitable scalings of the least energy solutions converge in $C^1_{\mathrm{loc}}(\mathbb{R}^2)$, as $p \to +\infty$, to the function U in (4.64) which is a regular solution of the famous Liouville problem (4.65) in the plane. They also showed that the L^∞-norm of the least energy solutions converge to $\sqrt{e}$, thus confirming a previous conjecture of [61].

Concerning sign changing solutions, the asymptotic analysis was started in [136] by considering a family of low-energy nodal solutions but adding the hypothesis that the minimum and the maximum of the solutions are "comparable" (see [136] for the precise definition). In this case, the positive and the negative part of the solutions separate, each one having the limit profile (after scaling) of the regular solution U of the Liouville problem (4.65). However, as shown in [137], this is not the case of nodal radial solutions in the ball.

The newelty, and somehow surprising result, of [137] is that the limit profiles of the positive and negative parts of the least energy nodal radial solutions u_p are different. Indeed, assuming $u_p(0) > 0$, by (4.62) we have that the rescaling z_p^+ converge to the regular solution of (4.65), while, by (4.63), the rescaling z_p^- converge to a singular radial solution Z_l (given by (4.66)) of the problem (4.68) in the plane. Hence, asymptotically, the solutions u_p look like a superposition of different bubbles, given by a regular and a singular solutions of (4.65) and (4.68). So again, we can say that the limit profile is a tower of two bubbles, but the bubbles are different, unlike the higher dimensional case. Moreover, it can be proved that each bubble carries a different energy which emphasizes the difference with the case $N \geq 3$.

The same kind of phenomenon is shown to happen in some symmetric domains by analysing the asymptotic behavior of low energy symmetric solutions (see [103]). In [102], similar results are shown for low Morse index symmetric solutions. However, as pointed out before, in this more general situations, precise estimates are not available.

As in higher dimensions, the asymptotic behavior of these bubble-tower solutions induces a peculiar blow-up (in time) phenomenon in the associated parabolic problem with initial data close to this kind of stationary solutions [100, 101, 109].

Finally, we observe that the Liouville problems (4.65) and (4.68) are important both in geometry and in physics. They arise in the study of surfaces with prescribed Gaussian curvature and in the study of vortices in the Chern–Simon theory [214].

4.3 Computation of the Morse index of radial solutions in dimension $N \geq 3$

At the end of Section 4.1, as a result of the spectral decomposition we were left out with the estimates of the first n eigenvalues $\tilde{\beta}_i^h$, $i = 1, \ldots, n$, of the linear weighted operator $\tilde{L}_{u,\mathrm{rad}}^h$ in the interval $(\frac{1}{h}, 1)$ (see (4.16)).

Since, as in Section 4.2, the results will depend on the number of nodal regions of the radial solutions, as well as on the exponent p, we will indicate explicitly the dependence on them. Therefore, the radial solutions with n nodal regions of (4.1) will be denoted by u_p^n and the operator in (4.16) and its eigenvalues will be denoted by $\tilde{L}_{p,\mathrm{rad}}^{h,n}$ and by $\tilde{\beta}_i^h(n,p)$.

The first result is about an estimate for the nth eigenvalue.

Proposition 4.12. *Let* $h \in \mathbb{N}^+$, $p \in (1, p_s)$ *and* $h_p'' = h_p''(n) = [\frac{1}{r_{1,p}^n}] + 1$, *where* $r_{1,p}^n$ *is the first nodal radius of* u_p^n, *as defined in* (4.33). *Then*

$$\tilde{\beta}_n^h(n,p) > -(N-1) \quad \text{for any } h \geq h_p''. \tag{4.71}$$

Proof. Let $\eta(r) = \frac{du_p^n(r)}{dr}$, then, by the choice of h_p'' it follows that for any $h \geq h_p''$ it holds $\frac{1}{h} < r_{1,p}^n$, so that the function η satisfies

$$\begin{cases} \tilde{L}_{p,\mathrm{rad}}^{h,n}\eta = -(N-1)\eta & \text{in } \left(\frac{1}{h}, 1\right) \\ \eta\left(\frac{1}{h}\right) < 0 \\ \eta(1) < 0 \text{ if } n \text{ is odd or } \eta(1) > 0 \text{ if } n \text{ is even,} \end{cases}$$

the last alternative depending on the assumption $u_p^h(0) > 0$ and the sign are strict, by the Hopf lemma. Moreover, by the behavior of u_p^n, we know that, for $h \geq h_p''$, the function η has exactly $(n-1)$ zeros in the interval $(\frac{1}{h}, 1)$, given (if $n \geq 2$) by the points $s_{i,p}^n$, $i = 1, \ldots, n-1$, which satisfy (4.42).

Let w be an eigenfunction of $\tilde{L}_{p,\mathrm{rad}}^{h,n}$ associated to the eigenvalue $\tilde{\beta}_n^h(n,p)$, i.e.,

$$\begin{cases} \tilde{L}_{p,\mathrm{rad}}^{h,n}w = \tilde{\beta}_n^h(n,p)w & \text{in } \left(\frac{1}{n}, 1\right) \\ w\left(\frac{1}{n}\right) = w(1) = 0 \end{cases}$$

It is well known that w has exactly n nodal regions. Then, arguing by contradiction, let us assume that $\tilde{\beta}_n^h(n, p) \leq -(N-1)$. If $\tilde{\beta}_n^h(n, p) = -(N-1)$ then η and w are two solutions of the same Sturm–Liouville equation

$$(r^{N-1}v')' + \left[p|u_p(r)|^{p-1}r^{N-1} + \frac{\tilde{\beta}_n^h(n, p)}{r^{3-N}} \right] v = 0, \quad r \in \left(\frac{1}{n}, 1 \right)$$

and they are linearly independent, because $\eta(1) \neq 0 = w(1)$. As a consequence of the Sturm separation theorem, the zeros of η and w must alternate. Since η has $(n-1)$ zeros, then w must have $(n-1)$ nodal regions and this gives a contradiction. If $\tilde{\beta}_n^h(n, p) < -(N-1)$, then by the Sturm comparison theorem, η must have a zero between any two consecutive zeros of w. As a consequence, since we know that w has $(n-1)$ zeros in $(\frac{1}{h}, 1)$ and also $w(\frac{1}{h}) = w(1) = 0$, then η must have n zeros in $(\frac{1}{h}, 1)$, which gives again a contradiction. $\qquad\square$

Now we should study the eigenvalues $\tilde{\beta}_i^h(n, p)$, for $i = 1, \ldots, n-1$, of the linear operator $\tilde{L}_{p,\mathrm{rad}}^{h,n}$ (see (4.16)) which is defined in the annulus (4.9). It is convenient to choose h depending on p (and n) as follows:

$$h_p^n = \max\{h_p', h_p'', (M_{0,p}^n)^{p-1} + 1\} \tag{4.72}$$

where h_p' is as in Proposition (4.6), h_p'' as in Proposition (4.12) and $M_{0,p}^n$ is defined in (4.36).

Then we consider the eigenvalues

$$\tilde{\beta}_i(n, p) = \tilde{\beta}_i^h(n, p) \quad \text{for } h = h_p^n, \; i \in \mathbb{N}^+ \tag{4.73}$$

so that we can drop the dependence on h. By the choice of h_p^n and the previous results, we have that $\tilde{\beta}_i(n, p) < 0$ for $i = 1, \ldots, n-1$ and for every $p \in (1, p_s)$.

In the same way we shorten the notation for the linear operator, setting

$$\tilde{L}_{p,\mathrm{rad}}^n = \tilde{L}_{p,\mathrm{rad}}^{h,n} \quad \text{taking} \quad h = h_p^n.$$

The crucial result to estimate the eigenvalues $\tilde{\beta}_i(n, p)$, $i = 1, \ldots, n-1$, when p is close to p_s is the following.

Proposition 4.13. *For any* $n \in \mathbb{N}^+$,

$$\liminf_{p \to p_s} \tilde{\beta}_1(n, p) \geq -(N-1), \tag{4.74}$$

and hence

$$\liminf_{p \to p_s} \tilde{\beta}_i(n, p) \geq -(N-1) \quad \text{for } i = 1, \ldots, n-1. \tag{4.75}$$

Obviously, (4.75) follows from (4.74). It will be later proved that indeed equality holds in (4.74) and (4.75).

The proof of Proposition 4.13 is very long and requires several nontrivial estimates, therefore, we will only explain its strategy, referring to [106] for all details.

To get (4.74), it is natural to study the normalized eigenfunction associated to $\tilde{\beta}_1(n,p)$, for any $p \in (1, p_s)$, which is radial and positive. Let us denote it by φ_p^n. Hence it satisfies

$$
\begin{cases}
-\varphi_p^{n\prime\prime} - \dfrac{N-1}{r}\varphi_p^{n\prime} - p|u_p^n|^{p-1}\varphi_p^n = \tilde{\beta}_1(n,p)\dfrac{\varphi_p^n}{r^2} & \text{in } \left(\dfrac{1}{h_p^n}, 1\right) \\[2mm]
\varphi_p^n\left(\dfrac{1}{h_p^n}\right) = \varphi_p^n(1) = 0
\end{cases}
\tag{4.76}
$$

To obtain (4.74), one would like to pass to the limit into (4.75) and study the limit eigenvalue problem. Since the term $p|u_p^n|^{p-1}$ in (4.76) is not bounded, this cannot be done straightforwardly but a scaling procedure should be used in order to get something in the limit. Usually these scalings are done using as parameters the unbounded quantity given by $p|u_p^n|^{p-1}$. However, since u_p^n has several nodal regions and blows up in each of them at a different rate, is convenient to use various rescalings of the eigenfunction φ_p^n with different parameters. More precisely, we define

$$
\widehat{\varphi}_p^{n,i} = \frac{1}{(M_{i,p}^n)^{\frac{(p-1)(N-2)}{4}}}\varphi_p^n\left(\frac{|x|}{(M_{i,p}^n)^{\frac{p-1}{2}}}\right), \quad i = 0,\ldots,n-1
\tag{4.77}
$$

where $x \in \widehat{A}_p^{n,i} = (M_{i,p}^n)^{\frac{p-1}{2}}A_p^n$ and

$$
A_p^n = \left\{ y \in \mathbb{R}^N : \frac{1}{h_p^n} < |y| < 1 \right\}
$$

where $M_{i,p}^n$ are the L^∞-norm of u_p^n in the corresponding ith nodal region, defined in (4.36).

The functions $\widehat{\varphi}_p^{n,i}$ satisfy the eigenvalue problem:

$$
\begin{cases}
-\Delta\widehat{\varphi}_p^{n,i} - V_{i,p}^n(x)\widehat{\varphi}_p^{n,i} = \tilde{\beta}_1(n,p)\dfrac{\widehat{\varphi}_p^{n,i}}{|x|^2} & x \in \widehat{A}_p^{n,i} \\[2mm]
\widehat{\varphi}_p^{n,i} = 0 & \text{on } \partial\widehat{A}_p^{n,i}
\end{cases}
\tag{4.78}
$$

where

$$
V_{i,p}^n(x) = p\frac{1}{(M_{i,p}^n)^{p-1}}\left|u_p^n\left(\frac{|x|}{(M_{i,p}^n)^{\frac{p-1}{2}}}\right)\right|^{p-1}.
$$

Note that by (4.41) and the fact that (4.44) and (4.45) imply that $\dfrac{M_{i,p}^n}{M_{i+1,p}^n} \to +\infty$, as $p \to p_s$, $\forall i = 0,\ldots,n-1$, it follows that the limit domain of $\widehat{A}_p^{n,i}$ is $\mathbb{R}^N \setminus \{0\}$. This suggests that a limit eigenvalue problem, obtained by considering the limits of the potential $V_{i,p}^n$ should play a role to evaluate the limit of the eigenvalues $\tilde{\beta}_1(n,p)$ as $p \to p_s$. This limit eigenvalue problem is the one for the weighted linear operator:

$$\tilde{L}^* v = |x|^2 [-\Delta v - V(x)v], \quad x \in \mathbb{R}^N \tag{4.79}$$

where

$$V(x) = p_s U(x)^{p_s-1} = \frac{N+2}{N-2} \left(\frac{N(N-2)}{N(N-2) + |x|^2} \right)^2 \tag{4.80}$$

and we recall that U, defined in (4.29), is the solution of (4.28) satisfying (4.30).

Then the corresponding eigenvalue problem is

$$-\Delta \eta - V(x)\eta = \lambda \frac{\eta}{|x|^2}, \quad x \in \mathbb{R}^N \setminus \{0\} \tag{4.81}$$

One of the main steps in computing the Morse index of u_p^n is to show that the first eigenvalue $\tilde{\beta}^*$ of (4.81) is exactly

$$\tilde{\beta}^* = -(N-1). \tag{4.82}$$

This result has its own interest independent of the computation of the Morse index of u_p^n. It also holds in dimension $N = 2$ with an appropriate definition of the potential $V(x)$ and will play a crucial role in the computation of the Morse index of least energy nodal radial solution of (4.1) in the 2-dimensional ball that we describe in the next section. Therefore, we will prove (4.82) in Section 4.5.

Going back to (4.78), let us observe that, passing to the limit as $p \to p_s$, it could happen that the rescaled eigenfunctions $\widehat{\varphi}_p^{n,i}$ all converge to zero, for $i = 1, \ldots, n - 1$, in which case nothing is obtained in the limit. Since this cannot be excluded, the asymptotic analysis of the eigenvalues $\tilde{\beta}_1(n,p)$ is done in [106] by evaluating with great accuracy the contribution to the limit of $\tilde{\beta}_1(n,p)$ given by each nodal region of u_p^n.

Once Proposition 4.13 is obtained, we can prove the following exact formula for the Morse index of the solution u_p^n.

Theorem 4.14. *Let $N \geq 3$ and u_p^n be a radial solution of (4.1) with $n \in \mathbb{N}^+$ nodal domains. Then*

$$m(u_p^n) = n + N(n-1) \tag{4.83}$$

for p sufficiently close to p_s.

Proof. We recall that we have approximated the ball B by an annulus A_h, choosing $h = h_p^n$ as in (4.72). Then we have considered the radial weighted linear operators $\tilde{L}_{p,\mathrm{rad}}^n$ whose eigenvalue we have denoted by $\tilde{\beta}_i(n,p)$, $i \in \mathbb{N}^+$ (see (4.73)).

To simplify the notation, we set $\tilde{L}_p^n = \tilde{L}_{u_p^n}^h$ for $h = h_p^n$, where $\tilde{L}_{u_p^n}^h$ is defined in (4.15) and we denote its eigenvalues by

$$\tilde{\mu}_i(n,p) = \tilde{\mu}_i^h(n,p), \quad \text{for } h = h_p^n, \quad i \in \mathbb{N}^+$$

emphasizing the dependence on n and p. The number of negative eigenvalues of $\tilde{L}_p^h$ is denoted by $\tilde{k}_p(n) = \tilde{k}_p^h(n)$.

We have seen, by Proposition 4.6, that determining the Morse index $m(u_p^n)$ is equivalent to counting the number $\tilde{k}_p(n)$ of negative eigenvalues $\tilde{\mu}_i(n,p)$ of the operator $\tilde{L}_p^n$. Hence we should show that

$$\tilde{k}_p(n) = n + N(n-1) \quad \text{for } p \text{ close to } p_s. \tag{4.84}$$

By (4.21), we have that

$$\tilde{\mu}_j(n,p) = \tilde{\beta}_i(n,p) + \lambda_k, \quad \text{for } i,j \in \mathbb{N}^+, k \in \mathbb{N} \tag{4.85}$$

where λ_k are the eigenvalues of the Laplace–Beltrami operator $-\Delta_{S^{N-1}}$ on the unit sphere S^{N-1}, $N \geq 3$. As we already mentioned in (4.20),

$$\lambda_k = k(k+N-2) \quad (\geq 0), \quad k = 0,1,\dots$$

with multiplicity (see [34])

$$N_k - N_{k-2} \tag{4.86}$$

where N_s, $s \in \mathbb{Z}$, is defined in (4.23).

By (4.27), we already know that

$$\tilde{k}_p(n) \geq n + N(n-1) \tag{4.87}$$

and that

$$\tilde{\beta}_1(n,p) \leq \cdots \leq \tilde{\beta}_n(n,p) < 0 \leq \tilde{\beta}_{n+1}(n,p) \leq \dots \tag{4.88}$$

By (4.88), since $\lambda_k \geq 0$, it immediately follows that

$$\tilde{\beta}_i(n,p) + \lambda_k \geq 0 \quad \forall i \geq n+1, \forall k \geq 0 \tag{4.89}$$

so that *all the eigenvalues $\tilde{\beta}_i(n,p)$ with $i \geq n+1$ cannot produce any negative eigenvalue $\tilde{\mu}_j(n,p)$* by the formula (4.85).

Next, we analyze the contribution given by the last negative eigenvalues $\tilde{\beta}_n(n,p)$. Observe that $\lambda_1 = N-1$ and, by Proposition 4.12, $\tilde{\beta}_n(n,p) > -(N-1)$, hence we get

$$\tilde{\beta}_n(n,p) + \lambda_k > 0, \quad \forall k \geq 1. \tag{4.90}$$

On the other side, from (4.88) and observing that $\lambda_0 = 0$, we have that

$$\tilde{\beta}_n(n,p) + \lambda_0 = \tilde{\beta}_n(n,p) < 0. \tag{4.91}$$

Hence, by (4.85) and (4.91) we have that $\tilde{\beta}_n(n,p)$ is a *negative eigenvalue of $\tilde{L}_p^n$, which is radial and simple*, since by (4.86) it follows that λ_0 has multiplicity one. In fact, the eigenfunctions corresponding to λ_0 are just constants. Furthermore, because of (4.90), this eigenvalue is *the only* negative eigenvalue obtained by summing $\tilde{\beta}_n(n,p)$ with the eigenvalues of $-\Delta_{S^{N-1}}$.

Then (4.83) is obviously proven in the case $n = 1$.

In the case $n \geq 2$, we need to study the remaining negative eigenvalues $\tilde{\beta}_i(n,p)$, $i = 1,\ldots,n-1$ and, since there are exactly n radial simple negative eigenvalues of $\tilde{L}_p^n$, we have to prove that they produce exactly $N(n-1)$ negative *nonradial* eigenvalues $\tilde{\mu}_j(n,p)$ by the formula (4.85) (counted with their multiplicity).

Since by Proposition (4.13), we have

$$\liminf_{p \to p_s} \tilde{\beta}_i(n,p) \geq -(N-1), \quad \text{for any } i = 1,\ldots,n-1 \tag{4.92}$$

and observing that $\lambda_k \geq 2N > N-1$ for all $k \geq 2$, it follows that for p sufficiently close to p_s

$$\tilde{\beta}_i(n,p) + \lambda_k > 0, \quad \text{for any } i = 1,\ldots,n-1, \text{ for all } k \geq 2. \tag{4.93}$$

By (4.93) and the estimate (4.87), we immediately have that for p close to p_s,

$$\tilde{\beta}_i(n,p) + \lambda_1 < 0, \quad \text{for any } i = 1,\ldots,n-1. \tag{4.94}$$

Indeed, since there are exactly n radial simple negative eigenvalues of $\tilde{L}_p^n$, by (4.87) there must be at least $N(n-1)$ negative *nonradial* eigenvalues of $\tilde{L}_p^n$ (counted with their multiplicity). By (4.93), for p close to p_s, these nonradial eigenvalues must be obtained by the formula (4.85) for $i = 1,\ldots,n-1$ and $k = 1$ (for $k = 0$ only radial eigenvalues may be constructed). Hence, observing that the multiplicity of λ_1 is N (by (4.86)), we deduce that, if (4.94) does not hold, then (4.87) cannot be true.

In conclusion, by (4.93) and (4.94) we get that, for p close to p_s there are exactly $N(n-1)$ negative nonradial eigenvalues of $\tilde{L}_p^n$ counted with their multiplicity, given by the formula (4.94).

This proves the assertion (4.83). $\qquad\qquad\qquad\qquad\qquad\qquad\qquad\qquad\qquad\square$

We end this section by a few comments about the formula (4.83). It can be written as

$$m(u_p^n) = n(N+1) - N$$

which shows that the Morse index $m(u_p^n)$ grows linearly with respect to the number n of nodal domains, which, in turns, corresponds to the number of negative radial eigenvalues of the linearized operator $L_{u_p^n}$. This is somehow surprising since, in general, one would expect *many more* negative nonradial eigenvalues than the negative radial ones. Indeed, if we look at the distribution of the radial and nonradial eigenvalues of the Laplace operator $(-\Delta)$ in $H_0^1(B)$, we observe that:

(i) on one side by results of Brüning-Heintze and Donnelly [51, 52, 111] we get that

$$\lambda_{r,n} \sim Cn^2 \quad \text{as } n \to +\infty$$

where $\lambda_{r,n}$ is the nth radial eigenvalues of $(-\Delta)$, which implies that the number $k_r(n^2)$ of the radial eigenvalues of $(-\Delta)$ bounded by n^2 is n, more precisely

$$k_r(n^2) \sim n \quad \text{as } n \to +\infty$$

(ii) on the other side by the classical Weil law (see, e. g., [210]):

$$k(n^2) \sim Cn^N \quad \text{as } n \to +\infty \quad (N \text{ is the dimension})$$

where $k(n^2)$ is the number of all eigenvalues of $(-\Delta)$ in $H_0^1(B)$ less than or equal to n^2.

In an equivalent way, we can observe that if we consider a radial eigenfunction of $(-\Delta)$ in $H_0^1(B)$ with n nodal regions, i. e., corresponding to the eigenvalue $\lambda_{r,n}$, then its *Morse index* is just the number of the eigenvalues less that $\lambda_{r,n}$ which, by (i) and (ii), grows at a rate of order n^N and so faster than n (if $N \geq 2$) as $n \to +\infty$.

So $L_{u_p^n}$ represents an example of a linear, Schrödinger-type, operator determined by the potential $V_p^n(x) = p|u_p^n(x)|^{p-1}$, for p approaching p_s, for which (i) and (ii) do not hold, at least for negative eigenvalues.

Another interesting consequence could be derived studying (4.1) as $p \to 1$. In this case, it is reasonable to conjecture the convergence of the Morse index $m(u_p^n)$ to the Morse index of the Dirichlet radial eigenfunction of $(-\Delta)$ with n nodal regions (i. e., the eigenfunction corresponding to the radial eigenvalue $\lambda_{r,n}$) possibly augmented by the multiplicity of $\lambda_{r,n}$ which is 1. Indeed, suitable normalizations of solutions of (4.1) converge to eigenfunctions of the Laplacian as $p \to 1$ (see [37, 135]). Therefore, the previous considerations indicate that for large n the Morse index $m(u_p^n)$ for p close to 1 is of order n^N, hence it is much larger that $n + N(n - 1)$, which is, by (4.83), the Morse index of u_p^n for p close to p_s. So bifurcations from u_p^n should appear (as for the problems described in Chapter 5), as p ranges from 1 to p_s, showing that the structure of the solution set of (4.1) is richer than one could imagine.

Finally, we would like to point out another interesting fact: the formula (4.83) *does not hold in dimension $N = 2$, as $p \to p_s = +\infty$*, as we will show in the next section.

4.4 Computation of the Morse index in dimension $N = 2$

As in Section 4.2.2, we only consider the nodal radial solution of (4.1) with the least energy that we denote by u_p and assume $u_p(0) > 0$.

By the results of Section 4.1, since u_p has only two nodal regions and its radial Morse index is two, we have to estimate the first two eigenvalues $\tilde{\beta}_i^h$, $i = 1, 2$, of the linear weighted operator $\tilde{L}_{u_p,\mathrm{rad}}^h$ in the interval $(\frac{1}{h}, 1)$ (see (4.16)). Making explicit the dependence on p we denote by $\tilde{\beta}_i^h(p)$ the eigenvalues $\tilde{\beta}_i^h$, $i = 1, 2$.

The first estimate which holds for $\tilde{\beta}_2^h(p)$ is the following.

Proposition 4.15. *Let $h \in \mathbb{N}^+$, $p \in (1, +\infty)$. Then there exists $h_p'' \in \mathbb{N}^+$ such that*

$$\tilde{\beta}_2^h(p) > -1, \quad \text{for any } h \geq h_p''.$$

The proof is the same as that of Proposition 4.12, a bit simpler because there is only one nodal radius.

As for the case $N \geq 3$, it is convenient to choose h depending on p as follows:

$$h_p = \max\{h_p', h_p'', [(\epsilon_p^+)^{-2}] + 1\} \tag{4.95}$$

where h_p' is as in Proposition 4.6, h_p'' as in Proposition 4.15 and $(\epsilon_p^+)^{-2}$ as in (4.58).

Since Proposition 4.15 gives an estimate for $\tilde{\beta}_2^h(p)$, we only have to study the eigenvalues $\tilde{\beta}_1^{h_p}(p)$ that we denote by $\tilde{\beta}_1(p)$, having fixed $h = h_p$. The corresponding linear operator will be denoted by $\tilde{L}_{p,\mathrm{rad}}$ instead of $\tilde{L}_{u_p,\mathrm{rad}}^h$. The crucial result to estimate these eigenvalues for large values of p is the following.

Proposition 4.16. *We have*

$$\lim_{p \to +\infty} \tilde{\beta}_1(p) = -\frac{l^2 + 2}{2} \approx -26.9 \tag{4.96}$$

where the number l is defined in (4.67).

Let us immediately observe that (4.96) is different from the estimate (4.74) which holds in the higher dimensional case $N \geq 3$. This is due to the different asymptotic behavior of the solutions u_p, as $p \to +\infty$, with respect to the case $N \geq 3$ where $p \to p_s = \frac{N+2}{N-2}$, as explained in Section 4.2.2.

The proof of Proposition 4.16 is very long and difficult, therefore, we indicate its main steps, referring to [105] for the details.

The strategy is the same as for the case $N \geq 3$, i.e., we consider the normalized eigenfunction φ_p corresponding to $\tilde{\beta}_1(p)$, $p \in (1, +\infty)$, which is radial and positive. The convenient normalization is $\|\frac{\varphi_p}{|y|}\|_{L^2(A_p)} = 1$ where A_p is the annulus

$$A_p = A_{h_p} = \left\{ y \in \mathbb{R}^2 : \frac{1}{h_p} < |y| < 1 \right\}. \tag{4.97}$$

It satisfies

$$\begin{cases} -\varphi_p'' - \dfrac{N-1}{r}\varphi_p' - p|u_p|^{p-1}\varphi_p = \tilde{\beta}_1(p)\dfrac{\varphi_p}{r^2} & \text{in } \left(\dfrac{1}{h_p}, 1\right) \\[2ex] \varphi_p\left(\dfrac{1}{h_p}\right) = \varphi_p(1) = 0 \end{cases}$$

(4.98)

Again, we cannot pass to the limit, as $p \to +\infty$, in (4.98) since the term $p|u_p|^{p-1}$ is not bounded. Hence it is convenient to rescale φ_p. The solutions u_p have two nodal domains and we have seen in Section 4.2, Proposition 4.11, that the asymptotic behavior of the rescaling of the positive part u_p^+ and of the negative part u_p^- (i. e., z_p^+ and z_p^- defined in (4.59) and (4.60)) with the parameters $\epsilon_p^\pm$ defined in (4.58) are different. In particular, their limits are different (see (4.62), (4.63)).

Hence we have two possible scalings for the eigenfunction φ_p:

$$\widehat{\varphi}_p^\pm(x) = \varphi_p(\epsilon_p^\pm x)$$

(4.99)

which satisfy

$$\begin{cases} -\Delta\widehat{\varphi}_p^\pm(x) - V_p^\pm(x)\widehat{\varphi}_p^\pm(x) = \tilde{\beta}_1(p)\dfrac{\widehat{\varphi}_p^\pm(x)}{|x|^2} & \text{in } \dfrac{A_p}{\epsilon_p^\pm} \\[2ex] \widehat{\varphi}_p^\pm = 0 & \text{on } \partial\left(\dfrac{A_p}{\epsilon_p^\pm}\right) \end{cases}$$

(4.100)

where

$$V_p^+(x) = \left|\frac{u_p(\epsilon_p^+ x)}{u_p(0)}\right|^{p-1}, \qquad V_p^-(x) = \left|\frac{u_p(\epsilon_p^- x)}{u_p(s_p)}\right|^{p-1}$$

(4.101)

where s_p is defined in (4.57).

For both rescalings, we have by (4.69) and (4.95) that the limit domain is $\mathbb{R}^2 \setminus \{0\}$, as $p \to +\infty$. Moreover, by (4.62) and (4.63) we have

$$V_p^+ = \left|1 + \frac{z_p^+}{p}\right|^{p-1} \longrightarrow V^+ = e^U \quad \text{in } C_{\text{loc}}^0(\mathbb{R}^2),$$

(4.102)

$$V_p^- = \left|1 + \frac{z_p^-}{p}\right|^{p-1} \longrightarrow V^- = e^{Z_l} \quad \text{in } C_{\text{loc}}^0(\mathbb{R}^2 \setminus \{0\}).$$

(4.103)

This, together with bounds of $\widehat{\varphi}_p^\pm$ (see [105]) allows to pass to the limit in (4.100). However, the difficulty underlined in the case $N \geq 3$ in Section 4.3 could appear, i. e., both functions $\widehat{\varphi}_p^\pm$ could vanish in the limit and nothing can be deduced by the limit eigenvalue problem. Here is the main difference with respect to the higher dimensional case, since in [105] it is proved that $\widehat{\varphi}_p^-$ does not converge to zero. Indeed, it holds

$$\liminf_{p \to +\infty} \int_{\{\frac{1}{k}\leq|x|\leq k\}} \frac{\widehat{\varphi}_p^-(x)^2}{|x|^2}\, dx > 0$$

(4.104)

for some $k > 1$ (see [105], Proposition 6.6).

The proof of (4.104) is nontrivial and requires several delicate estimates. Now we are ready to outline the proof of (4.96).

Sketch of proof of Proposition 4.16. Let us define the Hilbert space

$$L^2_{\frac{1}{|x|}}(\mathbb{R}^N) = \left\{ v \in \mathbb{R}^N \longrightarrow \mathbb{R} : \frac{v}{|x|} \in L^2(\mathbb{R}^N) \right\}, \quad N \geq 2 \tag{4.105}$$

endowed with the scalar product $(u, v) = \int_{\mathbb{R}^N} \frac{u(x)v(x)}{|x|^2} \, dx$. Then we define the space

$$D_{\mathrm{rad}}(\mathbb{R}^N) = D^{1,2}_{\mathrm{rad}}(\mathbb{R}^N) \cap L^2_{\frac{1}{|x|}}(\mathbb{R}^N) \tag{4.106}$$

where $D^{1,2}_{\mathrm{rad}}(\mathbb{R}^N)$ is the subspace of the radial functions in $D^{1,2}(\mathbb{R}^N)$ which, in turn, is the closure of $C^\infty_c(\mathbb{R}^N)$ with respect to the Dirichlet norm $|v|_{D^{1,2}(\mathbb{R}^N)} = (\int_{\mathbb{R}^N} |\nabla v(x)|^2 \, dx)^{\frac{1}{2}}$.

It is not difficult to see that the sequence $\{\widehat{\varphi}_p\}$ is bounded in $D_{\mathrm{rad}}(\mathbb{R}^2)$, and hence, up to a subsequence, weakly converge to a nonnegative function $\widehat{\varphi}$ in $D_{\mathrm{rad}}(\mathbb{R}^2)$, as $p \to +\infty$.

By (4.104), it is possible to prove that $\widehat{\varphi} \neq 0$. This allows to pass to the limit in (4.100), proving that there exists $\tilde{\beta}_1 < 0$ such that

$$\int_{\mathbb{R}^2 \setminus \{0\}} \nabla\widehat{\varphi}(x)\nabla\rho(x) \, dx - \int_{\mathbb{R}^2} V^-(x)\widehat{\varphi}(x)\rho(x) \, dx$$

$$= \tilde{\beta}_1 \int_{\mathbb{R}^2 \setminus \{0\}} \frac{\varphi(x)\rho(x)}{|x|^2} \, dx \tag{4.107}$$

where $V^-(x) = e^{Z_l}$, Z_l as in (4.66), (4.67) and ρ is any test function in $C^\infty_0(\mathbb{R}^2 \setminus \{0\})$. Namely, $\widehat{\varphi}$ is a weak (and also classical) nontrivial nonnegative solution to the limit equation

$$-\widehat{\varphi}''(s) - \frac{\widehat{\varphi}'(s)}{s} - V^-(s)\widehat{\varphi}(s) = \tilde{\beta}_1 \frac{\widehat{\varphi}(s)}{s^2}, \quad s \in (0, +\infty). \tag{4.108}$$

As in [128], it is convenient to set $\eta(s) = \widehat{\varphi}(\delta(\frac{s}{2\sqrt{2}})^{\frac{2}{2+\gamma}})$, δ as in (4.67) so that by (4.108) η satisfies

$$-\eta'' - \frac{\eta'}{s} - V(s)\eta = \frac{4\tilde{\beta}_1}{(\gamma + 2)^2} \frac{\eta}{s^2} \quad \text{in } (0, +\infty) \tag{4.109}$$

with $\frac{4\tilde{\beta}_1}{(\gamma+2)^2} < 0$ and $V(s) = (\frac{1}{1+\frac{1}{8}s^2})^2$. This means that $\frac{4\tilde{\beta}_1}{(\gamma+2)^2}$ is the first eigenvalue $\tilde{\beta}_*$ for the problem

$$-\Delta\eta - V(|x|)\eta = \lambda \frac{\eta}{|x|^2} \quad \text{in } \mathbb{R}^2 \setminus \{0\}. \tag{4.110}$$

To know the value of $\tilde{\beta}^*$ is then crucial to compute $\tilde{\beta}_1$, and hence the Morse index of u_p as shown in the next theorem.

In Section 4.5, we will show that

$$\tilde{\beta}^* = -1. \tag{4.111}$$

Then using (4.111) we have, by (4.109), that

$$\frac{4\tilde{\beta}_1}{(y+2)^2} = -1$$

which, by the definition of y in (4.67), implies (4.96) since $l \approx 7.1979$ (see (4.67)), $\qquad \square$

We can now prove the result on the Morse index of the solution u_p.

Theorem 4.17. *Let u_p be the least energy nodal radial solution of (4.1) with $u_p(0) > 0$. Then*

$$m(u_p) = 12 \quad \text{for } p \text{ sufficiently large}.$$

Proof. The proof follows the same procedure as in Theorem 4.14.

Let h_p be defined as in (4.95). By Proposition 4.6, we know that determining the Morse index $m(u_p)$ is equivalent to counting the number $\tilde{k}_p = \tilde{k}_p^{h_p}$ of the negative eigenvalues $\tilde{\mu}_i(p) = \tilde{\mu}_i^{h_p}$, $i \in \mathbb{N}^+$, of the operator $\tilde{L}_p = \tilde{L}_{u_p}^{h_p}$, defined in (4.15).

Hence we have to show that

$$\tilde{k}_p = 12, \quad \text{for } p \text{ sufficiently large}. \tag{4.112}$$

By (4.21), we have that

$$\tilde{\mu}_j(p) = \tilde{\beta}_i(p) + \lambda_k, \quad \text{for } i, j \in \mathbb{N}^+ \text{ and } k \in \mathbb{N} \tag{4.113}$$

where $\lambda_k = k^2$, by (4.20).

Moreover, by (4.22), (4.23) the eigenspace associated to λ_0 has dimension one, while the eigenspace associated to each λ_k has dimension two, for $k \in \mathbb{N}^+$.

By (4.27), we already know that the only negative eigenvalues $\tilde{\beta}_i(p)$ are $\tilde{\beta}_1(p) \leq \tilde{\beta}_2(p) < 0$, so that these are the only eigenvalues which can make $\tilde{\mu}_j(p) < 0$ in (4.113).

By Proposition 4.15, we know that $\tilde{\beta}_2(p) > -1$ and this implies that

$$\tilde{\beta}_2(p) + \lambda_k > 0, \quad \text{for } k > 0 \tag{4.114}$$

while

$$\tilde{\beta}_2(p) + \lambda_0 = \tilde{\beta}(p) < 0 \tag{4.115}$$

This gives one negative eigenvalue of $\tilde{L}_p$ which is a radial one.

Let us now consider $\tilde{\beta}_1(p)$. By Proposition 4.15, we know exactly its limit which is given in (4.96). Therefore, for p large we get

$$-\lambda_6 = -36 < \tilde{\beta}_1(p) < -25 = \lambda_5$$

so that

$$\tilde{\beta}_1(p) + \lambda_k > 0 \quad \text{for } k \geq 6$$

while

$$\tilde{\beta}_1(p) + \lambda_k < 0 \quad \text{for } k \leq 5. \tag{4.116}$$

Remembering that λ_k has multiplicity two for $k \geq 1$ and λ_0 has multiplicity one for $k = 0$, from (4.115) and (4.116) we get exactly 12 negative eigenvalues of the linear operator $\tilde{L}_p$. Hence (4.112) is proved. $\qquad\square$

It is interesting to observe that Theorem 4.17 shows that the formula (4.83) for the Morse index of radial solutions of (4.1) in dimension $N \geq 3$, as $p \to p_s$, does not hold in dimension $N = 2$ as $p \to +\infty$. Indeed, since the least energy radial solution to (4.1) has two nodal regions inserting $N = 2$ and $n = 2$ in (4.83), one would get $m(u_p) = 4$ while Theorem 4.17 gives $m(u_p) = 12$.

The reason for this difference relies on the behavior of the radial solutions of (4.1) in dimension $N \geq 3$ which is different from that in dimension $N = 2$. Indeed, while in dimension $N \geq 3$ the rescalings of the solutions in each nodal region converge to the same function U defined in (4.29), in dimension $N = 2$ it happens that the rescalings of the negative and positive parts of u_p converge to two different functions, namely the ones defined in (4.62) and (4.63). In particular, the rescaling z_p^- of u_p^- converge to a solution Z_l of the singular Liouville problem in $\mathbb{R}^2$ (see Proposition 4.11). The proof of Theorem 4.17, by means of Proposition 4.16, shows clearly that the main contribution to the Morse index of u_p comes from the nodal domain where u_p is negative.

4.5 A weighted eigenvalue problem in $\mathbb{R}^N$

In this section, we compute the first eigenvalue of the weighted linear operator

$$\tilde{L}^*(v) = |x|^2(-\Delta v - V(x)v), \quad x \in \mathbb{R}^N, \ N \geq 2 \tag{4.117}$$

where the potential V is defined as

$$V(x) = \begin{cases} e^{U(x)} = \left(\dfrac{1}{1 + \frac{1}{8}|x|^2}\right)^2 & \text{if } N = 2 \\[4mm] p_s U^{p_s - 1}(x) = \dfrac{N+2}{N-2}\left(\dfrac{N(N-2)}{N(N-2) + |x|^2}\right)^{\frac{N-2}{2}} & \text{if } N \geq 3 \end{cases} \tag{4.118}$$

where $U(x)'$ is defined in (4.29) for $N \geq 3$ and (4.64) for $N = 2$.

As in Section 4.4, we consider the Hilbert space $D_{\mathrm{rad}}(\mathbb{R}^N)$ defined in (4.106) endowed with the scalar product

$$(u, v) = \int_{\mathbb{R}^N} \nabla u(x) \nabla v(x)\, dx + \int_{\mathbb{R}^N} \frac{u(x)v(x)}{|x|^2}\, dx.$$

Obviously, $D_{\mathrm{rad}}(\mathbb{R}^N)$ is continuously embedded both in $D^{1,2}_{\mathrm{rad}}(\mathbb{R}^N)$ and in $L^2_{\frac{1}{|x|}}(\mathbb{R}^N)$ (see (4.106)). Moreover, by the Hardy inequality [139, 140, 184] $D_{\mathrm{rad}}(\mathbb{R}^N) = D^{1,2}_{\mathrm{rad}}(\mathbb{R}^N)$ when $N \geq 3$, while it is well known that $D_{\mathrm{rad}}(\mathbb{R}^2) \subsetneq D^{1,2}_{\mathrm{rad}}(\mathbb{R}^2)$.

Let us set

$$\tilde{\beta}^* = \inf_{\substack{v \in D_{\mathrm{rad}}(\mathbb{R}^N) \\ v \neq 0}} \tilde{R}^*(v) \tag{4.119}$$

where

$$\tilde{R}^*(v) = \frac{\tilde{Q}^*(v)}{\left\| \frac{v}{|x|} \right\|^2_{L^2(\mathbb{R}^N)}},$$

$$\tilde{Q}^*(v) = \int_{\mathbb{R}^N} \left(|\nabla v(x)|^2 - V(x)v(x)^2 \right) dx.$$

Since the function $V(x)|x|^2$ is bounded, $\tilde{Q}^*(v)$ and $\tilde{R}^*(v)$ are well-defined for $v \in D_{\mathrm{rad}}(\mathbb{R}^N)$. Indeed,

$$\int_{\mathbb{R}^N} V(x)v(x)^2\, dx \leq \sup_{\mathbb{R}^N}(V(x)|x|^2) \int_{\mathbb{R}^N} \frac{v(x)}{|x|^2}\, dx < +\infty.$$

The aim of this section is to compute exactly the value of the first eigenvalue $\tilde{\beta}^*$ of the operator $\tilde{L}^*$ defined in (4.117). To do this, an important step is the following result on the spectrum of $\tilde{L}^*$.

Proposition 4.18. *Let $\lambda \leq 0$ and $\eta \in C^2(\mathbb{R}^N \setminus \{0\}) \cap D_{\mathrm{rad}}(\mathbb{R}^N), \eta \geq 0, \eta \not\equiv 0$, be a radial solution to*

$$-\Delta\eta - V(x)\eta = \lambda \frac{\eta}{|x|^2}, \quad in\ \mathbb{R}^N \setminus \{0\} \tag{4.120}$$

Then

$$\lambda = -(N - 1).$$

The proof of Proposition 4.18 relies on some properties of regular functions in the space $D_{\mathrm{rad}}(\mathbb{R}^N)$ which are stated in the following lemmas.

Lemma 4.19. *Let $N \geq 3$ and $\eta \in C^2(\mathbb{R}^N \setminus \{0\}) \cap D_{\mathrm{rad}}(\mathbb{R}^N)$, then*

$$|x|^{N-1}\eta(x) \to 0 \quad \text{as } |x| \to 0 \quad \text{and} \quad \frac{\eta(x)}{|x|} \to 0 \quad \text{as } |x| \to +\infty.$$

Lemma 4.20. *Let $f \in L^\infty(\mathbb{R}^2)$, $f \geq 0$ be such that $\frac{1}{|x|^4}f(\frac{x}{|x|^2}) \in L^\infty(\mathbb{R}^2)$ and let $\alpha \geq 0$. If $\eta \in C^2(\mathbb{R}^2 \setminus \{0\}) \cap D_{\mathrm{rad}}(\mathbb{R}^2)$, $\eta \geq 0$ is a nontrivial radial solution of*

$$-\Delta\eta - f(x)\eta = -\alpha^2 \frac{\eta}{|x|^2} \quad \text{in } \mathbb{R}^2 \setminus \{0\} \tag{4.121}$$

then

$$|x|\eta(x) \to 0 \quad \text{as } |x| \to 0 \quad \text{and} \quad \frac{\eta(x)}{|x|} \to 0 \quad \text{as } |x| \to +\infty.$$

The proof of Lemma 4.19 and Lemma 4.20 are contained in [105] (see Lemma 8.1 and Lemma 8.2 therein).

Proof of Proposition 4.18. It is easy to see that the function

$$\eta_1(x) = \begin{cases} \dfrac{|x|}{1 + \frac{1}{8}|x|^2} & \text{if } N = 2 \\[2em] \dfrac{|x|}{\left(1 + \frac{|x|^2}{N(N-2)}\right)^{\frac{N}{2}}} & \text{if } N \geq 3 \end{cases} \tag{4.122}$$

satisfies the equation

$$-\Delta\eta_1 - V(x)\eta_1 = \lambda_1 \frac{\eta_1}{|x|^2} \quad \text{in } \mathbb{R}^N \setminus \{0\} \tag{4.123}$$

with $\lambda_1 = -(N-1)$ and $V(x)$ as in (4.118).

Let us assume that there exists a function $\eta_2 \in C^2(\mathbb{R}^N \setminus \{0\}) \cap D_{\mathrm{rad}}(\mathbb{R}^N)$ radial and nonnegative for which

$$-\Delta\eta_2 - V(x)\eta_2 = \lambda_2 \frac{\eta_2}{|x|^2} \quad \text{in } \mathbb{R}^N \setminus \{0\} \tag{4.124}$$

for some $\lambda_2 \leq 0$.

Since $\int_{\mathbb{R}^N} |\nabla\eta_2|^2 \, dx < +\infty$, then there exist two sequences of radii $r_n \to 0$ and $R_n \to +\infty$ such that

$$r_n^N |\nabla\eta_2(r_n)|^2 \longrightarrow 0 \quad \text{and} \quad R_n^N |\nabla\eta_2(R_n)|^2 \longrightarrow 0 \quad \text{as } n \to +\infty.$$

As a consequence,

$$r_n^N |\nabla\eta_2(r_n)| \longrightarrow 0 \quad \text{and} \quad R_n^N |\nabla\eta_2(R_n)| \longrightarrow 0 \quad \text{as } n \to +\infty. \tag{4.125}$$

Applying Lemma 4.19 and Lemma 4.20, we get that

$$r_n^{N-1}\eta_2(r_n) \to 0 \quad \text{and} \quad \frac{\eta_2(R_n)}{R_n} \to 0, \quad \text{as } n \to +\infty. \tag{4.126}$$

Next, multiplying (4.123) by η_2 and (4.124) by η_1, adding them and integrating over $B_{R_n}(0) \setminus B_{r_n}(0)$ we get

$$(\lambda_1 - \lambda_2) \underbrace{\int_{B_{R_n}(0) \setminus B_{r_n}(0)} \frac{\eta_1(x)\eta_2(x)}{|x|^2} \, dx = \int_{\partial B_{R_n}(0)} \eta_1 \nabla \eta_2 \cdot v \, ds}_{=:A_n}$$

$$+ \underbrace{\int_{\partial B_{R_n}(0)} \eta_2 \nabla \eta_1 \cdot v \, dS}_{=:B_n} - \underbrace{\int_{\partial B_{r_n}(0)} \eta_1 \nabla \eta_2 \cdot v \, ds}_{=:C_n}$$

$$- \underbrace{\int_{\partial B_{r_n}(0)} \eta_2 \nabla \eta_1 \cdot v \, dS}_{=:D_n} \tag{4.127}$$

where v is the outer normal to $\partial B_{R_n}(0)$. Then using the explicit expression of η_1 and (4.125), (4.126), we can estimate A_n, B_n, C_n and D_n as follows:

$$|A_n| \le c_n R_n^{N-1} \eta_1(R_n)|\nabla \eta_2(R_n)| \le c_N R_n^{N-1} R_n^{-(N-1)}|\nabla \eta_2(R_n)| \longrightarrow 0,$$

$$|B_n| \le c_n R_n^{N-1} \eta_2(R_n)|\nabla \eta_1(R_n)| \le c_N R_n^{N-1} \eta_2(R_n) R_n^{N-1} = \frac{c_N \eta_2(R_n)}{R_n} \longrightarrow 0,$$

$$|C_n| \le c_N r_n^{N-1} \eta_1(r_n)|\nabla \eta_2(r_n)| \le c_N r_n^{N-1} r_n|\nabla \eta_2(r_n)| \longrightarrow 0,$$

$$|D_n| \le c_N r_n^{N-1} \eta_2(r_n)|\nabla \eta_1(r_n)| \le c_N r_n^{N-1} \eta_2(r_n) \longrightarrow 0,$$

where in the above estimates the limits are for $n \to +\infty$ and we have denoted by c_N a generic constant depending only on N.

Thus, passing to the limit in (4.127), we get

$$(\lambda_1 - \lambda_2) \int_{\mathbb{R}^N} \frac{\eta_1(x)\eta_2(x)}{|x|^2} \, dx = 0,$$

which implies that $\lambda_2 = \lambda_1 = -(N-1)$ because $\eta_1 > 0$ in $\mathbb{R}^N \setminus \{0\}$ and, by assumption, $\eta_2 \ge 0$, $\eta_2 \not\equiv 0$. $\qquad\square$

A first estimate for the first eigenvalue $\tilde{\beta}^*$ is easily obtained by using the function η_1 defined in (4.122) to evaluate the infimum in (4.119).

Indeed, since $\eta_1 \in D_{\mathrm{rad}}(\mathbb{R}^N)$ and satisfies (4.123), multiplying both sides of (4.123) by η_1 and integrating we get

$$\tilde{R}^*(\eta_1) = -(N-1).$$

Hence we have

$$\tilde{\beta}^* \leq -(N-1). \tag{4.128}$$

We will show that equality holds in (4.128). Indeed, we have the following.

Theorem 4.21. *For any $N \geq 2$,*

$$\tilde{\beta}^* = -(N-1)$$

and it is achieved by the function η_1 defined in (4.122).

Proof. We first show a coercivity property. For any $v \in D_{\mathrm{rad}}(\mathbb{R}^N)$ which satisfies $\int_{\mathbb{R}^N} \frac{v(x)^2}{|x|^2}\, dx = 1$, one has

$$
\begin{aligned}
\tilde{Q}^*(v) &= \int_{\mathbb{R}^N} |\nabla v(x)|^2\, dx - \int_{\mathbb{R}^N} V(x)|x|^2 \frac{v(x)^2}{|x|^2}\, dx \\
&\geq \int_{\mathbb{R}^N} |\nabla v(x)|^2\, dx - \sup_{\mathbb{R}^N}(V(x)|x|^2) \int_{\mathbb{R}^N} \frac{v(x)^2}{|x|^2}\, dx \\
&= \int_{\mathbb{R}^N} |\nabla v(x)|^2\, dx - C,
\end{aligned}
\tag{4.129}
$$

where $(0 <) \; C := \sup_{\mathbb{R}^N}(V(x)|x|^2) < +\infty$.

Since one can easily show that

$$\tilde{\beta}^* = \inf_{\substack{v \in D_{\mathrm{rad}}(\mathbb{R}^N) \\ \|\frac{v}{|x|}\|^2_{L^2(\mathbb{R}^N)}=1}} \tilde{Q}^*(v),$$

then clearly (4.129) implies that $\tilde{\beta}^* > -\infty$.

Let $\{v_n\}_n \subset D_{\mathrm{rad}}(\mathbb{R}^N)$ be a minimizing sequence for (4.119) with $\|\frac{v_n}{|x|}\|_{L^2(\mathbb{R}^N)} = 1$. We can assume without loss of generality that $v_n \geq 0$ (because otherwise we could consider $|v_n|$). By the coercivity property (4.129), it follows that v_n is bounded in $D^{1,2}_{\mathrm{rad}}(\mathbb{R}^N)$, and hence in $D_{\mathrm{rad}}(\mathbb{R}^N)$, because $\|\frac{v_n}{|x|}\|_{L^2(\mathbb{R}^N)} = 1$. Therefore, by the reflexivity of $D_{\mathrm{rad}}(\mathbb{R}^N)$, there exists $v \in D_{\mathrm{rad}}(\mathbb{R}^N)$ such that, up to a subsequence:

$$v_n \rightharpoonup v \quad \text{in } D_{\mathrm{rad}}(\mathbb{R}^N),$$

$$v_n \to v \quad \text{in } L^q(B_R),\, 1 < q < +\infty \text{ if } N = 2;\, 1 < q < \frac{2N}{N-2} \text{ if } N \geq 3,$$

$$v_n \rightharpoonup v \quad \text{in } D^{1,2}_{\mathrm{rad}}(\mathbb{R}^N) \text{ by the embedding } D_{\mathrm{rad}}(\mathbb{R}^N) \subset D^{1,2}_{\mathrm{rad}}(\mathbb{R}^N),$$

$$v_n \rightharpoonup v \quad \text{in } L^2_{\frac{1}{|x|}}(\mathbb{R}^N) \text{ by the embedding of } D_{\mathrm{rad}}(\mathbb{R}^N) \subset L^2_{\frac{1}{|x|}}(\mathbb{R}^N),$$

$$v_n \to v \quad \text{a. e. in } \mathbb{R}^N.$$

Hence $v \geq 0$,

$$\|\nabla v\|_{L^2(\mathbb{R}^N)} \leq \liminf_{n \to +\infty} \|\nabla v_n\|_{L^2(\mathbb{R}^N)} \tag{4.130}$$

and

$$\left\|\frac{v}{|x|}\right\|_{L^2(\mathbb{R}^N)} \leq \liminf_{n \to +\infty} \left\|\frac{v_n}{|x|}\right\|_{L^2(\mathbb{R}^N)} = 1. \tag{4.131}$$

Next, we show that

$$\int_{\mathbb{R}^N} V(x)v_n(x)^2 \, dx \longrightarrow \int_{\mathbb{R}^N} V(x)v(x)^2 \, dx \quad \text{as } n \to +\infty. \tag{4.132}$$

Let us fix $\epsilon > 0$ then

$$\left| \int_{\{|x|>R\}} V(x)(v_n(x)^2 - v(x)^2) \, dx \right|$$

$$\leq \sup_{|x|>R}(V(x)|x|^2)\left[\int_{\{|x|>R\}} \frac{v_n(x)^2}{|x|^2} \, dx + \int_{\{|x|>R\}} \frac{v(x)^2}{|x|^2} \, dx \right]$$

$$\leq \frac{C}{R^2} < \frac{\epsilon}{2}$$

choosing R sufficiently large.

On the other hand, fixing the same R, since $v_n \to v$ in $L^2(B_R)$, also

$$V^{\frac{1}{2}}v_n \to V^{\frac{1}{2}}v \quad \text{in } L^2(B_R),$$

and hence

$$\int_{B_R} V(x)v_n(x)^2 \, dx \longrightarrow \int_{B_R} V(x)v(x)^2 \, dx.$$

Therefore, for n large

$$\left| \int_{B_R} V(x)v_n(x)^2 \, dx - \int_{B_R} V(x)v(x)^2 \right| < \frac{\epsilon}{2},$$

thus proving (4.132).

By (4.130), (4.132) and (4.127), it follows that

$$\tilde{Q}^*(v) = \int_{\mathbb{R}^N} \left(|\nabla v(x)|^2 - V(x)v(x)^2\right) dx$$

$$\leq \liminf_n \int_{\mathbb{R}^N} \left(|\nabla v_n(x)|^2 - V(x)v_n(x)^2\right) dx$$

$$= \tilde{\beta}^* \leq -(N-1) < 0, \tag{4.133}$$

in particular, $\tilde{Q}^*(v) < 0$ and so $v \neq 0$.

Next, we show that

$$\left\| \frac{v}{|x|} \right\|_{L^2(\mathbb{R}^N)} = 1. \tag{4.134}$$

By the definition of $\tilde{\beta}^*$ and (4.133), we have

$$\tilde{\beta}^* \le \tilde{R}^*(v) = \frac{\tilde{Q}^*(v)}{\left\| \frac{v}{|x|} \right\|^2_{L^2(\mathbb{R}^N)}} \le \frac{\tilde{\beta}^*(v)}{\left\| \frac{v}{|x|} \right\|^2_{L^2(\mathbb{R}^N)}}. \tag{4.135}$$

Since $\tilde{\beta}^* < 0$, then necessarily

$$\left\| \frac{v}{|x|} \right\|_{L^2(\mathbb{R}^N)} \ge 1,$$

which together with (4.131) gives (4.134). As a consequence from (4.135), we get

$$\tilde{R}^*(v) = \tilde{\beta}^*,$$

namely the infimum in (4.119) is attained at v.

Finally, since $v \ge 0$, $v \ne 0$, is a radial solution to

$$-\Delta v(x) - V(x)v(x) = \tilde{\beta}^* \frac{v(x)}{|x|^2}, \quad x \in \mathbb{R}^N,$$

with $\tilde{\beta}^* < 0$, we can apply Proposition 4.18 obtaining that $\tilde{\beta}^* = -(N-1)$. $\qquad\square$

5 Bifurcation from radial solutions

In this chapter, we show how Morse theory can be used to get some bifurcation results. After recalling preliminary theorems in Section 5.1, we will focus on bifurcation from radial positive solutions of the Lane–Emden Dirichlet problem (3.17) in an annulus.

We will consider two cases. In the first one, we will take the exponent of the nonlinearity as a bifurcation parameter and will show the existence of many "branches" of nonradial solutions. In the second case, we will study the bifurcation from radial solutions, as the radii of the annulus vary.

Most of the results we describe are taken from [129] where the more general case of a linear perturbation added to the nonlinear term is considered. As compared with [129], we will present simplified or more detailed proofs.

Finally, let us mention that bifurcation results can also be obtained when the domain is a ball or for sign changing solutions, again using Morse index estimates (see [10, 11, 130]).

5.1 Preliminaries

We consider the Lane–Emden Dirichlet problem that we rewrite in the case when the domain is an annulus and for positive solutions:

$$\begin{cases} -\Delta u = u^p & \text{in } A \\ u > 0 & \text{in } A \\ u = 0 & \text{on } \partial A \end{cases} \tag{5.1}$$

where $p > 1$ and $A = \{ x \in \mathbb{R}^n, a < |x| < b \}, b > a > 0, N \geq 2$.

We recall that for any $p > 1$, problem (5.1) has only one radial solution u (see [148, 182]) which is also nondegenerate in the space of radial functions. This means that if we denote, as in (3.21), by L_u the linearized operator at u:

$$L_u = -\Delta - pu^{p-1} \tag{5.2}$$

then zero is not an eigenvalue of L_u to which a radial eigenfunction corresponds.

A direct easy proof of this property of u is given in the next proposition.

Proposition 5.1. *Let u be the radial solution of* (5.1). *Then the linearized problem at u:*

$$\begin{cases} -\Delta v - pu^{p-1}v = 0 & \text{in } A \\ v = 0 & \text{in } A \end{cases} \tag{5.3}$$

does not admit any nontrivial radial solution.

https://doi.org/10.1515/9783110538243-005

Proof. Arguing by contradiction, let us assume that $\bar{v} \neq 0$ is a radial solution of (5.3).

Since the positive radial solution of (5.1) is unique, it is obvious that u is the least energy radial solution, and hence it has radial Morse index one, as shown in Section 3.2.2 (arguing in the space $H^1_{0,\mathrm{rad}}(A)$). Then, denoting by β_i, $i \in \mathbb{N}^+$, the radial eigenvalues of L_u, as in Section 4.1, we have that the contradiction assumption implies that $\beta_2 = 0$. Thus the corresponding eigenfunction φ_2 has only two nodal regions, by Courant's nodal domain theorem. We denote them by $A_1 = \left\{ x \in \mathbb{R}^N : a < |x| < d \right\}$ and $A_2 = \left\{ x \in \mathbb{R}^N : d < |x| < b \right\}$, $d \in (a,b)$, and in each of them the first eigenvalue of the linearized operator L_u is zero.

It is easy to see that the radial function

$$z = x \cdot \nabla u + \frac{2}{p-1} u$$

satisfies

$$\begin{cases} -\Delta z - pu^{p-1}z = 0 & \text{in } A \\ z > 0 & \text{if } |x| = a \\ z < 0 & \text{if } |x| = b \end{cases}$$

Hence z changes sign in A. We claim that z has only two nodal domains which are annuli. Indeed if it had more than two nodal regions then, by the boundary behavior it should have at least four nodal regions B_j, $j = 1,\ldots,k$, $k \geq 4$. Thus, for at least one region B_j it should happen that $B_j \subset A_i$, for some $i \in \{1,2\}$, and obviously $z = 0$ on ∂B_j. This implies that

$$0 = \lambda_1(L_u, B_j) > \lambda_1(L_u, A_i) = 0$$

which is a contradiction (here $\lambda_1(L_u, X)$ denotes the first eigenvalue of the operator L_u in the domain X).

Hence the function z has only two nodal regions B_1 and B_2 and, necessarily, $A_i \subseteq B_j$, for some $i,j = 1$ or 2. Using z as a test function, it is easy to see that $\lambda_1(L_u, B_j) > 0$ against the fact that $\lambda_1(L_u, B_j) \leq \lambda_1(L_u, A_i) = 0$. $\qquad \square$

In order to study possible bifurcations from the radial solution u, as described in Section 5.2 and Section 5.3, we need to analyze the spectrum of the linearized operator L_u to detect the values of the exponent p or of the radii of A at which L_u admits zero as an eigenvalue, in the whole space $H^1_0(A)$.

To do this, we use the same spectral decomposition performed in Section 4.1, for the case of radial nodal solutions in a ball. However, the procedure is a bit easier when the domain is an annulus since we do not have to worry about the singularity at the origin of the auxiliary weighted eigenvalue problem.

Let us summarize what is needed in the next sections.

We consider the weighted linear operator $\tilde{L}_u : H^2(A) \cap H_0^1(A) \longrightarrow L^2(A)$

$$\tilde{L}_u(v) := |x|^2(-\Delta v - pu(x)^{p-1}v), \quad x \in A \tag{5.4}$$

and the associated radial operator $\tilde{L}_{u,\mathrm{rad}} : H^2((a,b)) \cap H_0^1((a,b)) \longrightarrow L^2((a,b))$

$$\tilde{L}_{u,\mathrm{rad}}(v) := r^2\left(-v'' - \frac{N-1}{r}v' - pu^{p-1}(r)v\right), \quad r \in (a,b). \tag{5.5}$$

By Lemma 4.4, we have the spectral decomposition

$$\sigma(\tilde{L}_u) = \sigma(\tilde{L}_{u,\mathrm{rad}}) + \sigma(-\Delta_{S^{N-1}}) \tag{5.6}$$

and, by Lemma 4.3, we have that the Morse index $m(u)$ of the solution u is equal to the number of the negative eigenvalues of $\tilde{L}_u$, while the radial Morse index $m_{\mathrm{rad}}(u)$ is equal to the number of negative eigenvalues of the operator $\tilde{L}_{u,\mathrm{rad}}$.

To get bifurcation results, we need to understand whether the linearized operator L_u can have zero as an eigenvalue.

Denoting by $\tilde{\mu}_i$ the eigenvalues of $\tilde{L}_u$, by $\tilde{\beta}_i$ those of $\tilde{L}_{u,\mathrm{rad}}$, $i \in \mathbb{N}^+$, and by λ_k the eigenvalues of $\Delta_{S^{N-1}}$, $k \in \mathbb{N}$, we get the following.

Lemma 5.2. *Let u be the radial solution of (5.1). The linearized operator L_u admits zero as an eigenvalue if and only if there exists $k \geq 1$ such that*

$$\tilde{\beta}_1 + \lambda_k = 0. \tag{5.7}$$

Proof. We start observing that a function v is a nontrivial solution of (5.3) if and only if it satisfies $\tilde{L}_u(v) = 0$ and $v = 0$ on ∂A. Hence zero is an eigenvalue of L_u if and only if zero is an eigenvalue of $\tilde{L}_u$.

By (5.6), all eigenvalues $\tilde{\mu}_j$ of $\tilde{L}_u$, $j \in \mathbb{N}^+$ are given by

$$\tilde{\mu}_j = \tilde{\beta}_i + \lambda_k, \quad i,j \in \mathbb{N}^+, \; k \in \mathbb{N}. \tag{5.8}$$

Thus we have to check for what i and k the eigenvalue $\tilde{\mu}_j$ is zero in (5.8).

Since $\lambda_k \geq 0$ (by (4.20)), we have to analyze when $\tilde{\beta}_i \leq 0$, for $i \in \mathbb{N}^+$. By Lemma 4.3, the number of the negative eigenvalues of $\tilde{L}_{u,\mathrm{rad}}$ is the same as the number of the negative radial eigenvalues of L_u. We have already observed in the proof of Proposition 5.1 that only the first radial eigenvalue $\tilde{\beta}_1$ of L_u is negative. Hence $\tilde{\beta}_i < 0$ only for $i = 1$.

On the other side, some $\tilde{\beta}_i$ can be zero if and only if (5.3) has a nontrivial radial solution, which is excluded by Proposition 5.1. Thus the assertion is proved because $\lambda_0 = 0$ (by (4.20)). $\qquad\square$

Note that in (5.7) λ_k depends only on the dimension N, while $\tilde{\beta}_1 = \tilde{\beta}_1(u)$ depends on the solution u which, in turn, depends on the data of the Dirichlet problem (5.1) which are the exponent $p > 1$ and the annulus A.

In the next sections, we analyze the behavior of $\tilde{\beta}_1$ with respect to p and to A in order to deduce the bifurcation results.

5.2 Bifurcation with respect to the exponent p

In this section, we consider the exponent p in (5.1) as a bifurcation parameter, while the annulus A is fixed.

To emphasize the dependence on p of the unique radial solution of (5.1), we will denote it by u_p. For the same reason, the first eigenvalue $\tilde{\beta}_1$ of the operator $\tilde{L}_{u_p,\mathrm{rad}}$ which appears in (5.7) will be denoted by $\tilde{\beta}_1(p)$.

To solve the equation (5.7), we need to analyze the behavior of $\tilde{\beta}_1(p)$, as p varies, which, by the definition of the operator $\tilde{L}_{u_p,\mathrm{rad}}$, depends on the behavior of the radial solution u_p, with respect to p.

5.2.1 Asymptotic analysis of the radial solution

The exponent p of the nonlinear term in (5.1) belongs to the interval $(1, +\infty)$ and we wish to analyze the behavior of the corresponding radial solution u_p, as $p \to 1$ or $p \to +\infty$.

The first case is easier.

Proposition 5.3. *Let u_p be the radial solution of (5.1). Then*

$$\lim_{p\to 1}\|u_p\|_\infty^{p-1} = \lambda_1$$

where λ_1 is the first Dirichlet eigenvalue of $-\Delta$ in A.

Proof. Let us start by showing that $\|u_p\|_\infty^{p-1}$ is bounded, as $p \to 1$. To do this, we can follow the proof of Lemma 3.9, for $\lambda = 0$, arguing by contradiction and so assuming that for a sequence $p_n \searrow 1$ it holds

$$M_n^{p_n-1} = \|u_n\|_\infty^{p_n-1} = \left[u_{p_n}(x_{p_n})\right]^{p_n-1} \longrightarrow +\infty.$$

Then, if x_{p_n} stay away from ∂A as in the case of Lemma 3.9, where the domain is supposed to be convex, then we reach a contradiction exactly as in the proof of Lemma 3.9.

If instead the distance d_n of x_{p_n} to the boundary ∂A tends to zero, as $n \to +\infty$, then we distinguish two cases:

$$\text{either} \quad \frac{d_n}{(M_n)^{p_n-1}} \longrightarrow +\infty, \quad \text{or} \quad \frac{d_n}{(M_n)^{p_n-1}} \longrightarrow \delta \geq 0.$$

In the first case, considering the rescaled function $\tilde{u}_n = \frac{1}{M_n}u_n\left(\frac{x}{M_n^{\frac{p_n-1}{2}}} + x_p\right)$ it is easy to see that it converges locally uniformly to a function $\tilde{u}$ which is a nontrivial solution of the same linear problem in $\mathbb{R}^N$ as in Lemma 3.9 and then a contradiction is reached in the same way.

In the second case, exactly as in the proof of Case 2 of Theorem 1.1 of [125] we can prove that $\delta > 0$, so that the rescaled functions $\tilde{u}_n$ converge locally uniformly to a function $\tilde{u}$ which solves

$$\begin{cases} -\Delta \tilde{u} = \tilde{u} & \text{in } H = \{\, x = (x_1, \ldots, x_N), x_N > -\delta \,\} \\ \tilde{u} > 0 & \text{in } H \\ \tilde{u}(0) = 1 \\ \tilde{u} = 0 & \text{on } \partial H \end{cases} \tag{5.9}$$

This is not possible by Theorem 2 in [87] which shows that a solution of (5.9) must be strictly monotone increasing in the x_N direction.

Hence $\|u_p\|_\infty^{p-1}$ is bounded, as $p \to 1$ and we denote by μ its limit. Then, exactly as in the proof of Lemma 3.9 (with $\lambda = 0$), we obtain that $\mu = \lambda_1$. $\qquad\square$

Now we pass to the study of u_p, as $p \to +\infty$.

This is much more difficult, and we state below the results obtained in [134].

Proposition 5.4. *Let u_p be the radial solution of (5.1). Then, as $p \to \infty$,*

$$u_p(|x|) \longrightarrow w(|x|) \quad \text{in } C^0(\overline{A}), \tag{5.10}$$

where, if $N \geq 3$,

$$w(|x|) = \frac{2}{a^{2-N} - b^{2-N}} \begin{cases} a^{2-N} - |x|^{2-N} & \text{for } a \leq |x| \leq r_0 \\ |x|^{2-N} - b^{2-N} & \text{for } r_0 \leq |x| \leq b \end{cases}$$

while, if $N = 2$

$$w(|x|) = \frac{2}{\log b - \log a} \begin{cases} \log|x| - \log a & \text{for } a \leq |x| \leq r_0 \\ \log b - \log|x| & \text{for } r_0 \leq |x| \leq b \end{cases}$$

with r_0 given by

$$\begin{cases} \left(\dfrac{a^{2-N} + b^{2-N}}{2} \right)^{\frac{1}{2-N}} & \text{if } N \geq 3 \\ \sqrt{ab} & \text{if } N = 2 \end{cases}$$

and we recall that a, b are the radii of the annulus A.

Moreover, there exists $C > 0$ such that

$$|\nabla u_p| \leq C \quad \forall p > 1. \tag{5.11}$$

Proof. For (5.10), we refer to Theorem 1.1 of [134] and for (5.11) to Proposition 2.4 of [134]. $\qquad\square$

To study the local behavior of u_p at the radius of A where it achieves its maximum we consider the rescalings, in polar coordinates:

$$\tilde{u}_p(r) = \frac{p}{\|u_p\|_\infty} \left(u_p(\epsilon_p r + r_p) - \|u_p\|_\infty \right) \tag{5.12}$$

where

$$\epsilon_p^{-2} = p\|u_p\|_\infty^{p-1}, \quad r = |x| \quad \text{and} \quad u_p(r_p) = \|u_p\|_\infty.$$

Note that ϵ_p is defined analogously as in (4.58).

Proposition 5.5. *We have*

$$\tilde{u}_p \longrightarrow U \quad \text{in } C^1_{loc}(\mathbb{R})$$

where

$$U(r) = \log \frac{4e^{\sqrt{2}r}}{(1 + e^{\sqrt{2}r})^2} \tag{5.13}$$

is the only solution of the ODE problem

$$\begin{cases} -z'' = e^z & \text{in } \mathbb{R} \\ z(0) = z'(0) = 0 \end{cases} \tag{5.14}$$

Proof. See Proposition 3.2 in [134]. $\qquad\qquad\square$

The last asymptotic estimate we need is the following one.

Proposition 5.6. *Let u_p be the radial solution of (5.1). Then, for p sufficiently large and some positive constants c_1, c_2,*

$$\tilde{u}_p \leq c_1 U + c_2 \tag{5.15}$$

in the interval $[a_p, b_p]$ with $a_p = \frac{a - r_p}{\epsilon_p}$, $b_p = \frac{b - r_p}{\epsilon_p}$ and the function U is as in (5.12).

Proof. See Lemma 4.2 of [129]. $\qquad\qquad\square$

5.2.2 Asymptotic behavior of the radial eigenvalue $\tilde{\beta}_1(p)$

We start by showing the following result.

Proposition 5.7. *The map $p \longmapsto \tilde{\beta}_1(p)$ is real analytic.*

Proof. The real analyticity of $\tilde{\beta}_1(p)$ follows by a result of Kato [147], once we prove that u_p is real analytic in p.

Let φ_1 be the first positive Dirichlet normalized eigenfunction of $-\Delta$ in A. We define

$$C_{\varphi_1} = \left\{ u \in C_0(A) : u \text{ is radial and } \left\| \frac{u}{\varphi_1} \right\|_\infty \leq k, \text{ for some } k > 0 \right\}.$$

It was proved by Dancer in [88] that if p_0 and u_0 are such that there exists $\mu > 0$ for which $u \geq \mu\varphi_1$, then the map

$$(p, u) \in (1, +\infty) \times C_{\varphi_1} \longrightarrow (-\Delta)^{-1}(u^p) \in C_{\varphi_1}$$

is analytic near (p_0, u_0).

We claim that

$$\forall p > 1, \text{ the function } u_p \text{ belongs to } C_{\varphi_1} \text{ and}$$
$$\exists \mu > 0 \text{ such that } u_p \geq \mu\varphi_1 \text{ in } A. \tag{5.16}$$

Let us assume that (5.16) holds. Then the function u_p is the only solution of the equation $F(p, u) = 0$ where $F: (1, +\infty) \times C_{\varphi_1} \longrightarrow C_{\varphi_1}$ is defined as $F(p, u) = u - (-\Delta)^{-1}(u^p)$. We observe that, by the result of [88] quoted before, the map F is analytic in a neighborhood of (p_0, u_{p_0}) for any $p_0 > 1$. Moreover, the map $F_u(p_0, u_{p_0})$ is an isomorphism since u_{p_0} is nondegenerate in the space of radial functions, for any $p_0 > 1$. Thus, from the analytic version of the implicit function theorem, it follows that the map $p \longmapsto u_p$ is real analytic.

It remains to prove the claim (5.16).

By the symmetry of the domain, we know that the first eigenfunction φ_1 is radial. We set $\psi = \frac{u_p}{\varphi_1} > 0$, in A. By the Hopf boundary lemma, we have that $\lim_{r \to a} \psi(r) = \frac{u_p'(a)}{\varphi_1'(a)} > 0$ and the same holds at b. This implies that there exist constants $c, C > 0$ such that $C \geq \psi(r) \geq c > 0$ in a neighborhood of ∂A and the same holds (possibly changing the constants) in the interior of A.

This completes the proof. $\qquad\qquad\square$

In the next proposition, we study the behavior of $\tilde{\beta}_1(p)$, as $p \to 1$.

Proposition 5.8. *We have*

$$\tilde{\beta}_1(p) \longrightarrow 0, \quad as\ p \to 1. \tag{5.17}$$

Proof. Let us consider the positive eigenfunction $w_{1,p}$ corresponding to the eigenvalue $\tilde{\beta}_1(p)$, with $\|w_{1,p}\|_\infty = 1$. It satisfies

$$\begin{cases} -w_{1,p}'' - \dfrac{N-1}{r} w_{1,p}' - p u_p^{p-1} w_{1,p} = \dfrac{\tilde{\beta}_{1,p}}{r^2} w_{1,p} & \text{in } (a, b) \\ w_{1,p}(a) = w_{1,p}(b) = 0 \end{cases} \tag{5.18}$$

Moreover, by the variational characterization of $\tilde{\beta}_1(p)$ we have

$$\tilde{\beta}_1(p) = \frac{\int_a^b (w_{1,p}')^2 r^{N-1}\, dr - p\int_a^b u_p^{p-1} w_{1,p}^2 r^{N-1}\, dr}{\int_a^b w_{1,p}^2 r^{N-3}\, dr} \geq$$

$$\geq \frac{-p\|u_p\|_\infty^{p-1} \int_a^b w_{1,p}^2 r^{N-1}\, dr}{\int_a^b w_{1,p}^2 r^{N-3}\, dr} \geq -Cp \tag{5.19}$$

so that $\tilde{\beta}_1(p)$ is bounded, as $p \to 1$, and hence, up to a sequence, $\tilde{\beta}_1(p) \longrightarrow \tilde{\beta}_1 \leq 0$, as $p \to 1$.

By standard regularity theorem we have that $w_{1,p} \longrightarrow w_1$, as $p \to 1$ in $C^2(A)$. Therefore, passing to the limit in (5.18) and using Proposition 5.3 we get that w_1 satisfies

$$\begin{cases} -w_1'' - \dfrac{N-1}{r} w_1' = \lambda_1 w_1 + \dfrac{\tilde{\beta}_1}{r^2} w_1 & \text{in } (a,b) \\[2mm] w_1 > 0 & \text{in } (a,b) \\[2mm] w_1(a) = w_1(b) = 0 \end{cases}$$

i. e.,

$$\begin{cases} -\Delta w_1 - \lambda_1 w_1 = \dfrac{\tilde{\beta}_1}{|x|^2} & \text{in } A \\[2mm] w_1 = 0 & \text{on } \partial A \end{cases} \tag{5.20}$$

Then, multiplying by φ_1, which is the first positive normalized eigenfunction of $-\Delta$ in A, and integrating we get

$$\int_A \nabla w_1 \nabla \varphi_1 - \lambda_1 \int_A w_1 \varphi_1 = \tilde{\beta}_1 \int_A \frac{w_1 \varphi_1}{|x|^2}. \tag{5.21}$$

But, by the definition of φ_1, using w_1 as a test function, we have that

$$\int_A \nabla w_1 \nabla \varphi_1 - \lambda_1 \int_A w_1 \varphi_1 = 0.$$

Hence, from (5.21) we deduce that $\tilde{\beta}_1 = 0$ and then $w_1 = \varphi_1$, by (5.20). □

Now we prove an asymptotic expansion of $\tilde{\beta}_1(p)$, as $p \to +\infty$.

Proposition 5.9. *We have*

$$\lim_{p \to +\infty} \frac{\|u_p\|_\infty^p}{p} = \lim_{r \to r_0^-} \frac{\omega'(r)^2}{2} = M > 0 \tag{5.22}$$

and

$$\tilde{\beta}_1(p) = -\frac{1}{2} M r_0^2 p^2 + o(p^2), \quad \text{as } p \to +\infty \tag{5.23}$$

where ω and r_0 are defined as in Proposition 5.4.

Proof. Let us start by proving (5.22).

It is convenient to write the equation for u_p in radial coordinates, i. e.,

$$-u_p'' - \frac{N-1}{r} u_p' = u_p^p \quad \text{in } (a,b). \tag{5.24}$$

Denoting again by r_p the radius where u_p achieves its maximum, multiplying (5.24) by u_p' and integrating on the interval (a, r_p) we get

$$-\int_a^{r_p} u_p'' u_p' \, dr - (N-1) \int_a^{r_p} \frac{(u_p')^2}{r} \, dr = \int_a^{r_p} u^p u' \, dr.$$

This implies

$$\frac{u_p'(a)^2}{2} - (N-1) \int_a^{r_p} \frac{(u_p')^2}{r} \, dr = \frac{u^{p+1}(r_p)}{p+1} = \frac{\|u_p\|_\infty^{p+1}}{p+1}.$$

By Proposition 5.4, we have that $\|u_p\|_\infty \longrightarrow 1$, as $p \to +\infty$ and (5.11) holds. Hence we can pass to the limit obtaining

$$\lim_{p \to +\infty} \frac{\|u_p\|_\infty^p}{p} = \lim_{p \to +\infty} \frac{\|u_p\|_\infty^{p+1}}{p+1} = \frac{1}{2} w'(a)^2 - (N-1) \int_a^{r_0} \frac{(w')^2}{r} \, dr. \tag{5.25}$$

On the other side, the function w satisfies

$$\begin{cases} -w'' - \dfrac{N-1}{r} w' = 0 & \text{in } (a, r_0) \\ w(a) = 0, \ w(r_0) = 1 \end{cases} \tag{5.26}$$

as well as an analogous problem in (r_0, b).

Thus, multiplying the equation (5.26) by w' and integrating on the interval (a, r_0) we get

$$-\frac{w_-'(r_0)^2}{2} + \frac{w'(a)^2}{2} - (N-1) \int_a^{r_0} \frac{(w')^2}{r} \, dr = 0 \tag{5.27}$$

where $w_-'(r_0) = \lim_{r \to r_0^-} w'(r)$.

Combining (5.27) and (5.25), we obtain (5.22).

To prove (5.23), we consider, as in Proposition 5.8, the positive, L^∞-normalized eigenfunction $w_{1,p}$, corresponding to the eigenvalue $\tilde{\beta}_1(p)$. We recall that it satisfies (5.18) and the variational characterization (5.19) holds.

Then, by (5.22) we get

$$\tilde{\beta}_1(p) \geq \frac{-p\|u_p\|_\infty^{p-1} \int_a^b w_{1,p}^2 r^{N-1} \, dr}{\int_a^b w_{1,p}^2 r^{N-3} \, dr} \geq -Cp^2 \tag{5.28}$$

for a constant $C \geq 0$.

Now we define $\tilde{w}_{1,p}(t) = w_{1,p}(\epsilon_p t + r_p)$, with ϵ_p as in (5.12), which solves

$$-\tilde{w}_{1,p}'' - \epsilon_p \frac{N-1}{\epsilon_p s + r_p} \tilde{w}_{1,p}' = \left(1 + \frac{\tilde{u}_p}{p}\right)^p \tilde{w}_{1,p} + \frac{\epsilon_p^2 \tilde{\beta}_1(p)}{(\epsilon_p s + r_p)^2} \tilde{w}_{1,p}. \tag{5.29}$$

Since $\|\tilde{w}_{1,p}\|_\infty = 1$, we have that $\tilde{w}_{1,p} \longrightarrow \tilde{\psi}_1 \geq 0$, as $p \to +\infty$ uniformly on compact sets of $\mathbb{R}$.

Passing to the limit in (5.29), by Proposition 5.5, we get

$$\begin{cases} -\tilde{\psi}_1 - e^U \tilde{\psi}_1 = \tilde{\beta}_1 \tilde{\psi}_1 & \text{in } \mathbb{R} \\ \tilde{\psi}_1 \geq 0 \end{cases} \tag{5.30}$$

where U is defined in (5.13) and

$$\tilde{\beta}_1 = \lim_{p \to +\infty} \frac{\epsilon_p^2 \tilde{\beta}_1(p)}{r_p^2}. \tag{5.31}$$

Note that the limit in (5.31) is finite by (5.22), (5.28) and the definition of ϵ_p.

We would like to prove that $\tilde{\psi}_1 \neq 0$ so that it is a first eigenfunction for (5.30).

To do this, we proceed as follows.

With the change of variable $r = \sqrt{2}\log s$ and $Z(s) = \varphi_1(r)$, we get that the function Z satisfies

$$\begin{cases} -Z'' - \dfrac{Z'}{s} = \dfrac{8}{(1+s^2)^2}Z + \dfrac{2\tilde{\beta}_1}{s^2}Z & \text{in } (0, +\infty) \\ \|Z\|_\infty \leq 1 \text{ and } Z \geq 0 \end{cases} \tag{5.32}$$

To show that $Z \neq 0$, we denote by η_p the point where $w_{1,p}$ achieves its maximum, i. e.,

$$w_{1,p}(\eta_p) = 1, \quad w_{1,p}'(\eta_p) = 0, \quad w_{1,p}''(\eta_p) \leq 0$$

and set $\tilde{\eta}_p = \frac{\eta_p - r_p}{\epsilon_p}$. Then

$$\tilde{w}_{1,p}(\tilde{\eta}_p) = 1, \quad \tilde{w}_{1,p}'(\tilde{\eta}_p) = 0, \quad \tilde{w}_{1,p}''(\tilde{\eta}_p) \leq 0.$$

From (5.29), we deduce

$$-\tilde{w}_{1,p}''(\tilde{\eta}_p) = \left(1 + \frac{\tilde{u}_p(\tilde{\eta}_p)}{p}\right)^p + \epsilon_p^2 \tilde{\beta}_1(p)\frac{1}{\eta_p^2}. \tag{5.33}$$

We claim that

$$\tilde{\eta}_p \longrightarrow \eta_0 \in \mathbb{R}.$$

By contradiction, let us assume that $\tilde{\eta}_p \longrightarrow +\infty$ (the case when $\tilde{\eta}_p \longrightarrow -\infty$ is similar). By (5.15), we get

$$\left(1 + \frac{\tilde{u}_p(\tilde{\eta}_p)}{p}\right)^p \le Ce^{\tilde{u}_p(\tilde{\eta}_p)} \le Ce^{-C\tilde{\eta}_p} \longrightarrow 0.$$

Observing that $\epsilon_p^2 \tilde{\beta}_1(p) \longrightarrow \tilde{\beta}_1 r_0^2 < 0$, as $p \to +\infty$ and $\eta_p \in (a, b)$, from (5.33) it follows

$$-\tilde{w}_{1,p}''(\tilde{\eta}_p) < 0$$

which is not possible.

Hence $\tilde{\eta}_p \longrightarrow \eta_0 \in \mathbb{R}$ and $Z(\eta_0) = 1$, so that $Z \ne 0$.

Therefore, Z is a first eigenfunction for (5.32), and hence, by Theorem 4.21 of Section 4.5 (the case $N = 2$) we have that $2\tilde{\beta}_1 = \tilde{\beta}^* = -1$.

Since $\tilde{\beta}_1 = -\frac{1}{2}$, by the limit (5.31) and (5.22) we deduce

$$-\frac{1}{2} + o(1) = \frac{\epsilon_p^2 \tilde{\beta}_1(p)}{r_p^2} = \frac{\tilde{\beta}_1(p)}{p\|u_p\|_\infty^{p-1} r_p^2} = \frac{\tilde{\beta}_1(p)}{p^2(M + o(1))(r_0^2 + o(1))}$$

which gives (5.23). $\qquad\square$

5.2.3 Bifurcation results

We start with a definition (see [66, 150, 195] and the references therein for classical results in bifurcation theory).

Definition 5.10. Let $\bar{p} > 1$ and let $u_{\bar{p}}$ be the radial solution of (5.1) in the annulus A, for $p = \bar{p}$. We say that a **nonradial bifurcation** occurs at $(u_{\bar{p}}, \bar{p})$ if in every neighborhood of $(u_{\bar{p}}, \bar{p})$ in $C^{1,\alpha}(\overline{A}) \times (1, +\infty)$ there exists a point (v_p, p) where v_p is a nonradial solution of (5.1).

Definition 5.11. If, in the previous definition, it happens that for any neighborhood of $(u_{\bar{p}}, \bar{p})$ in $C^{1,\alpha}(\overline{A}) \times (1, +\infty)$ there exist at least m distinct points (u_{p_i}, p_i), $i = 1, \dots, m$ with u_{p_i} nonradial solution of (5.1), then we say that a m-**nonradial bifurcation** occurs at $(u_{\bar{p}}, \bar{p})$.

The bifurcation results we present are the following.

Theorem 5.12. *There exists a sequence of exponents p_k, $k \in \mathbb{N}^+$, such that $p_k \longrightarrow +\infty$, as $k \to +\infty$, and a nonradial bifurcation occurs at (u_{p_k}, p_k). Moreover, for k even, a $[\frac{N}{2}]$-nonradial bifurcation occurs at (u_{p_k}, p_k) ($[\frac{N}{2}]$ is the integer part of $\frac{N}{2}$).*

To prove this theorem, we go back to equation (5.7) and show the following result.

Proposition 5.13. *For any $k \geq 1$, there exists an exponent p_k such that*

$$\tilde{\beta}_1(p_k) + \lambda_k = 0. \tag{5.34}$$

Moreover, $p_k \longrightarrow +\infty$ as $k \to +\infty$.

Proof. It follows by Proposition 5.7, Proposition 5.8 and Proposition 5.9. $\qquad\square$

As a consequence of (5.23) and using Lemma 4.3 and Lemma 4.5 (where A_h is just A), we easily get the following.

Proposition 5.14. *Let u_p be the radial solution of (5.1). The Morse index $m(u_p)$ goes to $+\infty$, as $p \to +\infty$.*

Now we observe that the equation in (5.1) can be written as $u = T_p(u)$, where $T_p(w) = (-\Delta)^{-1}(|u|^{p-1}u)$ and T_p is a compact operator from $C_0^{1,\alpha}(\overline{A})$ into $C_0^{1,\alpha}(\overline{A})$.

If D is an open bounded set in $C_0^{1,\alpha}(\overline{A})$ such that $I - T_p \neq 0$ on ∂D, then the Leray–Schauder degree for the map $I - T_p$, i. e., $\deg(I - T_p, D, 0)$ is well-defined.

Now let us observe that, by Proposition 5.7, the set of numbers p which satisfy (5.7), for a given $k \geq 1$, is at most finite. Hence the exponents p_k, defined by (5.34) are isolated and so, for p in a suitable neighborhood of p_k the operator $I - T_p'(u_p)$ is invertible. Hence for $p \neq p_k$, we have

$$\deg(I - T_p, B, 0) = \deg(I - T_p'(u_p), D, 0) = (-1)^{m_p} \tag{5.35}$$

if B is a neighborhood of u_p in $C_0^{1,\alpha}(\overline{A})$ such that u_p is the only solution of $(I - T_p)(v) = 0$ in $\overline{B}$ and $m_p = m(u_p)$ is the Morse index of u_p.

Proof of Theorem 5.12. We start by pointing out that the equation (5.1) is invariant with respect to the action of the orthogonal group $O(N)$ and its subgroups. Then we consider the subspace $\mathcal{F}$ of $C^{1,\alpha}(\overline{A})$ given by

$$\mathcal{F} := \left\{ v \in C^{1,\alpha}(\overline{A}) : v(x',x_N) = v(g(x'),x_N), \forall g \in O(N-1) \right\}$$

where $x' = (x_1,\ldots,x_{N-1})$ and $O(N-1)$ is the orthogonal group in $\mathbb{R}^{N-1}$. By a result of [204] (Proposition 5.2 therein) we have that, for any $k \geq 1$, the eigenspace V_k spanned by the eigenfunctions corresponding to the eigenvalue λ_k of the Laplace–Beltrami operator on S^{N-1}, which are $O(N-1)$ invariant, is one-dimensional. Then let $\{p_k\}$ be the sequence of exponents defined in Proposition 5.13. Since the Morse index of u_{p_k} tends to $+\infty$, by Proposition 5.14, we have that, up to a subsequence, that we still denote by $\{p_k\}$, at each p_k the number of the negative eigenvalues of the linearized operator L_{u_p} increases, and in the space $\mathcal{F}$ can only increase by one because of (5.6) and the fact that the eigenfunctions of $\tilde{L}_{u_p}$ are a product of the eigenfunction of $\tilde{L}_{u,\mathrm{rad}}$ by an eigenfunction of $-\Delta_{S^{N-1}}$.

Hence the Morse index of u_p, in the space $\mathcal{F}$, increases by one, crossing p_k, i. e., $m(u_{p_k+\epsilon}) = m(u_{p_k-\epsilon}) + 1$, if ϵ is small enough.

Then

$$\deg(I - T_{p_k-\epsilon}, B_{p_k-\epsilon}, 0) = (-1)\deg(I - T_{p_k+\epsilon}, B_{p_k+\epsilon}, 0)$$

where B_p is a neighborhood of u_p in the space $\mathcal{F}$.

This implies that there is a change in the degree at the point (u_{p_k}, p_k) and then a bifurcation must occur, by the classical results in bifurcation theory (see, e. g., [150]). Moreover, the bifurcating solutions are nonradial since the radial solution u_p is nondegenerate in the space of radial functions. Finally, by the maximum principle, the bifurcating solutions are positive.

To get a $[\frac{N}{2}]$-nonradial bifurcation result when k is even, we consider the subgroup od $O(N)$ defined by

$$\mathcal{G}_h = O(h) \times O(N - h), \quad \text{for } 1 \le h \le \left[\frac{N}{2}\right].$$

In [205], it is shown that if k is even then the eigenspace relative to the eigenvalue λ_k of $-\Delta_{S^{N-1}}$, restricted to the functions invariant by the action of $\mathcal{G}_h$, has dimension one.

Moreover, any solution u of (5.1) that is invariant with respect to the action of both $\mathcal{G}_{h_1}$ and $\mathcal{G}_{h_2}$ with $h_1 \ne h_2$, must be radial. Hence nonradial solutions which are invariant for the action of different groups $\mathcal{G}_h$ are actually distinct.

So if we repeat the previous proof considering, instead of $\mathcal{F}$ the subspaces of functions which are invariant with respect to the action of $\tilde{G}_h$, for $1 \le h \le [\frac{N}{2}]$, we get the existence of $[\frac{N}{2}]$ distinct nonradial positive solutions of (5.1) bifurcating from u_{p_k}, for k even. $\qquad\square$

5.3 Bifurcation with respect to the radius of the annulus

In this section, we consider the case of expanding annuli $A_R := \{x \in \mathbb{R}^N, R < |x| < R + 1\}$, $R > 0$ and we will consider the radius R as the bifurcation parameter. To stress the dependence on R, we will denote by u_R the unique radial solution (5.1) in A_R and by $\tilde{\beta}_1(R)$ the first eigenvalue of the operator $\tilde{L}_{u_R,\text{rad}}$ which appears in (5.7).

As in Section 5.2, we have to analyze the behavior of u_R and of $\tilde{\beta}_1(R)$, as $R \to +\infty$.

5.3.1 Asymptotic estimates for the radial solution

The radial positive solution u_R satisfies the following ODE problem:

$$\begin{cases} -u_R'' - \dfrac{N-1}{r}u_R' = u_R^p & \text{in } (R, R+1) \\ u_R > 0 & \text{in } (R, R+1) \\ u_R(R) = u_R(R+1) = 0 \end{cases} \tag{5.36}$$

for any fixed $p > 1$.

As recalled in Section 5.1, there exists only one solution of (5.36) and it is nondegenerate in the space of radial functions (Proposition 5.1).

To study the behavior of u_R, as $R \to +\infty$, we consider the function

$$\tilde{u}_R(t) = u_R(t + R), \quad t \in [0, 1] \tag{5.37}$$

for which we prove the following results.

Proposition 5.15. *We have:*

(i) $\tilde{u}_R \longrightarrow \tilde{u}_0$ *uniformly in* $(0, 1)$ *where* $\tilde{u}_0$ *is the unique positive solution of*

$$\begin{cases} -u'' = u^p & \text{in } (0, 1) \\ u > 0 & \text{in } (0, 1) \\ u(0) = u(1) = 0 \end{cases} \tag{5.38}$$

(ii) *the function* $R \longmapsto \tilde{u}_R(t)$ *is continuously differentiable with respect to R, for any* $t \in (0, 1)$ *and it holds*

$$\lim_{R \to \infty} R^q \int_0^1 \left| \frac{\partial \tilde{u}_R}{\partial R} \right|^q dt = 0, \quad \forall q > 1. \tag{5.39}$$

Proof. Since u_R is the unique positive solution of (5.1) or (5.36) it is easy to see, by constrained minimization, that it achieves the following infimum:

$$\beta_R = \inf_{\substack{u \in H_0^1(R, R+1) \\ u \neq 0}} \frac{\int_R^{R+1} (u')^2 r^{N-1}\, dr}{\left(\int_R^{R+1} |u|^{p+1} r^{N-1}\, dr \right)^{\frac{2}{p+1}}}.$$

Taking a function $\varphi \in C_0^\infty((R, R+1))$ and making a change of variable, we get

$$\beta_R \leq \frac{\int_R^{R+1} (\varphi')^2 r^{N-1}\, dr}{\left(\int_R^{R+1} \varphi^{p+1} r^{N-1}\, dr \right)^{\frac{2}{p+1}}} \leq \frac{\int_0^1 (\varphi')^2 (R+t)^{N-1}\, dr}{\left(\int_0^1 \varphi^{p+1} (R+t)^{N-1}\, dr \right)^{\frac{2}{p+1}}}$$

$$\leq C R^{(N-1)(1 - \frac{2}{p+1})}.$$

Thus, from the equation (5.36) we get

$$\int_R^{R+1} u_R^{p+1} r^{N-1}\, dr = \beta_R^{\frac{p+1}{p-1}} \leq C R^{N-1}.$$

Then for the function $\tilde{u}_R(t)$ defined in (5.37) we obtain

$$\int_0^1 \tilde{u}_R^{p+1}\, dt = \int_R^{R+1} u_R^{p+1}\, dr \leq \frac{1}{R^{N-1}} \int_R^{R+1} u_R^{p+1} r^{N-1}\, dr \leq C. \tag{5.40}$$

On the other side, from (5.36), multiplying by u_R and integrating we get

$$\int_R^{R+1} (u_R')^2 r^{N-1}\, dr = \int_R^{R+1} u_R^{p+1} r^{N-1}\, dr. \tag{5.41}$$

By (5.40) and (5.41), we derive

$$\int_0^1 (\tilde{u}_R')^2\, dt \le \frac{1}{R^{N-1}} \int_0^1 (\tilde{u}_R')^2 (t+R)^{N-1}\, dt =$$
$$= \frac{1}{R^{N-1}} \int_R^{R+1} (u_R')^2 r^{N-1}\, dt \le C. \tag{5.42}$$

Now we observe that $\tilde{u}_R$ satisfies

$$\begin{cases} -\tilde{u}_R'' - \dfrac{N-1}{t+R}\tilde{u}_R' = \tilde{u}_R^p & \text{in } (0,1) \\[2mm] \tilde{u}_R > 0 & \text{in } (0,1) \\[2mm] \tilde{u}_R(0) = \tilde{u}_R(1) = 0 \end{cases} \tag{5.43}$$

Thus, since, by (5.42), $\tilde{u}_R$ is bounded in $H_0^1(0,1)$, it is also bounded in $C^2(0,1)\cap L^\infty(0,1)$, and hence $\tilde{u}_R \longrightarrow \tilde{u}_0$ uniformly, as $R \to +\infty$, where $\tilde{u}_0$ satisfies (5.38). Note that $u_0 \ne 0$ because $\|\tilde{u}_R\|_\infty = \|u_R\|_\infty \ge \alpha > 0$ where α is a constant independent of R.

Indeed multiplying the equation in (5.36) by the first eigenfunction $\psi_{1,R}$ of $-\Delta$ in A_R, with corresponding eigenvalue $\mu_1(R)$, and integrating we get

$$\int_{A_R} \nabla u_R \nabla \psi_{1,R}\, dx = \int_{A_R} u_R^p \psi_{1,R}\, dx = \mu_1(R) \int_{A_R} u_R \psi_{1,R}\, dx.$$

From this, we derive

$$\|u_R\|_\infty^{p-1} \int_{A_R} u_R \psi_{1,R}\, dx \ge \mu_1(R) \int_{A_R} u_R \psi_{1,R} 1,\, dx$$

so that

$$\|u_R\|_\infty \ge [\mu_1(R)]^{\frac{1}{p-1}}.$$

Since it is easy to see that $\mu_1(R) \longrightarrow \mu_1 > 0$, where μ_1 is the first eigenvalue of the problem

$$-u'' = \mu u \quad \text{in } (0,1), \quad u(0) = u(1) = 0$$

we get the bound from below for $\|u_R\|_\infty$. This proves assertion (i).

Since the solution u_R is nondegenerate in the space of radial functions, applying the implicit function theorem to the function

$$F(\psi, R) = \psi'' + \frac{N-1}{t+R}\psi' + \psi^p$$

it is easy to see that $\tilde{u}_R$ is continuously differentiable with respect to R. The function $V(\cdot, R) = \frac{\partial \tilde{u}_R}{\partial R}$ satisfies

$$\begin{cases} -V'' - \dfrac{N-1}{t+R}V' + \dfrac{N-1}{(t+R)^2}\tilde{u}_R' = p\tilde{u}_R^{p-1}V & \text{in } (0,1) \\ V(0) = V(1) = 0 \end{cases}$$

We claim that

$$R\|V(\cdot, R)\|_{H_0^1((0,1))} \leq C \tag{5.44}$$

for some constant $C > 0$.

Indeed, if (5.44) does not hold, then, for a sequence $R_n \to +\infty$ we have

$$R_n\|V(\cdot, R_n)\|_{H_0^1((0,1))} \longrightarrow +\infty, \quad \text{as } n \to +\infty.$$

The function $z_n = \dfrac{V(\cdot, R_n)}{\|V(\cdot, R_n)\|_{H_0^1((0,1))}}$ satisfies

$$\begin{cases} -z_n'' - \dfrac{N-1}{t+R_n}z_n' + \dfrac{(N-1)R_n\tilde{u}_{R_n}'}{(t+R_n)^2 R_n\|V(\cdot, R_n)\|_{H_0^1((0,1))}} = p\tilde{u}_{R_n}^{p-1}z_n \\ z_n(0) = z_n(1) = 0 \end{cases} \tag{5.45}$$

and it converges weakly in $H_0^1((0,1))$ and strongly in $L^q((0,1))$, for any $q > 1$ to a function z_0. Moreover, $\tilde{u}_{R_n}' = u_{R_n}'$ is bounded in $L^\infty((0,1))$. Indeed by the equation in (5.36), integrating between the point $a_R \in (R, R+1)$ where u_R achieves its maximum and r, we get

$$-u_R'(r)r^{N-1} = -\int_{a_r}^{r}(u_R' r^{N-1})'\, dr = \int_{a_r}^{r} t_R^p r^{N-1}\, dr \leq CR^{N-1}$$

by (5.42).

Hence

$$|u_R'(r)|R^{N-1} \leq CR^{N-1}.$$

So, passing to the limit in (5.45) we get

$$\begin{cases} -z_0'' = p\tilde{u}_0^{p-1}z_0 & \text{in } (0,1) \\ z(0) = z(1) = 0 \end{cases}$$

where $\tilde{u}_0$ is the limit function defined in (i).

Then $z_0 \equiv 0$, because $\tilde{u}_0$ is a nondegenerate solution of (5.38) as it is easy to see, repeating, for example, the proof of Proposition 5.1. Indeed the unique solution $\tilde{u}_0$ of (5.38) has Morse index one and so the proof of Proposition 5.1 shows that the second eigenvalue of the linearized operator cannot be zero.

On the other side, z_0 cannot vanish. Indeed, multiplying the equation for z_n in (5.45) by z_n and integrating we have

$$1 = \lim_{n \to +\infty} \int_0^1 (z_n')^2 \, dr = \lim_{n \to +\infty} p \int_0^1 \tilde{u}_{R_n}^{p-1} \cdot z_n^2 \, dr = p \int_0^1 \tilde{u}_0^{p-1} z_0^2 \, dr.$$

So we have reached a contradiction which shows that (5.44) holds. This implies that the functions $RV(\cdot, R)$ converge weakly in $H_0^1((0,1))$ and strongly in $L^q((0,1))$, for any $q > 1$ as $R \to +\infty$, to a function $\tilde{V}$ which solves as before the linearized problem

$$\begin{cases} -\tilde{V}'' = p\tilde{u}_0^{p-1}\tilde{V} & \text{in } (0,1) \\ \tilde{V}(0) = \tilde{V}(1) = 0 \end{cases}$$

We have just observed that this problem has only the solution $\tilde{V} \equiv 0$, so that (5.39) holds. $\qquad\square$

5.3.2 Asymptotic analysis of the eigenvalue $\tilde{\beta}_1(R)$

We start with an expansion of $\tilde{\beta}_1(R)$.

Proposition 5.16. *We have*

$$\tilde{\beta}_1(R) = \tilde{\beta}_1 R^2 + o(R^2) \quad as \ R \to +\infty \tag{5.46}$$

where $\tilde{\beta}_1$ is the first eigenvalue for the linear problem

$$\begin{cases} -v'' - p\tilde{u}_0^{p-1}b = \beta v & in \ (0,1) \\ v(0) = v(1) = 0 \end{cases}$$

and $\tilde{u}_0$ is the limit of $\tilde{u}_R$, as before.

Proof. To estimate $\tilde{\beta}_1(R)$, we use, as in Section 5.2.2, its variational characterization, i. e.,

$$\tilde{\beta}_1(R) = \inf_{\substack{v \in H_0^1((0,1)) \\ v \neq 0}} \frac{\int_R^{R+1} (v')^2 r^{N-1} \, dr - p \int_R^{R+1} u_R^{p-1} v^2 r^{N-1} \, dr}{\int_R^{R+1} v^2 r^{N-3} \, dr}. \tag{5.47}$$

Inserting a test function $\varphi \in C_0^\infty((R, R+1))$, $\varphi \geq 0$ and making a change of variable we have

$$\tilde{\beta}_1(R) \leq \frac{\int_0^1 (\varphi')^2 (R+t)^{N-1} \, dt - p \int_0^1 \tilde{u}_R^{p-1} \varphi^2 (R+t)^{N-1} \, dt}{\int_0^1 \tilde{\varphi}^2 (R+t)^{N-3} \, dt} \leq CR^2. \tag{5.48}$$

In order to get the reverse inequality, let us denote by $w_{1,R}$ the eigenfunction associated to $\tilde{\beta}_1(R)$ with $\|w_{1,R}\|_\infty = 1$. Plugging $w_{1,R}$ into (5.47) we have, for a positive constant $C = C(N,p)$,

$$\tilde{\beta}_1(R) = \frac{\int_R^{R+1}(w_{1,R}')^2 r^{N-1}\,dr - p\int_R^{R+1} u_R^{p-1} w_{1,R}^2 r^{N-1}\,dr}{\int_R^{R+1} w_{1,R}^2 R^{N-3}\,dr} \geq$$

$$\geq \frac{-p\|u_R\|_\infty^{p-1}\int_R^{R+1} w_{1,R}^2 r^{N-1}\,dr}{\int_R^{R+1} w_{1,R}^2 r^{N-3}\,dr} \geq -CR^2 \tag{5.49}$$

because we showed that $\|u_R\|_\infty \leq K$.

Now we define $\tilde{w}_{1,R}(t) = w_{1,R}(t + R)$ in $(0,1)$ which solves

$$\begin{cases} -\tilde{w}_{1,R}'' - \dfrac{N-1}{t+R}\tilde{w}_{1,R}' - p\tilde{u}_R^{p-1}\tilde{w}_{1,R} = \tilde{\beta}_1(R)\dfrac{\tilde{w}_{1,R}}{(t+R)^2} & \text{in } (0,1) \\ \tilde{w}_{1,R}(0) = \tilde{w}_{1,R}(1) = 0 \end{cases} \tag{5.50}$$

Since $\|\tilde{w}_{1,R}\|_\infty = 1$, we have that $\tilde{w}_{1,R} \longrightarrow \varphi_1 \neq 0$, as $R \to \infty$, uniformly in $(0,1)$, and $\varphi_1 \geq 0$ solves

$$\begin{cases} -\varphi_1'' - p\tilde{u}_0^{p-1}\varphi_1 = \tilde{\beta}_1\varphi_1 & \text{in } (0,1) \\ \varphi_1(0) = \varphi_1(1) = 0 \end{cases} \tag{5.51}$$

where $\tilde{\beta}_1 = \lim\frac{\tilde{\beta}_1(R)}{R^2}$, by (5.48) and (5.49). Hence $\tilde{\beta}_1$ is the first eigenvalue for the problem (5.51) and obviously $\tilde{\beta}_1 < 0$. Then we have, with a change of variable, from (5.49),

$$\tilde{\beta}_1(R) = \frac{\int_0^1(\tilde{w}_{1,R}')^2(t+R)^{N-1}\,dt - p\int_0^1 \tilde{u}_R^{p-1}\tilde{w}_{1,R}^2(t+R)^{N-1}\,dt}{\int_R^{R+1}\tilde{w}_{1,R}^2(t+R)^{N-3}\,dt}.$$

Thus passing to the limit, as $R \to +\infty$, we get

$$\frac{\tilde{\beta}_1(R)}{R^2} = \frac{\int_0^1(\varphi')^2\,dt - p\int_0^1 \tilde{u}_0^{p-1}\varphi_1^2\,dt}{\int_0^1 \varphi_1^2\,dt} + o(1)$$

which, together with (5.51) gives the assertion. $\qquad\square$

Now we make a deeper analysis on the behavior of $\tilde{\beta}_1(R)$ showing that, for large R, it is a strictly decreasing function of the radius R. More precisely, we show the following.

Proposition 5.17. *The eigenvalue $\tilde{\beta}_1(R)$ is differentiable with respect to R and*

$$\frac{\partial\tilde{\beta}_1(R)}{\partial R} = 2\tilde{\beta}_1(R) + o(R) \quad \text{as } R \to \infty \tag{5.52}$$

where $\tilde{\beta}_1$ is the first eigenvalue of the linear problem (5.51).

Proof. As in Proposition 5.16 we consider the eigenfunction $\tilde{w}_{1,R}$ corresponding to the eigenvalue $\tilde{\beta}_1(R)$ and recall that the functions $\tilde{w}_{1,R}$ converge, as $R \to +\infty$, to the function $\varphi_1 > 0$ which is a solution of (5.51). By results of Kato (see [147, p. 380]), we have that both $\tilde{w}_{1,R}$ and $\tilde{\beta}_1(R)$ depend analytically on R. Thus the function $\Phi = \Phi(t,R) = \frac{\partial \tilde{w}_{1,R}}{\partial R}$ satisfies

$$-\Phi'' - \frac{N-1}{t+R}\Phi' + \frac{N-1}{(t+R)^2}\tilde{w}'_{1,R} - p(p-1)\tilde{u}_R^{p-2}\frac{\partial \tilde{u}_R}{\partial R}\tilde{w}_{1,R} - p\tilde{u}_R^{p-1}\Phi$$

$$= \frac{\partial \tilde{\beta}_1(R)}{\partial R}\frac{1}{(t+R)^2}\tilde{w}_{1,R} + \frac{\tilde{\beta}_1(R)}{(t+R)^2}\Phi - 2\frac{\tilde{\beta}_1(R)}{(t+R)^3}\tilde{w}_{1,R}.$$

Multiplying this equation by $\tilde{w}_{1,R}$ and integrating, we get

$$\int_0^1 \Phi'\tilde{w}'_{1,R}(t+R)^{N-1}\,dt + (N-1)\int_0^1 \tilde{w}'_{1,R}\tilde{w}'_{1,R}(t+R)^{N-3}\,dt$$

$$- p(p-1)\int_0^1 \tilde{u}_R^{p-2}\frac{\partial \tilde{u}_R}{\partial R}\tilde{w}_{1,R}^2(t+R)^{N-1}\,dt$$

$$- p\int_0^1 \tilde{u}_R^{p-1}\tilde{w}_{1,R}\Phi(t+R)^{N-1}\,dt$$

$$= \frac{\partial \tilde{\beta}_1(R)}{\partial R}\int_0^1 \tilde{w}_{1,R}^2(t+R)^{N-3}\,dt + \tilde{\beta}_1(R)\int_0^1 \Phi\tilde{w}_{1,R}(t+R)^{N-3}\,dt$$

$$- 2\tilde{\beta}_1(R)\int_0^1 \tilde{w}_{1,R}^2(t+R)^{N-4}\,dt.$$

Multiplying (5.50) by Φ and integrating, we get

$$\int_0^1 \Phi'\tilde{w}'_{1,R}(t+R)^{N-1}\,dt - p\int_0^1 \tilde{u}_R^{p-1}\tilde{w}_{1,R}\Phi(t+R)^{N-1}\,dt$$

$$= \tilde{\beta}_1(R)\int_0^1 \Phi\tilde{w}_{1,R}(t+R)^{N-3}\,dt.$$

Subtracting the last two equations, we deduce

$$(N-1)\int_0^1 \tilde{w}'_{1,R}\tilde{w}_{1,R}(t+R)^{N-3}\,dt - p(p-1)\int_0^1 \tilde{u}_R^{p-2}\frac{\partial \tilde{u}_R}{\partial R}\tilde{w}_{1,R}^2(t+R)^{N-1}\,dt$$

$$= \frac{\partial \tilde{\beta}_1(R)}{\partial R}\int_0^1 \tilde{w}_{1,R}^2(t+R)^{N-3}\,dt - 2\tilde{\beta}_1(R)\int_0^1 \tilde{w}_{1,R}^2(t+R)^{N-4}\,dt.$$

Since

$$(N-1) \int_0^1 \tilde{w}'_{1,R} \tilde{w}_{1,R} (t+R)^{N-3}\, dt = -\frac{(N-1)(N-3)}{2} \int_0^1 \tilde{w}^2_{1,R}(t+R)^{N-4}\, dt = O(R^{N-4})$$

and by (5.39)

$$p(p-1) \int_0^1 \tilde{u}_R^{p-2}\left(R\frac{\partial \tilde{u}_R}{\partial R} \right) \tilde{w}^2_{1,R} \frac{(t+R)^{N-1}}{R}\, dt = o(R^{N-2})$$

we have

$$\frac{\partial \tilde{\beta}_1(R)}{\partial R} \int_0^1 \tilde{w}^2_{1,R} \frac{(t+R)^{N-3}}{R^{N-3}}\, dt = 2\tilde{\beta}_1(R) \int_0^1 \tilde{w}^2_{1,R} \frac{(t+R)^{N-4}}{R^{N-3}}\, dt + o(R).$$

Finally, using the convergence of $\tilde{w}_{1,R}$ to φ_1 as in (5.51) we get

$$\frac{\partial \tilde{\beta}_1(R)}{\partial R}\left(\int_0^1 \varphi_1^2\, dt + o(1) \right) = 2\tilde{\beta}_1 R\left(\int_0^1 \varphi_1^2\, dt + o(1) \right) + o(R)$$

so that (5.52) holds. $\qquad\square$

Hence, by Proposition 5.17 we have that $\tilde{\beta}_1(R)$ is strictly decreasing, as a function of the radius R, for R large. Thus we have the following.

Corollary 5.18. *There exists $\bar{k} \geq 1$ such that for any $k \geq \bar{k}$ there exists only one radius R_k such that $\tilde{\beta}_1(R_k)$ solves (5.7). Moreover, the sequence R_k behaves asymptotically as*

$$R_k = \sqrt{\frac{-k(k+N-2)}{\tilde{\beta}_1}} + o(1) \quad \text{as } k \to \infty. \tag{5.53}$$

As a consequence, the Morse index $m(u_R)$ tends to $+\infty$, as $R \to +\infty$.

Proof. As observed, by Proposition 5.17 we have the existence of $\bar{R} > 0$ such that $\tilde{\beta}_1(R)$ is strictly decreasing for $R > \bar{R}$. Then, by (5.46), there exists $\bar{k}$ such that, for any $k \geq \bar{k}$, the equation (5.7) has a solution $\beta = \tilde{\beta}_1(R_k)$ for only one radius R_k.

Moreover, by (5.46) we get

$$(\tilde{\beta}_1 + o(1))R_k^2 = -k(k+N-2)$$

which gives (5.53).

Finally (5.52), together with (5.7) and Lemma 4.3 implies that, as R crosses R_k, the Morse index of the radial solution u_R increases and diverges to $+\infty$ as $k \to +\infty$. $\qquad\square$

Remark 5.19. Let us point out that the radii R_k, which are the only ones for which the linearized operator L_{u_R} can be degenerate, are well distant from each other. Indeed, from (5.53) we derive

$$\tau := \lim_{k \to +\infty} (R_{k+1} - R_k) = \frac{1}{\sqrt{|\beta_1|}}.$$

5.3.3 Bifurcation results

In this section, we prove the existence of nonradial solutions, bifurcating from the radial solution u_R. Since the radial solutions, and hence the linearized operators L_{u_R} are defined in variable domains we first reduce the problems in a fixed annulus. This will allow to build a functional setting independent of R as in [159].

Let A be a fixed annulus with radii $0 < a < b$, as in (5.1). For any $R > 0$ we define the diffeomorphism $h_R : \overline{A}_R \longrightarrow \overline{A}$ defined by

$$h_R(r, \theta_1, \ldots, \theta_{N-1}) = (a + (b - a)(r - R), \theta_1, \ldots, \theta_{N-1})$$

(having used the polar coordinates in $\mathbb{R}^N$, $N \geq 2$).

The function h_R induces the map

$$h_R^* : C_0^{1,\alpha}(\overline{A}_R) \longrightarrow C_0^{1,\alpha}(\overline{A}), \quad h_R^*(u)(x) = u(h_R^{-1}(x)) \tag{5.54}$$

for any $x \in A$.

Then the equation (5.1), written in A_R, becomes

$$\begin{cases} D_R(u) = u^p & \text{in } A \\ u > 0 & \text{in } A \\ u = 0 & \text{on } \partial A \end{cases} \tag{5.55}$$

where $D_R(u) = h_R^*(-\Delta((h_R^*)^{-1}(u)))$.

Now we can give definitions analogous to those of Section 5.2.3.

Definition 5.20. Let $\bar{R} > 0$ and $u_{\bar{R}}$ be the radial solution of (5.1) in $A_{\bar{R}}$. We say that a nonradial bifurcation occurs at $(u_{\bar{R}}, \bar{R})$ if in every neighborhood of $(h_R^*(u_{\bar{R}}), \bar{R})$ in $C^{1,\alpha}(\overline{A}) \times (0, +\infty)$ a nonradial solution (w_R, R) of (5.55) exists.

Definition 5.21. If in the previous definitions it happens that for any neighborhood of $(h_{\bar{R}}^*(u_{\bar{R}}), \bar{R})$ in $C^{1,\alpha}(\overline{A}) \times (0, +\infty)$ at least m distinct nonradial solutions (w_{R^i}, R^i), $i = 1, \ldots, m$ of (5.55) exist, we say that an m-nonradial bifurcation occurs at $(u_{\bar{R}}, \bar{R})$.

The bifurcation results we obtain are as follows.

Theorem 5.22. *There exists a strictly increasing sequence of radii R_k, $k \in \mathbb{N}^+$, and $R_k \to +\infty$ such that a nonradial bifurcation occurs at (u_{R_k}, R_k). Moreover, if k is even and sufficiently large a $[\frac{N}{2}]$-nonradial bifurcation occurs at (u_{R_k}, R_k).*

To prove this theorem, we proceed as in Section 5.2.3. We start observing that the equation in (5.55) can be written as $w = T_R(w)$, with $T_R(w) = (D_R)^{-1}(|w|^{p-1}w)$. The operator T_R is well-defined from $C_0^{1,\alpha}(\overline{A})$ into $C_0^{1,\alpha}(\overline{A})$ and is a compact operator.

If D is an open bounded set in $C_0^{1,\alpha}(\overline{A})$ such that $I - T_R \neq 0$ on ∂D, then the Leray–Schauder degree for the map $I - T_R$, i. e., $\deg(I - T_R, D, 0)$ is well-defined. Applying a result of [183, Proposition 2, p. 243], we have that

$$\deg(I - T_R, D, 0) = \deg\left(I - P_R, \left(h_R^*\right)^{-1}(D), 0\right) \tag{5.56}$$

where $P_R(v) = (-\Delta)^{-1}(|v|^{p-1}v)$ and it is defined in $C_0^{1,\alpha}(\overline{A}_R)$. The operator P_R is differentiable at u_R and, by Corollary 5.18 and Remark 5.19 we have that the linearized operator $I - P_R'(u_R)$ is invertible, for any $R \neq R_k$, where $\{R_k\}$ is the sequence defined in Corollary 5.18. Hence, for any $R \neq R_k$,

$$\deg(I - P_R, B, 0) = \deg(I - P_R'(u_R), B, 0) = (-1)^{m_R} \tag{5.57}$$

where B is a neighborhood of u_R in $C_0^{1,\alpha}(\overline{A})$ such that u_R is the only solution of $(I - P_R)(v) = 0$ in $\overline{B}$ and $m_R = m(u_R)$ is the Morse index of u_R.

Proof of Theorem 5.22. It is similar to that of Theorem 5.12. Since we need to use the map h_R^*, we detail the main steps for the reader's convenience.

We consider the subspace $\mathcal{F}_R$ of $C^{1,\alpha}(\overline{A}_R)$ of the functions which are $O(N-1)$-invariant with respect to the first $(N-1)$ variables.

As in the proof of Theorem 5.12, we can invoke Proposition 5.2 of [204] to claim that the eigenspace V_k corresponding to the eigenvalue λ_k of $-\Delta_{S^{N-1}}$ and spanned by the $O(N-1)$-invariant eigenfunctions, is one-dimensional for any $k \geq 1$.

By Corollary 5.18, arguing as in the proof of Theorem 5.12 we then have that the Morse index $m(u_R)$, in the space $\mathcal{F}_R$ increases by one, as R crosses the values R_k, i. e., $m(u_{R_k+\epsilon}) = m(u_{R_k-\epsilon}) + 1$, if ϵ is small enough.

Then, by (5.56) and (5.57) we get

$$\deg(I - T_{R_k-\epsilon}, B_{R_k-\epsilon}, 0) = (-1)\deg(I - T_{R_k+\epsilon}, B_{R_k+\epsilon}, 0)$$

if B_R is a neighborhood of $w_R = h_R^*(u_R)$ in the space of functions which are $O(N-1)$-invariant in A.

The change of degree at the point (w_{R_k}, R_k) implies that a bifurcation occurs and the bifurcating solutions are nonradial since the solutions u_R are nondegenerate in the space of radial function. By maximum principle, the bifurcating solutions are positive.

The $[\frac{N}{2}]$-bifurcation, when k is even, is proved exactly as in Theorem 5.12. $\qquad\square$

We end mentioning that in [159] a nonradial bifurcation result is obtained for problem (5.1) in annuli with inner radius $a < 1$ and fixed outer radius $b = 1$. More precisely, the author proves the existence of $\tilde{k}$ such that, for any $k \geq \tilde{k}$ there exists a radius a_k for which nonradial bifurcation occurs.

The result of Theorem 5.22, which is taken from [129], is an improvement of that of [159] since it shows that for k large there is only one radius R_k at which a nonradial bifurcation occurs, so these radii R_k are the only ones for which bifurcation can take place.

6 Morse index and symmetry for semilinear elliptic equations in bounded domains

In this chapter, we show connections between the Morse index of solutions of semilinear elliptic equations and their symmetry properties. More precisely, we will prove that, under some convexity hypotheses on the nonlinearity, the solutions which have Morse index not greater than the dimension of the space have an axial symmetry property. This kind of results were obtained in [186, 190] and extended in [35] to a class of fully nonlinear equations. We describe them in an unified context simplifying some proofs and adding some recent developments obtained in [78] for nonlinear mixed boundary value problems. Finally, we mention that similar symmetry results can be obtained for variational problems by using some symmetrization and reflection methods (see [24, 48–50, 152, 161, 165–167, 172, 203, 220–223] and the survey paper [224]).

To point out the range of applications, we start by presenting in Section 6.1 the famous symmetry theorems of Gidas, Ni and Nirenberg [122] obtained by the method of moving planes, together with some counterexamples.

6.1 Symmetry and monotonicity of positive solutions

6.1.1 Moving planes and symmetry

In this section, we recall the famous result by Gidas, Ni and Nirenberg [122] about the radial symmetry of positive solutions of semilinear elliptic equations in balls. It is based on the Alexandrov–Serrin moving planes method that goes back to Alexandrov (see [8]) and has been introduced in the context of partial differential equations by Serrin in [199].

In the proof of next theorem and in the sequel, we will use mostly the Berestycki–Nirenberg version of the moving planes method (see [31]), based on the strong maximum principle together with the weak maximum principle in small domains proved in Chapter 1. This technique allows us to prove a variant of the method, known as rotating planes method (see Theorem 6.3) as well as analogous symmetry results for solutions of semilinear elliptic equations in cylindrically symmetric domains and for solutions of semilinear elliptic systems.

Nevertheless, we will also sketch the original proof in [122] (see Remark 6.2).

Theorem 6.1. *Let $B = B_R = \{x \in \mathbb{R}^N : |x| < R\}$, $R > 0$, be a ball centered at the origin and assume that $u \in H_0^1(B) \cap C^0(\overline{B})$ is a weak solution of the problem*

$$
\begin{cases}
-\Delta u = f(|x|, u) & \text{in } B \\
u > 0 & \text{in } B \\
u = 0 & \text{on } \partial B
\end{cases}
\tag{6.1}
$$

https://doi.org/10.1515/9783110538243-006

where $f = f(r,s) : [0,+\infty) \times [0,+\infty) \to \mathbb{R}$ is a continuous function which is locally Lipschitz continuous in the s-variable uniformly with respect to r, and nonincreasing in r for any fixed $s \in [0,+\infty)$. Then u is radial and radially decreasing, i.e., $u(x) = v(|x|)$ for some function $v = v(r) : [0,R] \to \mathbb{R}$ and $v'(r) < 0$.

Proof. Since f is continuous by standard regularity results, the solution belongs at least to $C^1(B) \cap C^0(\overline{B})$. We will prove symmetry with respect to every hyperplane orthogonal to any direction $e \in S^{N-1}$. For simplicity of notation, we fix the hyperplane orthogonal to the $e_1 = (1,0,\ldots,0)$ direction. The proof is the same for any other direction. We set, for $-R < \lambda \le 0$,

$$H_\lambda = \{x \in \mathbb{R}^N : x_1 = \lambda\}$$
$$B_\lambda = \{x = (x_1,\ldots,x_N) \in B : x_1 < \lambda\}$$
$$x_\lambda = R_\lambda(x) = R_\lambda(x_1,\ldots x_N) = (2\lambda - x_1,\ldots x_N)$$

i.e., R_λ is the reflection through the hyperplane H_λ. Then we define

$$u_\lambda(x) = u(x_\lambda) \text{ for } x \in B_\lambda$$

Note that u_λ satisfies the equation $-\Delta u_\lambda = f(|x_\lambda|, u_\lambda)$ in B_λ and, since $0 = u \le u_\lambda$ on $\partial B_\lambda \setminus H_\lambda$ and $u = u_\lambda$ on $H_\lambda \cap \overline{B}$, we have that $u \le u_\lambda$ on ∂B_λ. Moreover, since $f(|x|, u)$ is nonincreasing in $|x|$ and $|x| \ge |x_\lambda|$ when $\lambda \le 0$, we get that $f(|x_\lambda|, u_\lambda) \ge f(|x|, u_\lambda)$.

Summing up, we have, for $-R < \lambda < 0$:

$$\begin{cases} -\Delta u = f(|x|, u); -\Delta u_\lambda \ge f(|x|, u_\lambda) & \text{in } B_\lambda \\ u \le u_\lambda & \text{on } \partial B_\lambda \end{cases} \tag{6.2}$$

We will prove the following inequality:

$$u \le u_\lambda \quad \text{in } B_\lambda \text{ for any } \lambda \in (-R, 0) \tag{6.3}$$

Let us first show that if (6.3) holds then we also have

$$u < u_\lambda \quad \text{in } B_\lambda, \quad \forall \lambda \in (-R, 0) \tag{6.4}$$

and

$$u_{x_1}(x) > 0, \forall x \in H_\lambda \cap B \quad \text{and} \quad \forall \lambda \in (-R, 0) \tag{6.5}$$

To prove (6.4) and (6.5), let us observe that if $\lambda < 0$ then by (6.2) and (6.3) and by the strong comparison principle (Theorem 1.31) we get that $u < u_\lambda$ in B_λ. Indeed, when $\lambda < 0$, there are points on the boundary of the ball where the strict inequality $0 = u < u_\lambda$ holds (because these points are reflected inside the ball, where u is positive). This proves (6.4) and, again by Theorem 1.31, we get that $u_{x_1} > (u_\lambda)_{x_1} = -u_{x_1}$ on $H_\lambda \cap B$, since

$u = u_\lambda$ on $H_\lambda \cap B \subset \partial B_\lambda$ and e_1 is an *outer* direction with respect to the cap B_λ. Hence (6.5) follows.

Let us show that once (6.3) is proved we can conclude the proof of the theorem. Indeed, if (6.3) holds, by continuity we obtain that $u \leq u_0$ in $B_0 = [x_1 < 0]$. By considering the opposite direction $-e_1$, we deduce

$$u \equiv u_0 \text{ in } B_0 \tag{6.6}$$

Moreover, since (6.5) holds, it follows that

$$u_{x_1} > 0 \quad \text{in } B_0 \tag{6.7}$$

To get the radial symmetry, it is enough to repeat the argument leading to (6.6) and (6.7) for every direction $e \in S^{N-1}$.

Let us now prove (6.3) in two step.

Step 1, starting the moving planes:

Let us set $A = \|u\|_{L^\infty(B)}$. Then for any $\lambda \in (-R, 0)$ we have that

$$\|u\|_{L^\infty(B_\lambda)} \leq A, \quad \|u_\lambda\|_{L^\infty(B_\lambda)} \leq A$$

Let $\delta = \delta(A)$ as in Theorem 1.21 and observe that δ is independent of λ and u, u_λ satisfy (6.2).

If $\lambda > -R$ is close to $-R$, then the measure of B_λ can be made arbitrarily small, in particular meas $(B_\lambda) < \delta$, so that we get from (6.2) and Theorem 1.21 that $u \leq u_\lambda$ in B_λ. Hence we get that

$$\exists \alpha > 0 : u < u_\lambda \text{ in } B_\lambda \text{ and } u_{x_1} > 0 \text{ in } B_\lambda, \ \forall \lambda \in (-R, -R+\alpha)$$

Step 2, continuing the moving planes:

Let us set

$$\lambda_0 = \sup\{\lambda \in (-R, 0) : u \leq u_\mu \text{ in } B_\mu, \ \forall \mu \in (-R, \lambda)\}.$$

To prove (6.3), we have to prove that $\lambda_0 = 0$.

Suppose by contradiction that $\lambda_0 < 0$. Then by continuity $u \leq u_{\lambda_0}$ in B_{λ_0}, and by the strong comparison principle (Theorem 1.31) we get, as before, that $u < u_{\lambda_0}$ in B_{λ_0}. Take now a compact set $K \subset B_{\lambda_0}$ such that $|B_{\lambda_0} \setminus K| < \frac{\delta}{2}$, where δ is again as in Theorem 1.21, and observe that $\min_K(u_{\lambda_0} - u) = m > 0$ since $(u_{\lambda_0} - u)$ is continuous and positive in K. By continuity for $\lambda > \lambda_0$ and close to λ_0, we still have that $\lambda < 0$ and

$$|B_\lambda \setminus K| < \delta, \quad \min_K(u_\lambda - u) \geq \frac{m}{2} > 0$$

Moreover, setting $B_\lambda' = (B_\lambda \setminus K)$ we have that $u \leq u_\lambda$ on $\partial B_\lambda'$.

Applying again Theorem 1.21, we get that $u \leq u_\lambda$ in B_λ', and since $u \leq u_\lambda$ in K, we deduce that $u \leq u_\lambda$ in B_λ for any λ greater than λ_0 and sufficiently close to λ_0. This contradicts the definition of λ_0, and hence necessarily $\lambda_0 = 0$. $\qquad\square$

Remark 6.2. Let us sketch here the original proof in [122], which is entirely based on the strong maximum principle and Hopf's lemma. This proof works for a $C^2(\overline{B})$ solution and relies deeply on the smoothness of the domain B.

We also assume that

$$f(R, 0) \geq 0$$

(for the case when $f(R, 0) < 0$ a variant of Hopf's lemma can be used, and we refer to [122], or to [115] for a simple proof).

Since $|x| \leq R$ and $f(r, s)$ is nonincreasing in r, we get that u satisfies

$$-\Delta u = f(|x|, u) - f(|x|, 0) + f(|x|, 0) \geq f(|x|, u) - f(|x|, 0) + f(R, 0)$$
$$\geq f(|x|, u) - f(|x|, 0)$$

Therefore, $u \geq 0$ satisfies the inequality $-\Delta u + c(x)u \geq 0$, with

$$-c(x) = \begin{cases} \frac{f(|x|, u(x)) - f(|x|, 0)}{u(x)} & \text{if } u(x) \neq 0 \\ 0 & \text{if } u(x) = 0 \end{cases},$$

which is bounded since $f(r, s)$ is Lipschitz continuous in s.

Since $u = 0$ on ∂B and $e_1 = (1, \ldots, 0)$ is an inner normal in every point x of ∂B where $x_1 < 0$, by Hopf's lemma (Theorem 1.28) we get

$$\frac{\partial u}{\partial x_1}(x) > 0 \quad \text{on } \partial B \cap [x_1 \leq \lambda] \text{ for any } \lambda < 0 \tag{6.8}$$

By (6.8), we can start the moving planes procedure and get that

$$\exists \alpha > 0 : u < u_\lambda \text{ in } B_\lambda \text{ and } u_{x_1} > 0 \text{ in } B_\lambda, \quad \forall \lambda \in (-R, -R + \alpha)$$

Indeed, if this is not the case, there exists a sequence $\lambda_n \to -R$ and a sequence of points $y_n = ((y_n)_1, y'_n) \in B_{\lambda_n}$, such that, up to a subsequence, $y_n \to (-R, 0, \ldots, 0)$, and $u(y_n) \geq u_{\lambda_n}(y_n) = u((2\lambda_n - (y_n)_1, y'_n))$.

By the mean value theorem, there exists a sequence $z_n = ((z_n)_1, z'_n)$ with $(y_n)_1 < (z_n)_1 < 2\lambda_n - (y_n)_1$, $z'_n = y'_n$ (so that again up to a subsequence $z_n \to (-R, 0, \ldots, 0)$) and $u_{x_1}(z_n) \leq 0$. As $n \to \infty$ we get $u_{x_1}((-R, 0, \ldots, 0)) \leq 0$, which contradicts (6.8).

So the set $\{\lambda \in (-R, 0) : u \leq u_\mu \text{ in } B_\mu, \forall \mu \in (-R, \lambda)\}$ is not empty and contains an interval $(-R, -R + \alpha)$, $\alpha > 0$. Setting

$$\lambda_0 = \sup\{\lambda \in (-R, 0) : u \leq u_\mu \text{ in } B_\mu, \forall \mu \in (-R, \lambda)\}$$

we have to prove that $\lambda_0 = 0$. Let us first observe that, by continuity, $u \leq u_{\lambda_0}$ in B_{λ_0}.

Exactly as in the previous proof, using Theorem 1.31, we get that if we suppose, by contradiction, that $\lambda_0 < 0$, then

$$u < u_{\lambda_0} \text{ in } B_{\lambda_0} \quad \text{and} \quad u_{x_1} > 0 \text{ on } H_{\lambda_0} \cap B$$

This, together with (6.8), gives

$$u_{x_1} > 0 \quad \text{on } \partial B_{\lambda_0} \tag{6.9}$$

The inequality (6.9) in turn implies that there exists $\varepsilon > 0$ such that $\lambda_0 + \varepsilon < 0$ and the inequality (6.3) still holds for $\lambda_0 < \lambda < \lambda_0 + \varepsilon$, contradicting the definition of λ_0. Indeed, if this is not the case, there exists a sequence $\lambda_n = \lambda_0 + \varepsilon_n \to \lambda_0$, a sequence of points $y_n = ((y_n)_1, y_n') \in B_{\lambda_n}$ with $u(y_n) \geq u_{\lambda_n}(y_n) = u((2\lambda_n - (y_n)_1, y_n'))$ and a sequence z_n of points in the segment joining y_n and $(y_n)_{\lambda_n}$ with $u_{x_1}(z_n) \leq 0$. For a subsequence, we then have that $y_n \to y_0 \in \overline{B_0}$ and, by continuity, we get $u(y_0) \geq u_{\lambda_0}(y_0)$, so that $y_0 \in H_{\lambda_0} \cap \overline{B}$, because $u < u_{\lambda_0}$ in $B_{\lambda_0} \cup \partial B \setminus H_{\lambda_0}$.

Hence also $z_n \to y_0$ and by continuity $u_{x_1}(y_0) \leq 0$, contradicting (6.9). $\qquad\square$

A variant of the moving planes method is the so called *rotating planes method* (exploited, e. g., in [190],) which is similar to the one used in the proof of Theorem 6.1, but rotating the planes instead of moving them in a fixed direction.

Let Ω be a bounded rotationally symmetric domain in $\mathbb{R}^N$, i. e., a ball or an annulus. For a unit vector $e \in S^{N-1}$, we consider the hyperplane

$$H(e) = \{x \in \mathbb{R}^N : x \cdot e = 0\} \tag{6.10}$$

orthogonal to the direction e and the open half-domain

$$\Omega(e) = \{x \in \Omega : x \cdot e > 0\} \tag{6.11}$$

We then set

$$\sigma_e(x) = x - 2(x \cdot e)e \quad \text{for every } x \in \Omega \tag{6.12}$$

i. e., $\sigma_e : \Omega \to \Omega$ is the *reflection with respect to the hyperplane $H(e)$*. Note that

$$H(-e) = H(e) \quad \text{and} \quad \Omega(-e) = \sigma_e(\Omega(e)) = -\Omega(e) \quad \text{for every } e \in S^{N-1}. \tag{6.13}$$

Finally, if $u : \Omega \to \mathbb{R}$ is a continuous function we define the *reflected function $u^{\sigma(e)}$* : $\Omega \to \mathbb{R}$ by

$$u^{\sigma(e)}(x) = u(\sigma_e(x)) \tag{6.14}$$

Now let us consider solutions, possibly sign-changing, of the following problem:

$$\begin{cases} -\Delta u = f(|x|, u) & \text{in } \Omega \\ u = 0 & \text{on } \partial\Omega \end{cases} \tag{6.15}$$

where Ω is a bounded rotationally symmetric domain and $f = f(r, s) : [0, \infty) \times \mathbb{R} \to \mathbb{R}$ is a continuous function which is locally Lipschitz continuous in the variable $s \in \mathbb{R}$.

In the next theorem, we denote by e_ϑ the directions defined by

$$e_\vartheta = (\cos(\vartheta), \sin(\vartheta), 0, \ldots, 0), \quad \vartheta \in \mathbb{R}$$

Theorem 6.3 (Rotating planes method). *Let Ω and f be as in (6.15) and let $u \in H_0^1(\Omega) \cap C^0(\overline{\Omega})$ be a weak solution of (6.15). If there exists $\vartheta_0 \in \mathbb{R}$ such that $u < u^{\sigma(e_{\vartheta_0})}$ in $\Omega(e_{\vartheta_0})$, then there exists $\vartheta_1 > \vartheta_0$ such that*

$$u \equiv u^{\sigma(e_{\vartheta_1})} in\ \Omega(e_{\vartheta_1}), \quad u < u^{\sigma(e_\vartheta)} in\ \Omega(e_\vartheta), \ \forall \vartheta \in (\vartheta_0, \vartheta_1)$$

Proof. Let us observe that the functions $u^{\sigma(e_\vartheta)}$ satisfy the same equation as u, namely $-\Delta u^{\sigma(e_\vartheta)} = f(|x|, u^{\sigma(e_\vartheta)})$ in Ω, and both $\|u\|_{L^\infty(\Omega(e_\vartheta))}$ and $\|u^{\sigma(e_\vartheta)}\|_{L^\infty(\Omega(e_\vartheta))}$ are bounded by $\|u\|_{L^\infty(\Omega)} =: A$.

Let us fix $\delta = \delta(A)$ as in Theorem 1.21 and observe that δ is independent of ϑ, and the functions $u, u^{\sigma(e_\vartheta)}$ satisfy

$$\begin{cases} -\Delta u = f(|x|, u); -\Delta u^{\sigma(e_\vartheta)} = f(|x|, u^{\sigma(e_\vartheta)}) & \text{in } \Omega(e_\vartheta) \\ u = u^{\sigma(e_\vartheta)} & \text{on } \partial\Omega(e_\vartheta) \end{cases} \tag{6.16}$$

Let us set $\Theta = \{\vartheta \geq \vartheta_0 : u < u^{\sigma(e_{\vartheta'})}$ in $\Omega(e_{\vartheta'})\ \forall \vartheta' \in (\vartheta_0, \vartheta)\}$ and let us show that the set Θ is nonempty and contains an interval $[\vartheta_0, \vartheta_0 + \varepsilon)$ for $\varepsilon > 0$ sufficiently small. Indeed we can take a compact set $K \subset \Omega(e_{\vartheta_0})$ such that $|\Omega(e_{\vartheta_0}) \setminus K| \leq \frac{\delta}{2}$ and $m = \min_K(u^{\sigma(e_{\vartheta_0})} - u) > 0$. By continuity if ϑ is close to ϑ_0, we have that $K \subset \Omega(e_\vartheta)$, $(u^{\sigma(e_\vartheta)} - u) \geq \frac{m}{2} > 0$ in K, $|\Omega(e_\vartheta) \setminus K| \leq \delta$ and $(u^{\sigma(e_\vartheta)} - u) \geq 0$ on $\partial(\Omega(e_\vartheta) \setminus K)$. Then by the weak comparison principle in small domains (Theorem 1.21) we get that $u \leq u^{\sigma(e_\vartheta)}$ in $\Omega' = \Omega(e_\vartheta) \setminus K$, and hence in $\Omega(e_\vartheta)$. Moreover, $u < u^{\sigma(e_\vartheta)}$ in $\Omega(e_\vartheta)$ by the strong comparison principle (Theorem 1.31). So the set Θ is nonempty, and is bounded from above by $\vartheta_0 + \pi$, since, considering the opposite direction, the inequality between u and the reflected function gets reversed. Let us set $\vartheta_1 = \sup \Theta$.

We claim that $u \equiv u^{\sigma(e_{\vartheta_1})}$ in $\Omega(e_{\vartheta_1})$. Indeed, if this is not the case, we get $u < u^{\sigma(e_{\vartheta_1})}$ in $\Omega(e_{\vartheta_1})$ by the strong comparison principle (Theorem 1.31), since by continuity $u \leq u^{\sigma(e_{\vartheta_1})}$ in $\Omega(e_{\vartheta_1})$. Then, using again the weak comparison principle in small domains and the previous technique we get $u < u^{\sigma(e_\vartheta)}$ in $\Omega(e_\vartheta)$ for $\vartheta > \vartheta_1$ and close to ϑ_1, contradicting the definition of ϑ_1. $\qquad\square$

6.1.2 Monotonicity by the method of moving planes

The bibliography on the moving planes method is huge; we refer, e. g., to [28–31, 33, 47, 49, 53, 56, 68, 69, 75, 76, 80–85, 87, 94, 95, 118, 119, 122, 123, 155, 156, 199, 216, 217, 219] and the references therein for many applications to different elliptic problems.

Let us observe that the moving planes method yields not only symmetry results but also monotonicity results in general domains as we recall below.

Let us start with some notation. Let Ω be a bounded domain and e a direction in $\mathbb{R}^N$. For a real number λ, we define

$$H_\lambda^e = \{x \in \mathbb{R}^N : x \cdot e = \lambda\} \tag{6.17}$$

$$\Omega_\lambda^e = \{x \in \Omega : x \cdot e > \lambda\} \tag{6.18}$$

$$x_\lambda^e = R_\lambda^e(x) = x + 2(\lambda - x \cdot e)e, \quad x \in \mathbb{R}^N \tag{6.19}$$

(i. e., R_λ^e is the reflection through the hyperplane H_λ^e)

$$a(e) = \sup_{x \in \Omega} x \cdot e \tag{6.20}$$

If $\lambda < a(e)$, then Ω_λ^e is nonempty, thus we set

$$(\Omega_\lambda^e)' = R_\lambda^e(\Omega_\lambda^e) \tag{6.21}$$

If Ω is smooth and $\lambda < a(e)$, λ close to $a(e)$, then the reflected cap $(\Omega_\lambda^e)'$ is contained in Ω and will remain in it, at least until one of the following situations occurs:
(i) $(\Omega_\lambda^e)'$ becomes internally tangent to $\partial\Omega$ at some point not on H_λ^v
(ii) H_λ^e is orthogonal to $\partial\Omega$ at some point

Let $\Lambda_1(e)$ be the set of those $\lambda' < a(e)$ such that for each $\lambda \in (\lambda', a(e)]$ none of the conditions (i) and (ii) holds and define

$$\lambda_1(e) = \inf \Lambda_1(e) \tag{6.22}$$

In the terminology of [122], $\Omega_{\lambda_1(e)}^e$ is the *maximal cap*.

In some situations, e. g. in a rectangle, it may happen that for $\lambda < \lambda_1(e)$ the reflected cap is still contained in Ω, i. e., $(\Omega_\lambda^e)' \subset \Omega$. We consider the set $\Lambda_2(e)$ of those $\lambda < a(e)$ such that $(\Omega_\mu^e)' \subset \Omega$ for each $\mu \in (\lambda, a(e)]$ and define

$$\lambda_2(e) = \inf \Lambda_2(e) \tag{6.23}$$

In the terminology of [122], $\Omega_{\lambda_2(e)}^e$ is the *optimal* cap.
Let us consider the problem

$$\begin{cases} -\Delta u = f(x, u) & \text{in } \Omega \\ u > 0 & \text{in } \Omega \\ u = 0 & \text{on } \partial\Omega \end{cases} \tag{6.24}$$

One of the main results by Gidas, Ni and Nirenberg [122] is the following.

Theorem 6.4. *Let Ω be a bounded smooth domain, e a direction and $u \in C^2(\overline{\Omega})$ be a solution of (6.24), where $f = f(x, s)$ is a locally Lipschitz continuous function in $\overline{\Omega} \times [0, +\infty)$ and it is monotone in the direction e w. r. t. the first variable, in the sense that for any λ in the interval $(\lambda_1(e), a(e))$ it holds*

$$f(x, s) \leq f(x_\lambda^e, s) \ \text{if } x \in \Omega_\lambda^e, \ s \in [0, +\infty)$$

Then for λ in the interval $(\lambda_1(e), a(e))$ we have

$$u(x) \leq u(x_\lambda^e) \quad \forall x \in \Omega_\lambda^e \tag{6.25}$$

Moreover,

$$\frac{\partial u}{\partial e}(x) < 0 \quad \forall x \in \Omega_{\lambda_1(e)}^e \tag{6.26}$$

The proof is essentially the same sketched in Remark 6.2 where we have fixed the direction $e = -e_1$. The radiality of the solutions in balls is just a corollary of Theorem 6.4, which is however a monotonicity result, in particular in a neighborhood of the boundary of Ω.

It has been exploited in many situations, e. g., in *a priori estimates* using the *blow-up* technique (see [62, 71, 96, 124, 125] and the references therein).

The proof of Theorem 6.4 given in [122] fails for simple domains with corners as a rectangle, while the proof that we have presented in the case of the ball (Theorem 6.1), due to Berestycki and Nirenberg [31], also works in these cases (and also assuming that u is a $C^1(\overline{\Omega})$ weak solution). Moreover, (6.25) and (6.26) hold in the optimal cap defined by $\lambda_2(e)$. The proof is essentially the same as the one of Theorem 6.1 given above, where we have fixed the direction $e = -e_1$.

Note however that in [31] much more general equations are considered, in particular, the paper focuses on fully nonlinear elliptic equations.

6.1.3 Counterexamples to radial symmetry

If one of the hypotheses of Theorem 6.1 fails, or the ball is replaced by an annulus, the solutions need not to be radial anymore. In particular, this can happen when Ω is a rotationally symmetric bounded domain, e. g., a ball or an annulus, but either one of the following situations occurs:
a) the nonlinearity $f(r, s)$ is not locally Lipschitz continuous in the variable s
b) the nonlinearity $f(r, s)$ is not monotone decreasing in the variable r
c) the domain is an annulus
d) the solution changes sign in Ω

Let us discuss briefly, following the survey paper [187], some counterexamples to the radial symmetry in each of these cases.
a) In the case when the nonlinearity is not Lipschitz continuous, there can be some compactly supported positive solutions which exhibit a combination of "bumps." As an example, if $p > 2$ the function

$$w(x) = \begin{cases} (1 - |x|^2)^p & \text{if } |x| \leq 1 \\ 0 & \text{if } |x| > 1 \end{cases} \tag{6.27}$$

satisfies in $\mathbb{R}^N$ the equation

$$- \Delta w = f(w) \tag{6.28}$$

with the Hölder continuous function f with exponent $(1 - \frac{2}{p})$ defined by

$$f(s) = -2p(p-2)s^{1-\frac{2}{p}} + 2p(N + 2p - 2)s^{1-\frac{1}{p}}$$

If x_0 is any point with $|x_0| = 3$, the function $u(x) = w(x) + w(x - x_0)$ satisfies the Dirichlet problem associated to (6.28) in $\Omega = B_5$ but it is not radial.

b) The problem (6.1) when $f(|x|, u) = |x|^\alpha u^p$ is the Hénon problem (see [144]) already considered in Chapter 3. Note that if $\alpha > 0$ then the monotonicity hypothesis on the function $r \mapsto f(r, s)$ in Theorem 6.1 is not satisfied.
Let us assume that $\alpha > 0$ and $1 < p < +\infty$ if $N = 2$, $1 < p < \frac{N+2}{N-2}$ if $N \geq 3$.
A solution of (6.1) can be found using a constrained minimization method. Indeed, if we define

$$S_\alpha = \inf_{v \in H_0^1(B),\, v \neq 0} \frac{\int_B |\nabla v|^2\, dx}{\left(\int_B |x|^\alpha |v|^{p+1}\, dx \right)^{\frac{2}{p+1}}} \tag{6.29}$$

since the embedding of $H_0^1(B)$ in $L^{p+1}(B)$ is compact, we have that S_α is achieved by a function v that can be assumed to be nonnegative (taking its modulus) and by the strong maximum principle it follows that $v > 0$ in B.
In the same way, we can minimize among the radial function, i. e., if $H_{0,\mathrm{rad}}^1(B)$ is the subspace of the radial functions, we can consider the infimum

$$Z_\alpha = \inf_{v \in H_{0,\mathrm{rad}}^1(B),\, v \neq 0} \frac{\int_B |\nabla v|^2\, dx}{\left(\int_B |x|^\alpha |v|^{p+1}\, dx \right)^{\frac{2}{p+1}}} \tag{6.30}$$

As before, the infimum is achieved by a radial function that solves (6.1).
It is clear that $S_\alpha \leq Z_\alpha$ and it is proved in [202] that there exists $\alpha^* > 0$ such that $S_\alpha < Z_\alpha$ for any $\alpha > \alpha^*$. Obviously, for any such α there exists a nonradial solution of (6.1) in the ball.

c) An interesting counterexample was given in the classical paper [45] in the presence of a critical nonlinearity. Let us consider the problem

$$\begin{cases} -\Delta u = u^{\frac{N+2}{N-2}} + \lambda u & \text{in } \Omega \\ u > 0 & \text{in } \Omega \\ u = 0 & \text{on } \partial\Omega \end{cases} \tag{6.31}$$

where Ω is a bounded smooth domain in $\mathbb{R}^N$, $N \geq 4$ and $\lambda > 0$.

In [45], it is proved that for $\lambda \in (0, \lambda_1)$, where λ_1 is the first eigenvalue of the operator $-\Delta$ in $H_0^1(\Omega)$, there is a solution to problem (6.31). This is obtained by proving that the infimum

$$S_\lambda = \inf_{\substack{v \in H_0^1(\Omega) \\ \|v\|_{L^{2^*}(\Omega)} = 1}} \int_\Omega |\nabla v|^2 \, dx - \lambda \int_\Omega |v|^2 \, dx, \quad 2^* = \frac{2N}{N-2}$$

is achieved.

If Ω is an annulus $A = [r < |x| < R]$, we can consider also the infimum

$$Z_\lambda = \inf_{\substack{v \in H_{0,\mathrm{rad}}^1(\Omega) \\ \|v\|_{L^{2^*}(\Omega)} = 1}} \int_\Omega |\nabla v|^2 \, dx - \lambda \int_\Omega |v|^2 \, dx, \quad 2^* = \frac{2N}{N-2}$$

among the radial functions. By the compactness of the embedding of $H_{0,\mathrm{rad}}^1(A)$ in $L^{2^*}(A)$, the infimum Z_λ is also achieved when $\lambda = 0$, while the infimum S_0 is never achieved in any bounded domain. This remark is crucial in [45] to prove that $S_\lambda < Z_\lambda$ for λ close to zero, so that the least energy solution in an annulus is not radial.

d) It is very easy to construct a sign changing solution in a ball which is not radial. For example, the second eigenfunction of the Laplacian in a ball is an antisymmetric function with respect to a hyperplane passing through the origin.

More generally, if $f(u)$ is an odd function and u is a positive solution of the problem

$$\begin{cases} -\Delta u = f(u) & \text{in } B^- \\ u = 0 & \text{on } \partial B^- \end{cases} \tag{6.32}$$

in the half-ball $B^- = \{x = (x_1, \ldots, x_N) \in B_R : x_1 < 0\}$, then by reflecting u by oddness in $B^+ = \{x = (x_1, \ldots, x_N) \in B_R : x_1 > 0\}$ we get a sign-changing solution in B_R which is not radial.

6.2 Foliated Schwarz symmetry and related properties

Let us now define a particular kind of axial symmetry, known as *foliated Schwarz symmetry*. This name was introduced in [203] and refers to foliations by spheres with the same center. We will focus mainly on rotationally symmetric bounded domains Ω, i. e., either a ball or an annulus in $\mathbb{R}^N$ and that we always assume with the center at the origin. Nevertheless, the definition that we give and some of its properties that we discuss here also work for unbounded rotationally symmetric domains, such as the whole space $\mathbb{R}^N$ or the complement of a ball, as it will be sketched in Chapter 8.

Definition 6.5. Let Ω be a rotationally symmetric domain in $\mathbb{R}^N$, $N \geq 2$. We say that a function $v \in C^0(\overline{\Omega})$ is **foliated Schwarz symmetric** (briefly FSS) if there is a unit vector

$p \in \mathbb{R}^N$ such that $v(x)$ only depends on $r = |x|$ and (if $x \neq 0$) on $\theta := \arccos(\frac{x}{|x|} \cdot p)$, and v is nonincreasing in θ.

Remark 6.6. A foliated Schwarz symmetric function v is axially symmetric with respect to the axis containing the vector p. This means that v is symmetric with respect to every hyperplane orthogonal to any direction e which is orthogonal to p. Indeed, in this case, the points x and $\sigma_e(x)$ (σ_e defined in (6.12)) have the same modulus and the angle between x and p or $\sigma_e(x)$ and p is the same. Note however that besides the axial symmetry, Definition 6.5 requires the *monotonicity* with respect to the angular variable ϑ.

A radial function is a particular case of a FSS function. We will see that in general, e. g., for solutions of semilinear elliptic equations, a nonradial FSS function is actually strictly decreasing in the angle variable θ (see Remark 6.13, point 2).

Let us recall here some definitions from the previous section.

For a unit vector $e \in S = S^{N-1}$, we consider the hyperplane $H(e) = \{x \in \mathbb{R}^N : x \cdot e = 0\}$ orthogonal to the direction e and the open half-domain $\Omega(e) = \{x \in \Omega : x \cdot e > 0\}$. We then set $\sigma_e(x) = x - 2(x \cdot e)e$ for every $x \in \Omega$ i. e. $\sigma_e : \Omega \to \Omega$ is the *reflection with respect to the hyperplane $H(e)$*.

Finally, if $u : \Omega \to \mathbb{R}$ is a continuous function we define the *reflected function* $u^{\sigma(e)} : \Omega \to \mathbb{R}$ as $u^{\sigma(e)}(x) = u(\sigma_e(x))$

Let us now consider solutions of the following semilinear problem:

$$\begin{cases} -\Delta u = f(|x|, u) & \text{in } \Omega \\ u = 0 & \text{on } \partial\Omega \end{cases} \tag{6.33}$$

A sufficient condition for the foliated Schwarz symmetry of a solution of (6.33) is given in the following proposition, proved in [48] for the case of an annulus (but the proof carries over to the case of a ball) and also implicitly contained in [24].

Proposition 6.7. *Let Ω be a (bounded or unbounded) rotationally symmetric domain and $u \in H_0^1(\Omega) \cap C^0(\overline{\Omega})$ a weak solution of (6.33) where $f = f(r,s) : [0, \infty) \times \mathbb{R} \to \mathbb{R}$ is a continuous function which is locally Lipschitz continuous in the s-variable. Assume that for every unit vector $e \in S^{N-1}$ it holds:*

$$\text{either } u(x) \geq u(\sigma_e(x)) \; \forall x \in \Omega(e) \quad \text{or} \quad u(x) \leq u(\sigma_e(x)) \; \forall x \in \Omega(e) \tag{6.34}$$

Then u is foliated Schwarz symmetric.

Proof. Let $r > 0$ and $S_r = \{x \in \mathbb{R}^N : |x| = r\} \subset \Omega$, and let $p \in S^{N-1}$ be such that $u(rp) = \max_{S_r} u$.

We define $T_p^+ = \{e \in S^{N-1} : e \cdot p > 0\}$ and $T_p^- = \{e \in S^{N-1} : e \cdot p < 0\}$.

To prove that u is foliated Schwarz symmetric with respect to p, it is enough to show that

$$u(x) \geq u(\sigma_e(x)) \quad \text{for all } x \in \Omega(e) \tag{6.35}$$

whenever $e \in T_p^+$.

Indeed this would imply the reverse inequality for $e \in T_p^-$, so that if e is any direction orthogonal to p then $u(x) = u(\sigma_e(x))$, for all $x \in \Omega(e)$. In other words, u would be axially symmetric about the axis with direction p, and the angular monotonicity would follow also by the same inequalities.

So for fixed $e \in T_p^+$ we consider the difference $w = u - u^{\sigma(e)}$, and observe that w satisfies the linear equation

$$- \Delta w + c(x)w = 0 \quad \text{in } \Omega \tag{6.36}$$

with

$$-c(x) = \begin{cases} \frac{f(|x|,u(x))-f(|x|,u^{\sigma(e)}(x))}{u(x)-u^{\sigma(e)}(x)} & \text{if } u(x) \neq u^{\sigma(e)}(x) \\ 0 & \text{if } u(x) = u^{\sigma(e)}(x) \end{cases}$$

and obviously $c \in L^\infty(\Omega)$.

By assumption, either $w \geq 0$, or $w \leq 0$. In the first case, (6.35), is satisfied.

In the second case, since $w \leq 0$ satisfies a linear equation, by the strong maximum principle (Theorem 1.28), we get that either $w < 0$ or $w \equiv 0$ in $\Omega(e)$. The first case is not possible, since $rp \in \Omega(e)$ and $w(rp) \geq 0$ because $u(rp)$ is the maximum of u on S_r. Hence (if $w \leq 0$ then) $w \equiv 0$ in $\Omega(e)$ and (6.35) is satisfied. $\qquad\square$

In the rest of this section:

- Ω is either a ball or an annulus centered at the origin in $\mathbb{R}^N$, $N \geq 2$
- $u \in C^2(\overline{\Omega})$ is a classical solution of the semilinear problem (6.33), with $f \in C^1((0,+\infty) \times \mathbb{R})$.

Let $f'(|x|,s) = \frac{\partial f}{\partial s}(|x|,s)$ and let us set

$$V_u(x) = f'(|x|,u(x)), \quad x \in \Omega \tag{6.37}$$

We consider the quadratic form

$$Q_u(v;\Omega) = \int_\Omega \left(|\nabla v|^2 - V_u|v|^2\right) dx, \quad v \in H_0^1(\Omega) \tag{6.38}$$

corresponding to the linear operator $L_u = -\Delta - V_u$ and denote its eigenvalues and eigenfunctions by

$$\lambda_k = \lambda_k(-\Delta - V_u;\Omega); \quad \phi_k = \phi_k(-\Delta - V_u;\Omega) \tag{6.39}$$

If e is a direction, we also consider the corresponding quadratic form

$$Q_u(v;\Omega(e)) = \int_{\Omega(e)} \left(|\nabla v|^2 - V_u|v|^2\right) dx, \quad v \in H_0^1(\Omega(e)) \tag{6.40}$$

while we denote the eigenvalues and the eigenfunctions of the operator $-\Delta - V_u$ in the cap $\Omega(e)$ by

$$\lambda_k^e = \lambda_k(-\Delta - V_u; \Omega(e)); \quad \varphi_k^e = \varphi_k(-\Delta - V_u; \Omega(e)) \tag{6.41}$$

Observing that we can write

$$f(u(x)) - f(u^{\sigma(e)}(x)) = \int_0^1 [f'(|x|, tu(x) + (1-t)u^{\sigma(e)}(x))](u(x) - u^{\sigma(e)}(x))\, dt$$

we define, for any direction $e \in S^{N-1}$, the potential

$$V_e(x) = \int_0^1 f'(|x|, tu(x) + (1-t)u^{\sigma(e)}(x))\, dt$$

$$= \begin{cases} \frac{f(|x|,u(x))-f(|x|,u^{\sigma(e)}(x))}{u(x)-u^{\sigma(e)}(x)} & \text{if } u(x) \neq u^{\sigma(e)}(x) \\ f'(|x|, u(x)) & \text{if } u(x) = u^{\sigma(e)}(x) \end{cases}, \quad x \in \Omega(e) \tag{6.42}$$

Note that the difference $w^e(x) = u(x) - u^{\sigma(e)}(x)$ satisfies in $\Omega(e)$ the Dirichlet problem

$$\begin{cases} -\Delta w^e - V_e w^e = 0 & \text{in } \Omega(e) \\ w^e = 0 & \text{on } \partial\Omega(e) \end{cases} \tag{6.43}$$

Now we describe other sufficient conditions for the foliated Schwarz symmetry of a solution u of (6.33).

To this end, we begin with some geometric considerations about cylindrical coordinates with respect to a plane spanned by two orthogonal directions η_1, η_2. Using cylindrical coordinates with respect to this plane, we will write $x = (x_1, \ldots, x_N)$ as

$$x = r\cos(\vartheta)\eta_1 + r\sin(\vartheta)\eta_2 + \tilde{x} \tag{6.44}$$

where

$$r = r(\eta_1, \eta_2) = \sqrt{(x \cdot \eta_1)^2 + (x \cdot \eta_2)^2}, \quad \tilde{x} = \tilde{x}(\eta_1, \eta_2) = x - (x \cdot \eta_1)\eta_1 - (x \cdot \eta_2)\eta_2 \tag{6.45}$$

$$\vartheta = \vartheta(\eta_1, \eta_2) \in [0, 2\pi) : x \cdot \eta_1 = r\cos(\vartheta), \quad x \cdot \eta_2 = r\sin(\vartheta) \tag{6.46}$$

Let us denote by $u_\vartheta = u_{\vartheta(\eta_1,\eta_2)}$ the derivative of a function $u \in C^1(\overline{\Omega})$ whit respect to the angular coordinate ϑ, trivially extended where it is not well-defined, i. e., the function

$$u_\vartheta(r, \vartheta, \tilde{x}) = \begin{cases} \frac{\partial u}{\partial \vartheta}(r, \vartheta, \tilde{x}) & \text{if } r \neq 0 \\ 0 & \text{if } r = 0 \end{cases} \tag{6.47}$$

For simplicity, we omit the dependence on η_1, η_2 if the directions are fixed. Moreover, to simplify the notation, in the calculation that follows we will assume that $\eta_1 = e_1 = (1, 0, \ldots, 0)$, $\eta_2 = e_2 = (0, 1, 0, \ldots, 0)$.

Let $e_\beta = (\cos(\beta), \sin(\beta), 0, \ldots, 0)$ be a unit vector in the x_1-x_2 plane for $\beta \in \mathbb{R}$.

Then we consider as before the hyperplane $H(e_\beta)$ and the half-domains $\Omega(e_\beta)$ and, for simplicity, we will use the notation $\Omega_\beta = \Omega(e_\beta)$, $H_\beta = H(e_\beta)$, $\sigma_\beta = \sigma_{e_\beta}$ and u^{σ_β} for the reflection of a function u with respect to the hyperplane H_β.

It is easy to see that the reflection σ_β with respect to H_β can be written as

$$\sigma_\beta(r\cos\vartheta, r\sin\vartheta, \tilde{x}) = (r\cos(2\beta - \vartheta + \pi), r\sin(2\beta - \vartheta + \pi), \tilde{x})$$

$$= (r\cos(2\beta - \vartheta - \pi), r\sin(2\beta - \vartheta - \pi), \tilde{x}) \qquad (6.48)$$

since $2\beta - \vartheta + \pi = \vartheta + 2(\beta + \frac{\pi}{2} - \vartheta)$ and the angular variable is defined up to a multiple of 2π.

This can also be proved analytically writing the usual reflection in Cartesian coordinates and using simple trigonometric formulas. Let us set

$$h_\beta^{\pm}(\vartheta) = 2\beta - \vartheta \pm \pi \qquad (6.49)$$

Note that if we choose an interval $[\vartheta_0, \vartheta_0 + 2\pi)$ to which the angular coordinate ϑ belongs, the images $2\beta - \vartheta \pm \pi$ could not belong to the same interval. Nevertheless, we observe that for a fixed $\tilde{\beta} \in \mathbb{R}$, if we take $\vartheta \in [\tilde{\beta} - \frac{\pi}{2}, \tilde{\beta} + \frac{3}{2}\pi]$ then $h_{\tilde{\beta}}^+(\vartheta)$ belongs to the same interval, whereas if we take $\vartheta \in [\tilde{\beta} - \frac{3}{2}\pi, \tilde{\beta} + \frac{\pi}{2}]$ then $h_{\tilde{\beta}}^-(\vartheta)$ belongs to the same interval.

More precisely, we have

$$\vartheta \in \left[\tilde{\beta} - \frac{\pi}{2}, \tilde{\beta} + \frac{\pi}{2}\right] \Rightarrow 2\tilde{\beta} - \vartheta + \pi = h_{\tilde{\beta}}^+(\vartheta) \in \left[\tilde{\beta} + \frac{\pi}{2}, \tilde{\beta} + \frac{3}{2}\pi\right], \qquad (6.50)$$

$$\vartheta \in \left[\tilde{\beta} - \frac{3}{2}\pi, \tilde{\beta} - \frac{\pi}{2}\right] \Rightarrow 2\tilde{\beta} - \vartheta - \pi = h_{\tilde{\beta}}^-(\vartheta) \in \left[\tilde{\beta} - \frac{\pi}{2}, \tilde{\beta} + \frac{\pi}{2}\right]. \qquad (6.51)$$

This can be easily verified evaluating $h_{\tilde{\beta}}^{\pm}$ on the boundary of the intervals of definition, since the mappings $h_{\tilde{\beta}}^{\pm}$ are decreasing.

Proposition 6.8. *Let Ω be a rotationally symmetric domain in $\mathbb{R}^N$ centered at the origin, $\tilde{\beta} \in \mathbb{R}$, and assume that $u \in C^1(\overline{\Omega})$ satisfies:*
a)   *u is symmetric with respect to the hyperplane $H_{\tilde{\beta}}$.*
b)   *$u_\vartheta \geq 0$ in $\Omega_{\tilde{\beta}} = [x \cdot e_{\tilde{\beta}} > 0]$ where u_ϑ is as in (6.47)*

Then for any $\beta \in [\tilde{\beta} - \pi, \tilde{\beta}]$ and for any $x \in \Omega_\beta = [x \cdot e_\beta > 0]$ we have $u(x) \leq u(\sigma_\beta(x))$, while for every $\beta \in [\tilde{\beta}, \tilde{\beta} + \pi]$ we have $u(x) \geq u(\sigma_\beta(x))$ in Ω_β.

In other words, if a) and b) holds, then (6.34) holds for any direction e in the (x_1, x_2) plane that we consider.

To prove Proposition 6.8, we need the following simple lemma.

Lemma 6.9. *Suppose that the assumptions of Proposition 6.8 hold and let $\vartheta_0 \in (\tilde{\beta} - \frac{\pi}{2}, \tilde{\beta} + \frac{\pi}{2}]$. Then*

$$u(r, \vartheta', \tilde{x}) \geq u(r, \vartheta_0, \tilde{x}) \quad \forall \vartheta' \in [\vartheta_0, 2\tilde{\beta} - \vartheta_0 + \pi]. \tag{6.52}$$

Proof. Since u is symmetric, u_ϑ is antisymmetric with respect to $H_{\tilde{\beta}}$. By hypothesis $u_\vartheta \geq 0$ in $\Omega_{\tilde{\beta}} = [x \cdot e_{\tilde{\beta}} > 0] = \{(r, \vartheta, \tilde{x}) : \tilde{\beta} - \frac{\pi}{2} \leq \vartheta' \leq \tilde{\beta} + \frac{\pi}{2}\}$ so that $u_\vartheta(r, \vartheta', \tilde{x}) \leq 0$ if $\tilde{\beta} + \frac{\pi}{2} \leq \vartheta' \leq \tilde{\beta} + \frac{3}{2}\pi$. Moreover, by (6.50), if $\tilde{\beta} - \frac{\pi}{2} < \vartheta_0 < \tilde{\beta} + \frac{\pi}{2}$ then $h_{\tilde{\beta}}^+(\vartheta_0) = 2\tilde{\beta} - \vartheta_0 + \pi \in [\tilde{\beta} + \frac{\pi}{2}, \tilde{\beta} + \frac{3}{2}\pi]$. This means that $u(r, \cdot, \tilde{x})$ increases in $[\vartheta_0, \tilde{\beta} + \frac{\pi}{2}]$, then decreases in $[\tilde{\beta} + \frac{\pi}{2}, \tilde{\beta} + \frac{3}{2}\pi]$, in particular in $[\tilde{\beta} + \frac{\pi}{2}, 2\tilde{\beta} - \vartheta_0 + \pi]$. Since $u(r, \vartheta_0, \tilde{x}) = u(r, 2\tilde{\beta} - \vartheta_0 + \pi, \tilde{x})$, (6.52) follows. $\qquad\square$

Proof of Proposition 6.8. Let $\beta \in [\tilde{\beta} - \pi, \tilde{\beta}]$ and let $x \in \Omega_\beta = [x \cdot e_\beta > 0]$, equivalently $x = (r\cos\vartheta, r\sin\vartheta, \tilde{x})$ with $\beta - \frac{\pi}{2} < \vartheta < \beta + \frac{\pi}{2}$. We have to show that $u(x) \leq u^{\sigma_\beta}(x)$ if $\beta - \frac{\pi}{2} < \vartheta < \beta + \frac{\pi}{2}, \tilde{\beta} - \pi \leq \beta \leq \tilde{\beta}$.

Let us observe that, since u is symmetric with respect to $H_{\tilde{\beta}}$,

$$u^{\sigma_\beta}(x) = u(\sigma_{\tilde{\beta}}(\sigma_\beta(x))) = u(r\cos(\vartheta + 2(\tilde{\beta} - \beta)), r\sin(\vartheta + 2(\tilde{\beta} - \beta)), \tilde{x})$$

because $2\tilde{\beta} - (2\beta - \vartheta + \pi) + \pi = \vartheta + 2(\tilde{\beta} - \beta)$.

Let us first assume that $x \in \Omega_\beta \cap \Omega_{\tilde{\beta}}$, i.e. $x = (r\cos\vartheta, r\sin\vartheta, \tilde{x})$ with

$$\left(\beta - \frac{\pi}{2} \leq\right) \tilde{\beta} - \frac{\pi}{2} < \vartheta < \beta + \frac{\pi}{2} \left(\leq \tilde{\beta} + \frac{\pi}{2}\right).$$

Then we can apply Lemma 6.9 taking $\vartheta_0 = \vartheta$, $\vartheta' = \vartheta + 2(\tilde{\beta} - \beta)$, because we have that

$$\tilde{\beta} - \frac{\pi}{2} < \vartheta \leq \vartheta + 2(\tilde{\beta} - \beta) \leq 2\tilde{\beta} - \vartheta + \pi$$

Indeed $\tilde{\beta} - \beta \geq 0$, and the last equality is equivalent to $\vartheta < \beta + \frac{\pi}{2}$, which is true, since $x \in \Omega_\beta$.

So from Lemma 6.9 it follows that

$$u^{\sigma_\beta}(x) = u(r, \vartheta + 2(\tilde{\beta} - \beta), \tilde{x}) \geq u(r, \vartheta, \tilde{x}) = u(x).$$

Assume instead that $x = (r, \vartheta, \tilde{x}) \in \Omega_\beta \setminus \Omega_{\tilde{\beta}}$, i.e., $x = (r\cos\vartheta, r\sin\vartheta, \tilde{x})$ with

$$\left(\tilde{\beta} - \frac{3}{2}\pi \leq\right) \beta - \frac{\pi}{2} < \vartheta \leq \tilde{\beta} - \frac{\pi}{2} \left(\leq \beta + \frac{\pi}{2}\right).$$

Since $u \equiv u^{\sigma_{\tilde{\beta}}}$, we have that

$$u(x) = u(r, \vartheta, \tilde{x}) = u(r, 2\tilde{\beta} - \vartheta - \pi, \tilde{x})$$

while as before we get that

$$u^{\sigma_\beta}(x) = u(r, \vartheta + 2(\tilde{\beta} - \beta), \tilde{x})$$

Setting $\vartheta_0 = 2\tilde{\beta} - \vartheta - \pi$, $\vartheta' = \vartheta + 2(\tilde{\beta} - \beta)$, we have to prove the inequality

$$u(r, \vartheta', \tilde{x}) \geq u(r, \vartheta_0, \tilde{x}) = u(x)$$

This follows again by Lemma 6.9 provided we show that

$$\vartheta_0 := (2\tilde{\beta} - \vartheta - \pi) \in \left[\tilde{\beta} - \frac{\pi}{2}, \tilde{\beta} + \frac{\pi}{2}\right]$$

which follows from (6.51) (because $\tilde{\beta} - \frac{3}{2}\pi \leq \vartheta \leq \tilde{\beta} - \frac{\pi}{2}$), and

$$\vartheta_0 \leq \vartheta' \leq 2\tilde{\beta} - \vartheta_0 + \pi$$

i. e.

$$2\tilde{\beta} - \vartheta - \pi \leq \vartheta + 2(\tilde{\beta} - \beta) \leq 2\tilde{\beta} - \vartheta_0 + \pi = \vartheta + 2\pi$$

The inequality $\vartheta + 2(\tilde{\beta} - \beta) \leq \vartheta + 2\pi$ follows from the relation $\tilde{\beta} - \pi \leq \beta \leq \tilde{\beta}$, while $2\tilde{\beta} - \vartheta - \pi \leq \vartheta + 2(\tilde{\beta} - \beta)$ is equivalent to $\vartheta > \beta - \frac{\pi}{2}$, which is true since $x \in \Omega_\beta$. $\square$

A sufficient condition for the nonnegativity of the derivative u_ϑ in a symmetry position (which was one of the hypotheses of Proposition 6.8) is given in the following.

Proposition 6.10. *Let $\tilde{\beta} \in \mathbb{R}$ and assume that u is symmetric with respect to $H_{\tilde{\beta}}$. Assume further that there exists $\beta_1 < \tilde{\beta}$ such that for any $\beta \in (\beta_1, \tilde{\beta})$ we have $u \leq u^{\sigma_\beta}$ in Ω_β. Then $u_\vartheta = \frac{\partial u}{\partial \vartheta} \geq 0$ in $\Omega_{\tilde{\beta}}$.*

Proof. We can write the angular derivative as

$$u_\vartheta(r, \vartheta, \tilde{x}) = \lim_{\alpha \to 0^+} \frac{u(r, \vartheta + \alpha, \tilde{x}) - u(r, \vartheta, \tilde{x})}{\alpha}.$$

With the change of variable $\alpha = 2(\tilde{\beta} - \beta)$, $\beta = \tilde{\beta} - \frac{\alpha}{2}$, we have that $\beta \to \tilde{\beta}^-$. If α is small, then $\beta \in (\beta_1, \tilde{\beta})$, and, if $x \in \Omega_{\tilde{\beta}}$, then $x \in \Omega_\beta$ definitively for $\beta \to \tilde{\beta}^-$. Since, as observed, $u^{\sigma_\beta}(r, \vartheta, \tilde{x}) = u(r, \vartheta + 2(\tilde{\beta} - \beta), \tilde{x})$ we obtain

$$u_\vartheta(r, \vartheta, \tilde{x}) = \lim_{\beta \to \tilde{\beta}^-} \frac{u(r, \vartheta + 2(\tilde{\beta} - \beta), \tilde{x}) - u(r, \vartheta, \tilde{x})}{2(\tilde{\beta} - \beta)}$$

$$= \lim_{\beta \to \tilde{\beta}^-} \frac{u^{\sigma_\beta}(r, \vartheta, \tilde{x}) - u(r, \vartheta, \tilde{x})}{2(\tilde{\beta} - \beta)} \geq 0. \qquad \square$$

Other properties of the angular derivative are given in the following lemma.

Lemma 6.11. *Let Ω be a rotationally symmetric domain in $\mathbb{R}^N$, $N \geq 2$, $u \in C^2(\overline{\Omega})$ a solution of (6.33) and η_1, η_2 orthogonal directions. Let $u_\vartheta = u_{\vartheta(\eta_1, \eta_2)}$ be as in (6.47) the angular derivative with respect to cylindrical coordinates relative to this pair of directions. Then:*

(i) u_ϑ *weakly satisfies the Dirichlet problem*

$$\begin{cases} -\Delta u_\vartheta - V_u u_\vartheta = 0 & in\ \Omega \\ u_\vartheta = 0 & on\ \partial\Omega \end{cases} \tag{6.53}$$

where V_u is defined in (6.37).

(ii) *If $e \in span\,(\eta_1, \eta_2)$ is a direction of symmetry for the solution, i. e., $u \equiv u^{\sigma(e)}$ in Ω, then u_ϑ weakly satisfies in the half domain $\Omega(e)$ the Dirichlet problem*

$$\begin{cases} -\Delta u_\vartheta - V_u u_\vartheta = 0 & in\ \Omega(e) \\ u_\vartheta = 0 & on\ \partial\Omega(e) \end{cases} \tag{6.54}$$

Proof.

(i) Since the proof involves only two directions we assume for simplicity that $N = 2$. Writing the Laplacian in polar coordinate, we have that

$$\Delta u = u_{\rho\rho} + \rho^{-1} u_\rho + \rho^{-2} u_{\vartheta\vartheta}$$

and if the function u is sufficiently regular, say of class $C^3(\Omega)$, it is easy to check that the angular derivative satisfies the linearized equation $-\Delta u_\vartheta - V_u u_\vartheta = 0$ in Ω pointwisely. To see that if $u \in C^2(\overline{\Omega})$ then u weakly satisfies the equation let us consider the diffeomorphism $T = T(\rho, \vartheta) = (x(\rho, \vartheta), y(\rho, \vartheta))$ from $U = (0, +\infty) \times (0, 2\pi)$ and $V = \mathbb{R}^2 \setminus \{(x, y) \in \mathbb{R}^2 : y = 0, x \geq 0\}$ which defines the polar coordinates: $x = \rho \cos(\vartheta),\ y = \rho \sin(\vartheta)$.

Then the Jacobian matrix of T is given by

$$J_T(\rho, \vartheta) = \begin{pmatrix} x_\rho & x_\vartheta \\ y_\rho & y_\vartheta \end{pmatrix} = \begin{pmatrix} \cos(\vartheta) & -\rho \sin(\vartheta) \\ \sin(\vartheta) & \rho \cos(\vartheta) \end{pmatrix}$$

so that its inverse, as a function of (ρ, ϑ), is given by

$$J_{T^{-1}}(x(\rho, \vartheta),\, y(\rho, \vartheta)) = \begin{pmatrix} \rho_x & \rho_y \\ \vartheta_x & \vartheta_y \end{pmatrix} = \begin{pmatrix} \cos(\vartheta) & \sin(\vartheta) \\ -\rho^{-1} \sin(\vartheta) & \rho^{-1} \cos(\vartheta) \end{pmatrix}$$

Let $\varphi \in C_c^\infty(\Omega \setminus \{0\})$ be a test function, $\varphi_\vartheta = \frac{\partial\varphi}{\partial\vartheta}$ and assume that $u \in C^2(\overline{\Omega})$ is a solution of (6.33). Then we have

$$u_x = u_\rho \cos\vartheta - u_\vartheta \frac{\sin\vartheta}{\rho}, \quad u_y = u_\rho \sin\vartheta + u_\vartheta \frac{\cos\vartheta}{\rho},$$

$$(\varphi_\vartheta)_x = \varphi_{\rho\vartheta} \cos\vartheta - \varphi_{\vartheta\vartheta} \frac{\sin\vartheta}{\rho}, \quad (\varphi_\vartheta)_y = \varphi_{\rho\vartheta} \sin\vartheta + \varphi_{\vartheta\vartheta} \frac{\cos\vartheta}{\rho} \tag{6.55}$$

so that

$$\nabla u \cdot \nabla \varphi_\vartheta = \left(u_\rho \cos\vartheta - u_\vartheta \frac{\sin\vartheta}{\rho} \right) \left(\varphi_{\rho\vartheta} \cos\vartheta - \varphi_{\vartheta\vartheta} \frac{\sin\vartheta}{\rho} \right)$$

$$+ \left(u_\rho \sin\vartheta + u_\vartheta \frac{\cos\vartheta}{\rho} \right) \left(\varphi_{\rho\vartheta} \sin\vartheta + \varphi_{\vartheta\vartheta} \frac{\cos\vartheta}{\rho} \right) = u_\rho \varphi_{\rho\vartheta} + \frac{1}{\rho^2} u_\vartheta \varphi_{\vartheta\vartheta} \tag{6.56}$$

Exchanging the role of u and φ, we get

$$\nabla u \cdot \nabla \varphi_\vartheta = u_\rho \varphi_{\rho\vartheta} + \frac{1}{\rho^2} u_\vartheta \varphi_{\vartheta\vartheta}$$

$$\nabla u_\vartheta \cdot \nabla \varphi = u_{\rho\vartheta} \varphi_\rho + \frac{1}{\rho^2} u_{\vartheta\vartheta} \varphi_\vartheta \tag{6.57}$$

Taking $\psi = \varphi_\vartheta$ as a test function in (6.33) we have that

$$\int_\Omega \nabla u \cdot \nabla \varphi_\vartheta \, dx \, dy = \int_\Omega f(u)\varphi_\vartheta \, dx \, dy$$

Integrating in polar coordinates and using (6.57), we get

$$\int_\Omega \nabla u \cdot \nabla \varphi_\vartheta \, dx \, dy = \int_0^\infty \rho d\rho \int_0^{2\pi} d\vartheta \left(u_\rho \varphi_{\rho\vartheta} + \frac{1}{\rho^2} u_\vartheta \varphi_{\vartheta\vartheta} \right)$$

$$= - \int_0^\infty \rho d\rho \int_0^{2\pi} d\vartheta \left(u_{\rho\vartheta} \varphi_\rho + \frac{1}{\rho^2} u_{\vartheta\vartheta} \varphi_\vartheta \right) = - \int_\Omega \nabla u_\vartheta \cdot \nabla \varphi \, dx \, dy$$

and analogously $\int_\Omega f(u)\varphi_\vartheta \, dx \, dy = - \int_\Omega f'(|x|, u)u_\vartheta \varphi \, dx \, dy$ so that

$$\int_\Omega \nabla u_\vartheta \cdot \nabla \varphi \, dx \, dy = \int_\Omega f'(|x|, u)u_\vartheta \, \varphi \, dx \, dy \tag{6.58}$$

for any test function $\varphi \in C_c^\infty(\Omega \setminus \{0\})$.

By Proposition 1.46, we can extend (6.58) to all test functions $\varphi \in C_c^\infty(\Omega)$,[1] i. e., u_ϑ weakly solves $-\Delta u_\vartheta - V_u u_\vartheta = 0$ in Ω. Moreover, $u_\vartheta = 0$ on $\partial\Omega$.

ii) By the symmetry of u with respect to the hyperplane $H(e)$, we have that u_ϑ is antisymmetric with respect to $H(e)$ and, therefore, vanishes on $H(e)$. Since it vanishes on $\partial\Omega$ as well, it vanishes on $\partial\Omega(e)$, i. e., it satisfies the boundary condition in (6.54). □

Let us now prove some useful sufficient condition for the foliated Schwarz symmetry of a solution to the Dirichlet problem (6.33).

Theorem 6.12 (Sufficient conditions for FSS). *Let Ω be a bounded rotationally symmetric domain and $u \in C^2(\overline{\Omega})$ a solution of (6.33), where $f \in C^1([0, \infty) \times \mathbb{R})$. Then u is foliated Schwarz symmetric provided one of the following conditions holds:*

(i) *there exists a direction $e \in S^{N-1}$ such that $u \equiv u^{\sigma(e)}$ in Ω and $\lambda_1^e \geq 0$, with λ_1^e as in (6.41).*

(ii) *there exists a direction $e \in S^{N-1}$ such that $u < u^{\sigma(e)}$ or $u > u^{\sigma(e)}$ in $\Omega(e)$.*

---

[1] If $N \geq 3$, we start with test functions $\varphi \in C_c^\infty(\Omega \setminus F)$, with $F = \{x = (x_1, \ldots, x_n) \in \mathbb{R}^N : x_1 = x_2 = 0\}$, and then exploit Remark 1.49.

Proof. (i) For any direction η orthogonal to e the corresponding angular derivative $u_\vartheta = u_{\vartheta(e,\eta)}$ satisfies (6.54) in $\Omega(e)$.

Since $\lambda_1^e \geq 0$ either $\lambda_1^e > 0$ and then (by the maximum principle, which holds by Theorem 1.50) $u_\vartheta = 0$, or $\lambda_1^e = 0$ and then, if it does not vanish, u_ϑ is a first eigenfunction of the operator $-\Delta - V_u$ in $\Omega(e)$ with Dirichlet boundary condition and, therefore, it does not change sign. In any case by Proposition 6.8, we have that (6.34) holds for any direction in the plane spanned by e and η and since η is an arbitrary direction orthogonal to e we get that (6.34) holds for any direction and, by Proposition 6.7, u is foliated Schwarz symmetric.

(ii) If η is any direction orthogonal to e, by Theorem 6.3 we get that rotating the hyperplanes there exists a direction e' in the plane spanned by e and η such that $u \equiv u^{\sigma(e')}$ in $\Omega(e')$ and $u \leq u^{\sigma(e'')}$ for any direction e'' obtained while rotating e to reach e'. Therefore, the corresponding angular derivative (with respect to e and η) is nonnegative in $\Omega(e')$ by Proposition 6.10. Hence the assertion follows as in (i). $\qquad\square$

Remark 6.13.

1. In the proof of (ii) of Theorem 6.12, we can avoid using Proposition 6.10 by showing that $\lambda_1^{e'} = 0$ in the final direction of symmetry obtained by rotating the planes (see the proof of Theorem 6.12), and observing that the corresponding first eigenfunction u_ϑ does not change sign in $\Omega(e')$, so that the hypotheses of case (i) are satisfied.

   Indeed, using the notation of Theorem 6.3, we have that for every $\vartheta \in (\vartheta_0, \vartheta_1)$ the function $w^\vartheta = u - u^{\sigma(e_\vartheta)}$ is strictly negative in $\Omega(e_\vartheta)$ and, therefore, it is the first eigenfunction, with eigenvalue zero, of the operator $-\Delta - V_{e_\vartheta}$ in $\Omega(e_\vartheta)$. When $\vartheta \to \vartheta_1$, the potential $V_{e_\vartheta} = \frac{f(u)-f(u^{\sigma_{e_\vartheta}})}{u-u^{\sigma_{e_\vartheta}}}$ tends to the potential $V_u = f'(|x|, u)$, and, by continuity (see Theorem 1.44 v)), the first eigenvalue of $-\Delta - V_u$ in $\Omega(e_{\vartheta_1})$ is zero (with u_ϑ as a first eigenfunction).

2. Let u be a foliated Schwarz symmetric solution of (6.33) and e a direction such that $u \equiv u^{\sigma(e)}$ in Ω and $\lambda_1^e \geq 0$. As we have seen in the proof of Theorem 6.12 for any direction η orthogonal to e the corresponding angular derivative $u_\vartheta = u_{\vartheta(e,\eta)}$ (with respect to the plane spanned by e, η) either vanishes or is strictly positive (or negative) in $\Omega(e)$.

   In particular, if u is not radial and η is the direction of the vector p which appears in Definition 6.5 (i. e., the direction of the symmetry axis of u), then the derivative $u_{\vartheta(e,\eta)} = u_\vartheta^{e,\eta}$ is not zero, and it is strictly positive in $\Omega(e)$ since $u(rp) = \max_{S_r} u$ for $r > 0$ (see the proof of Proposition 6.7). Equivalently, interchanging the directions, we get that $u_{\vartheta(\eta,e)} < 0$ in $\Omega(e)$.

   Observe that if Π is the plane spanned by η and e then, for any $x \in \Pi \cap \Omega(e)$, the angular variable $\theta = \arccos(\frac{x}{|x|} \cdot p)$ which appears in Definition 6.5, coincides with the angular variable $\vartheta(\eta, e)$ in the plane Π. By the axial symmetry, the direction e can be substituted by any direction orthogonal to η, so that u is strictly decreasing with respect to the angular variable $\theta = \arccos(\frac{x}{|x|} \cdot p)$ if it is not radial.

6.3 Foliated Schwarz symmetry of low Morse index solutions of elliptic Dirichlet problems

In this section, we begin to study the relation between the Morse index of a solution to a semilinear elliptic problem and its symmetry.

6.3.1 Convex nonlinearities

We suppose that Ω is either a ball or an annulus centered at the origin in $\mathbb{R}^N$, $N \geq 2$, and $u \in C^2(\overline{\Omega})$ is a solution of the semilinear problem

$$\begin{cases} -\Delta u = f(|x|, u) & \text{in } \Omega \\ u = 0 & \text{on } \partial\Omega \end{cases} \tag{6.59}$$

where

$$f \in C^1((0, +\infty) \times \mathbb{R}) \tag{6.60}$$

We will use the definitions and notation (6.37)–(6.43).

The first result relating the Morse index of a solution to its FSS symmetry was proved in [186] and we discuss it below.

Theorem 6.14. *Suppose that the nonlinearity $f = f(|x|, s)$ in (6.59) is convex in the second variable. Then any solution $u \in C^2(\overline{\Omega})$ of (6.59) with Morse index $m(u) \leq 1$ is foliated Schwarz symmetric. Moreover, if u has Morse index zero, then it is radial.*

The proof is based on the following proposition, which relates the convexity of the function f to the Morse index of the solution in the caps $\Omega(e)$. It will be exploited again in the sequel.

Proposition 6.15. *Let the function $f = f(|x|, s)$ in (6.59) be convex in the second variable and let $u \in C^2(\overline{\Omega})$ be a solution of (6.59). If $e \in S^{N-1}$ is a direction such that $\lambda_1^e \geq 0$, then one of the following alternatives holds:*
1. $u \equiv u^{\sigma(e)}$ *in $\Omega(e)$*
2. $u < u^{\sigma(e)}$ *in $\Omega(e)$*
3. $u > u^{\sigma(e)}$ *in $\Omega(e)$*

If either f is strictly convex or if $\lambda_1^e > 0$ then (3) cannot hold.

Proof. Since f is convex, defining $v(x) = u^{\sigma(e)}(x)$ we have that

$$\text{if } u(x) > v(x) \text{ then } \frac{f(|x|, u(x)) - f(|x|, v(x))}{u(x) - v(x)} \leq f'(|x|, u(x)) \tag{6.61}$$

with strict sign if f is strictly convex. It follows that for any u, v

$$V_e\big((u-v)^+\big)^2 \le V_u\big((u-v)^+\big)^2 \tag{6.62}$$

with V_u and V_e as defined in (6.37) and (6.42) and the strict sign holds if f is strictly convex and $(u-v)^+$ does not vanish.

Testing the equation (6.43) with the function $w^+ = (w^e)^+ = (u - u^{\sigma(e)})^+$, we obtain

$$0 = \int_{\Omega(e)} \big(|\nabla w^+|^2 - V_e|w|^2\big)\,dx \ge \int_{\Omega(e)} \big(|\nabla w^+|^2 - V_u|w|^2\big)\,dx \tag{6.63}$$

Since $\lambda_1^e \ge 0$, we deduce that either $w^+ \equiv 0$ or w^+ is the first eigenfunction corresponding to the eigenvalue $0 = \lambda_1(-\Delta - V_u; \Omega(e))$.

Note that the second alternative cannot happen if f is strictly convex in the second variable or if $\lambda_1^e > 0$, in which case we deduce that $w^+ \equiv 0$.

In any case, if w^+ is positive in $\Omega(e)$, then $u > u^{\sigma(e)}$ in $\Omega(e)$, otherwise $w^+ \equiv 0$ so that $u \le u^{\sigma(e)}$ in $\Omega(e)$. In the latter case, since w satisfies (6.43), by the strong maximum principle either $u < u^{\sigma(e)}$ or $u \equiv u^{\sigma(e)}$ in $\Omega(e)$. $\qquad\square$

Corollary 6.16. *Under the hypotheses of Proposition 6.15 if, in addition, also $\lambda_1^{-e} \ge 0$, then $u \equiv u^{\sigma(e)}$ in $\Omega(e)$ provided one of the following holds:*
1. *$\lambda_1^e > 0, \lambda_1^{-e} > 0$*
2. *f is strictly convex in the second variable*

Remark 6.17. Proposition 6.15 and Corollary 6.16 hold more generally if Ω is a domain symmetric with respect to the direction e (see [186]).

Proof of Theorem 6.14. Let us assume that u is either positive or sign changing and let $x_0 \in \Omega$ be such that $u(x_0) = \max_\Omega u$ (if u is negative we take the minimum point and the same proof applies). Then we consider any direction e orthogonal to the direction p of the axis passing through the origin and the point x_0.

Since $m(u) \le 1$, by the variational characterization of the eigenvalues (Theorem 1.42), either $\lambda_1^e \ge 0$ or $\lambda_1^{-e} \ge 0$. Indeed, if this would not be the case, then the quadratic form $Q_u(v) = \int_\Omega (|\nabla v|^2 - V_u|v^2|)\,dx$ would be negative definite on the 2-dimensional space spanned by the trivial extensions of the eigenfunctions φ_1^e and φ_1^{-e}, and hence $m(u) \ge 2$. We can assume that $\lambda_1^e \ge 0$, otherwise we consider λ_1^{-e}. Since f is convex, by Proposition 6.15 either $u \equiv u^{\sigma(e)}$ or $u < u^{\sigma(e)}$ or $u > u^{\sigma(e)}$ in $\Omega(e)$. However, only the first possibility can hold, because if $w^e(x) = u(x) - u^{\sigma(e)}(x)$ had a strict sign in $\Omega(e)$, since it satisfies (6.43), by Hopf's lemma it would have a positive or negative normal derivative on the boundary. This is impossible, since $\nabla u(x_0) = 0$ because x_0 is the maximum of the function u. So $(u - u^{\sigma(e)}) \equiv 0$ in $\Omega(e)$ and $\lambda_1^e = \lambda_1^{-e} \ge 0$.

Since e is an arbitrary direction e orthogonal to p, we get that u is axially symmetric with respect to the axis passing through x_0 and by the sufficient condition given by Theorem 6.12 u is foliated Schwarz symmetric.

Finally, if the Morse index is zero then $\lambda_1(-\Delta - V_u; \Omega) \geq 0$, so that for any direction e we have that $\lambda_1^e = \lambda_1(-\Delta - V_u; \Omega(e)) > 0$. By Corollary 6.16, this implies that $w^e \equiv 0$ $\forall$ $e \in S^{N-1}$, i. e., u is radial. $\qquad\square$

Remark 6.18. If $m(u) = 0$, i.e. if u is a stable solution of (6.59), the radial symmetry actually follows as in Theorem 1.5. of [131], without requiring any convexity on f.

To generalize the result of Theorem 6.14 to higher Morse index solutions, we will use another, more indirect, method where the symmetry axis will be determined using also the rotating planes procedure. Moreover, we will exploit the well-known Borsuk–Ulam theorem. We recall here a version of it referring, e. g., to Corollary 4.2. in [107] for the proof.

Theorem 6.19 (Borsuk–Ulam theorem). *Let D be a bounded domain in $\mathbb{R}^N$ containing the origin and symmetric with respect to it, and let $g : \partial D \to \mathbb{R}^M$, with $M < N$, be a continuous function. Then $g(x) = g(-x)$ for some $x \in \partial D$.*

In particular, if $g : S^{N-1} \to \mathbb{R}^M$ is continuous and odd (i. e., $g(-x) = -g(x)$ for any $x \in S^{N-1}$) then it has a zero, i. e., there exists $e \in S^{N-1}$ such that $g(e) = 0 \in \mathbb{R}^M$.

Let us now prove the following generalization of Theorem 6.14 to the case of solutions with Morse index not exceeding the dimension N.

Theorem 6.20. *Let the function $f = f(|x|, s)$ in (6.59) be convex in the second variable. Then any solution $u \in C^2(\overline{\Omega})$ of (6.59) with Morse index $m(u) \leq N$ is foliated Schwarz symmetric.*

The proof will be based on the following.

Proposition 6.21. *Let u be a solution of problem (6.59) with Morse index $m(u) \leq N$. Then there exists a direction $e \in S^{N-1}$ such that $\lambda_1^e \geq 0$.*

Proof. The assertion is immediate if the Morse index of the solution satisfies $m(u) \leq 1$. Indeed, in this case for any direction e, at least one among λ_1^e and λ_1^{-e} must be non-negative, as observed at the beginning of the proof of Theorem 6.14. So let us assume that $2 \leq j = m(u) \leq N$.

Denote by Φ_k the $L^2(\Omega)$ normalized eigenfunctions of the operator $L_u = -\Delta - V_u$ in Ω, with Φ_1 positive in Ω, and for any direction $e \in S^{N-1}$ let us consider the function

$$\psi^e(x) = \begin{cases} \left(\dfrac{(\varphi_1^{-e}, \Phi_1)_{L^2(\Omega)}}{(\varphi_1^{e}, \Phi_1)_{L^2(\Omega)}}\right)^{\frac{1}{2}} \varphi_1^e(x) & \text{if } x \in \Omega(e) \\[2ex] -\left(\dfrac{(\varphi_1^{e}, \Phi_1)_{L^2(\Omega)}}{(\varphi_1^{-e}, \Phi_1)_{L^2(\Omega)}}\right)^{\frac{1}{2}} \varphi_1^{-e}(x) & \text{if } x \in \Omega(-e) \end{cases}$$

where φ_1^e is as in (6.41).

The mapping $e \mapsto \psi_e$ is odd and continuous from S^{N-1} to $H_0^1(\Omega)$ and, by construction,

$$(\psi^e, \Phi_1)_{L^2(\Omega)} = 0 \tag{6.64}$$

The function $h : S^{N-1} \to \mathbb{R}^{j-1}$ defined by

$$h(e) = ((\psi^e, \Phi_2)_{L^2(\Omega)}, \dots, (\psi^e, \Phi_j)_{L^2(\Omega)}) \tag{6.65}$$

is also odd and continuous. Since $(j-1) < N$, by the Borsuk–Ulam theorem it must have a zero. This means that there exists a direction $e \in S^{N-1}$ such that ψ^e is orthogonal to all the eigenfunctions $\Phi_1, \dots, \Phi_j$. Since $m(u) = j$, by Theorem 1.42 (ii) we deduce that $Q_u(\psi^e; \Omega) \geq 0$, which in turn implies that either $Q_u(\varphi_1^e; \Omega(e)) \geq 0$ or $Q_u(\varphi_1^{-e}; \Omega(-e)) \geq 0$, i. e. either λ_1^e or λ_1^{-e} is nonnegative, so the assertion is proved. $\qquad\square$

Proof of Theorem 6.20. Once we have a direction e such that $\lambda_1^e \geq 0$ we get from Proposition 6.15 that either $w^e > 0$, or $w^e < 0$ in $\Omega(e)$, or $w^e \equiv 0$ in $\Omega(e)$ (with $\lambda_1^e \geq 0$). Thus by the sufficient conditions given by Theorem 6.12 u is foliated Schwarz symmetric. $\qquad\square$

6.3.2 Nonlinearities with a convex derivative

The results in the previous section apply to several problems, in particular they apply to the model superlinear nonlinearity, $f(|x|, u) = g(|x|)u^p$, with $g(r) > 0$, $p > 1$, when the solutions are positive.

A corresponding model nonlinearity for sign changing solutions is the function $f(|x|, u) = g(|x|)|u|^{p-1}u$, $g > 0$, $p > 1$, which is not convex with respect to the second variable.

In this section, we study, following [190], another case, namely we suppose that the derivative $f'(|x|, s) = \frac{\partial f}{\partial s}(|x|, s)$ is convex.

In particular, if $f(|x|, s) = g(|x|)|s|^{p-1}s$ then the derivative is $f'(|x|, s) = g(|x|)p|s|^{p-1}$, which is convex when $p \geq 2$.

We assume, as in the previous section, that Ω is either a ball or an annulus centered at the origin in $\mathbb{R}^N$, $N \geq 2$, and $u \in C^2(\overline{\Omega})$ is a solution of the semilinear problem (6.59) where $f \in C^1((0, +\infty) \times \mathbb{R})$. We will use the definitions and notation of the previous section, in particular (6.37)–(6.43).

It is important now to introduce another potential for any direction $e \in S^{N-1}$, which is the e-symmetric part of the potential V_u:

$$V_{es}(x) = V_{(e,s)(u)}(x) = \frac{f'(|x|, u(x)) + f'(|x|, u(\sigma_e(x)))}{2} \tag{6.66}$$

and the corresponding quadratic forms

$$Q_{es}(v; \Omega) = \int_{\Omega} (|\nabla v|^2 - V_{es}|v|^2)\, dx, \quad v \in H_0^1(\Omega) \tag{6.67}$$

$$Q_{es}(v; \Omega(e)) = \int_{\Omega(e)} \left(|\nabla v|^2 - V_{es}|v|^2 \right) dx, \quad v \in H_0^1(\Omega(e)) \tag{6.68}$$

Let us denote by

$$\lambda_k(e, V_{es}) \tag{6.69}$$

the eigenvalues of the operator $-\Delta - V_{es}(x)$ in the half domain $B(e)$, with homogeneous Dirichlet boundary conditions.

Note that Q_{es} and Q_u coincide on functions which are either even or odd with respect to the reflection σ_e as it is easy to check with a change of variable in the integrals:

$$Q_{es}(v; \Omega) = Q_u(v; \Omega) \text{ if } v = \pm v^{\sigma(e)} \text{ in } \Omega \tag{6.70}$$

We will prove in this section the following result.

Theorem 6.22. *Assume that the nonlinearity $f = f(|x|, s)$ has a derivative $f'(|x|, s) = \frac{\partial f}{\partial s}(|x|, s)$ convex in the s-variable, for every $x \in \Omega$. Then every solution of (6.59) with Morse index $j \leq N$ is foliated Schwarz symmetric.*

Let us remark that in fact there are only two possibilities under the assumptions of Theorem 6.22: either the solution is radially symmetric or it is strictly monotone in the polar angle by Remark 6.13.

We also show an interesting consequence of Theorem 6.22 concerning the geometrical properties of the nodal set of a sign changing solution of (6.59) which relies on some results of [7].

Theorem 6.23. *Suppose that the nonlinearity f in (6.59) does not depend on x, i.e., $f = f(s)$ and its derivative f' is convex. Then the nodal set of any sign changing solutions of (6.59) with Morse index less than or equal to N intersects the boundary of Ω.*

Note that the assumptions of Theorem 6.23 are satisfied for the power type nonlinearity $f(s) = |s|^{p-1}s$, $p \geq 2$. In particular, under these assumptions, from Theorem 6.23 it follows that every sign changing solution u of (6.59) with Morse index less than or equal to N is nonradial. This was proved in [7] and we already discussed it in Chapter 3.

The proofs of Theorem 6.22 and Theorem 6.23 are postponed after the following result which gives a sufficient condition for the existence of a symmetry hyperplane for a solution of (6.59).

Proposition 6.24. *Assume that $f'(|x|, s)$ is convex in the s-variable. Let u be a solution of (6.59), such that there exists a direction $e \in S^{N-1}$ with $\lambda_1(e, V_{es}) \geq 0$. Then the following statements hold:*

(i) if either $\lambda_1(e, V_{es}) > 0$ or $f'(|x|, s)$ is strictly convex in s, then u is symmetric with respect to the hyperplane $H(e)$ and hence $\lambda_1(e, V_u) = \lambda_1(e, V_{es}) \geq 0$.

(ii) if $\lambda_1(e, V_{es}) = 0$, f' is only convex and u is not symmetric with respect to the hyperplane $H(e)$ then either $u < u^{\sigma(e)}$ or $u > u^{\sigma(e)}$.

Combining this proposition with Theorem 6.12, we immediately obtain the following.

Corollary 6.25. *Under the assumptions of Proposition 6.24, the solution u is foliated Schwarz symmetric.*

Proof of Proposition 6.24. Let us denote by w^e the difference between u and its reflection with respect to the hyperplane $H(e)$, i. e., $w^e(x) = u(x) - u(\sigma_e(x))$. It is easy to see that w^e solves the linear problem

$$\begin{cases} -\Delta w^e - V_e(x)w^e = 0 & \text{in } \Omega(e), \\ \qquad\qquad w^e = 0 & \text{on } \partial\Omega(e), \end{cases} \tag{6.71}$$

where

$$V_e(x) = \int_0^1 f'(|x|, tu(x) + (1-t)u(\sigma_e(x)))\, dt, \quad x \in \Omega(e). \tag{6.72}$$

Since f' is convex in the second variable, we have, for $x \in \Omega$,

$$V_e(x) \le \int_0^1 [tf'(|x|, u(x)) + (1-t)f'(|x|, u(\sigma_e(x)))]\, dt$$

$$= \frac{1}{2}[f'(|x|, u(x)) + f'(|x|, u(\sigma_e(x)))] = V_{es}(x) \tag{6.73}$$

Here, the strict inequality holds if f' is strictly convex and $u(x) \ne u(\sigma_e(x))$. Hence, denoting by $\lambda_k(e, V_e)$ the eigenvalues of the linear operator $-\Delta - V_e(x)$ in $\Omega(e)$ with homogeneous Dirichlet boundary conditions, we have, by (6.73) and Theorem 1.44(iv), that $\lambda_k(e, V_e) \ge \lambda_k(e, V_{es})$. In particular, $\lambda_1(e, V_e) \ge \lambda_1(e, V_{es}) \ge 0$ by hypothesis. If $\lambda_1(e, V_e) > 0$, then $w^e \equiv 0$ because it satisfies (6.71), and hence we get the symmetry of u with respect to the hyperplane $H(e)$. If $\lambda_1(e, V_e) = 0$, then $V_e \equiv V_{es}$ in Ω by the strict monotonicity of eigenvalues with respect to the potential, (Theorem 1.44(iv)). In the case when f' is strictly convex, this implies that $u(x) \equiv u(\sigma_e(x))$ for every $x \in \Omega$, so that again we get the symmetry of u with respect to the hyperplane $H(e)$.

It remains to consider the case when f' is only convex and $\lambda_1(e, V_e) = 0$. In this case, either $w^e \equiv 0$ or w^e does not change sign in $\Omega(e)$, and, by the strong maximum principle, $w^e > 0$ or $w^e < 0$ in $\Omega(e)$. $\qquad\square$

Before passing to the proof of Theorem 6.22, we recall that the quadratic forms Q_u and $Q_{e,s}$ are defined in (6.38) and (6.67), respectively.

Proof of Theorem 6.22. Let u be a solution of (6.59) with Morse index $m(u) = j \le N$. Thus, for the Dirichlet eigenvalues λ_k of the linearized operator $L = -\Delta - V_u(x)$ in Ω we have

$$\lambda_1 < 0, \dots, \lambda_j < 0 \quad \text{and} \quad \lambda_{j+1} \ge 0. \tag{6.74}$$

We distinguish two cases.

Case 1: $j \leq N - 1$

In this case, we obtain the foliated Schwarz symmetry of u, applying Corollary 6.25 in a simple way. Indeed, for any direction $e \in S$, let us denote by $g_e \in H_0^1(\Omega)$ the odd extension in Ω of the positive L^2-normalized eigenfunction of the operator $-\Delta - V_{es}(x)$ in the half domain $\Omega(e)$ corresponding to $\lambda_1(e, V_{es})$. It is easy to see that g_e depends continuously on e in the L^2-norm. Moreover, $g_{-e} = -g_e$ for every $e \in S$. Now we let $\phi_1, \phi_2, \ldots, \phi_j \in H_0^1(\Omega)$ denote L^2-orthonormal eigenfunctions of L_u corresponding to the eigenvalues $\lambda_1, \ldots, \lambda_j$. By Theorem 1.42, we have

$$\inf_{\substack{v \in H_0^1(\Omega) \setminus \{0\} \\ v \perp \phi_1, \ldots v \perp \phi_j}} \frac{Q_u(v)}{(v, v)_{L^2(\Omega)}} = \lambda_{j+1} \geq 0. \tag{6.75}$$

We consider the map

$$h : S = S^{N-1} \to \mathbb{R}^j, \quad h(e) = [(g_e, \phi_1)_{L^2(\Omega)}, \ldots, (g_e, \phi_j)_{L^2(\Omega)}]. \tag{6.76}$$

Since h is an odd and continuous map on the unit sphere $S \subset \mathbb{R}^N$ and $j \leq N - 1$, h must have a zero by the Borsuk–Ulam theorem. This means that there is a direction $e \in S$ such that g_e is L^2-orthogonal to all eigenfunctions $\phi_1, \ldots, \phi_j$. Thus $Q_u(g_e) \geq 0$ by (6.75). But, since g_e is an odd function, by (6.70)

$$Q_u(g_e) = Q_{es}(g_e) = 2\lambda_1(e, V_{es}),$$

which yields that $\lambda_1(e, V_{es}) \geq 0$. Having obtained a direction e for which $\lambda_1(e, V_{es})$ is nonnegative, Corollary 6.25 applies and we get the foliated Schwarz symmetry of u.

Case 2: $j = N$

The main difficulty in this case is that the map h, considered in (6.76), now goes from the $(N-1)$-dimensional sphere S into $\mathbb{R}^N$, so that the Borsuk–Ulam theorem does not apply. We therefore use a different and less direct argument to find a symmetry hyperplane $H(e')$ for u with $\lambda_1(e', V_u) \geq 0$.

Let $S_* \subset S$ be defined by

$$S_* = \{e \in S : w^e \equiv 0 \text{ in } \Omega \text{ and } \lambda_1(e, V_u) < 0\},$$

where, as before, w^e is given by $w^e(x) = u(x) - u(\sigma_e(x))$. Then S_* is a symmetric set. If $S_* \neq \varnothing$, we let k be the largest integer such that there exist k orthogonal directions $e_1, \ldots, e_k \in S_*$. We then observe that $k \in \{1, \ldots, N - 1\}$. Indeed, if we denote by V_0 the closed subspace of $L^2(\Omega)$ consisting of the functions which are symmetric with respect to the hyperplanes $H(e_1), \ldots, H(e_k)$, we have that $u \in V_0$ (by the very definition of S_*) and also $\phi_1 \in V_0$, where ϕ_1 denotes the unique positive L^2-normalized Dirichlet eigenfunction of the linearized operator $L_u = -\Delta - V_u(x)$ on Ω relative to the first eigenvalue

λ_1. Now we consider, for $l = 1, \ldots, k$, the functions $g_l \in H_0^1(\Omega)$ defined as the odd extensions in Ω of the unique positive L^2-normalized eigenfunction of the operator L_u in the half-domain $\Omega(e_l)$ corresponding to $\lambda_1(e_l, V_u) < 0$. Then, since u is symmetric with respect to $H(e_l)$, g_l is an eigenfunction of the operator $-\Delta - V_u(x)$ in the whole Ω corresponding to a certain eigenvalue $\lambda_m(= \lambda_1(e_l, V_u) < 0)$, $m \geq 2$. Since each g_l is odd with respect to the reflection at $H(e_l)$ and even with respect to the reflection at $H(e_m)$ for $m \neq l$, we have, letting l vary from 1 to k a set $\{g_1, \ldots, g_k\}$ of k orthogonal eigenfunctions of the operator $-\Delta - V_u(x)$, all corresponding to negative eigenvalues. Since the Morse index of u is equal to N and the first eigenfunction ϕ_1 is also orthogonal to each g_l we have that $k \leq N-1$ as we claimed. The same argument shows that if we denote by L_0 the self-adjoint operator which is the restriction of the operator L to the symmetric space V_0, i. e.,

$$L_0 : H_2(\Omega) \cap H_0^1(\Omega) \cap V_0 \to V_0, \quad L_0 v = -\Delta v - V_u(x)v$$

and we denote by m_0 the number of the negative eigenvalues of L_0 (counted with multiplicity), we have

$$1 \leq m_0 \leq N - k. \tag{6.77}$$

Note that the inequality on the left-hand side of (6.77) just follows from the fact that $\phi_1 \in V_0$ is an eigenfunction of L relative to a negative eigenvalue. Now we first assume that $S_* \neq \emptyset$ and consider the $(N - k - 1)$-dimensional sphere S^* defined by $S^* = S \cap H(e_1) \cap \cdots \cap H(e_k)$. By the maximality of k, we have $S^* \cap S_* = \emptyset$. We would like to prove that there exists a direction $e \in S^*$ such that

$$w^e(x) \geq 0 \quad \text{for every } x \in \Omega(e). \tag{6.78}$$

We argue by contradiction and suppose that w^e changes sign in $\Omega(e)$ for every $e \in S^*$. We consider the functions $w_1^e = (w^e)^+\chi_{\Omega(e)} + (w^e)^-\chi_{\Omega(-e)}$ and $w_2^e = (w^e)^-\chi_{\Omega(e)} + (w^e)^+\chi_{\Omega(-e)}$, where χ_A denotes the characteristic function of a set A. Since $u \in V_0$, we find that $w_1^e, w_2^e \in V_0 \cap H_0^1(\Omega) \setminus \{0\}$, and both functions are nonnegative and symmetric with respect to $H(e)$. Moreover,

$$w_1^{-e} = w_2^e \quad \text{and} \quad w_2^{-e} = w_1^e \quad \text{for every } e \in S^*. \tag{6.79}$$

Recalling the definition of V_e in (6.72), since w^e satisfies the linear equation $-\Delta w^e - V_e(x)w^e = 0$ in the whole domain Ω with $w^e = 0$ on $\partial\Omega$, multiplying by $(w^e)^+\chi_{B(e)} - (w^e)^-\chi_{\Omega(-e)}$ and integrating over Ω we obtain

$$0 = \int_\Omega \nabla w^e \cdot \nabla\left((w^e)^+\chi_{\Omega(e)} - (w^e)^-\chi_{\Omega(-e)}\right) dx$$

$$- \int_\Omega V_e(x)w^e\left((w^e)^+\chi_{\Omega(e)} - (w^e)^-\chi_{\Omega(-e)}\right) dx$$

$$= \int_{\Omega} \left(|\nabla((w^e)^+ \chi_{\Omega(e)})|^2 + |\nabla((w^e)^- \chi_{\Omega(-e)})|^2 \right) dx$$

$$- \int_{\Omega} V_e(x) [((w^e)^+ \chi_{\Omega(e)})^2 + ((w^e)^- \chi_{\Omega(e)})^2] \, dx$$

$$= \int_{\Omega} |\nabla w_1^e|^2 \, dx - \int_{\Omega} V_e(x)(w_1^e)^2 \, dx. \tag{6.80}$$

Now we can use the comparison between the potential $V_e(x)$ and $V_{es}(x)$, given by (6.73), and obtain from (6.80)

$$0 \geq \int_{\Omega} |\nabla w_1^e|^2 \, dx - \int_{\Omega} V_{es}(x)(w_1^e)^2 \, dx = Q_{es}(w_1^e) = Q_u(w_1^e), \tag{6.81}$$

because w_1^e is a symmetric function with respect to σ_e. Similarly, we can show

$$Q_u(w_2^e) \leq 0. \tag{6.82}$$

Now, for every $e \in S^*$, we let $\psi_e \in V_0 \cap H_0^1(\Omega)$ be defined by

$$\psi_e(x) = \left(\frac{(w_2^e, \phi_1)_{L^2(\Omega)}}{(w_1^e, \phi_1)_{L^2(\Omega)}} \right)^{1/2} w_1^e(x) - \left(\frac{(w_1^e, \phi_1)_{L^2(\Omega)}}{(w_2^e, \phi_1)_{L^2(\Omega)}} \right)^{1/2} w_2^e(x).$$

Using (6.79), it is easy to see that $e \mapsto \psi_e$ is an odd and continuous map from S^* to V_0. By construction, $\langle \psi_e, \phi_1 \rangle = 0$ for all $e \in S^*$. Moreover, since w_1^e and w_2^e have disjoint supports, (6.81) and (6.82) imply

$$Q_u(\psi_e) \leq 0 \quad \text{for all } e \in S^*. \tag{6.83}$$

Recalling (6.77), we first assume that $m_0 \geq 2$.

Let $\lambda_1, \tilde{\lambda}_2, \ldots, \tilde{\lambda}_{m_0}$ be the negative eigenvalues of the operator L_0 in increasing order, and let $\phi_1, \tilde{\phi}_2, \ldots, \tilde{\phi}_{m_0} \in V_0$ be corresponding L^2-orthonormal eigenfunctions. As in (6.75) we have

$$\inf_{\substack{v \in H_0^1(\Omega) \cap V_0, v \neq 0 \\ v \perp \phi_1, \ldots v \perp \tilde{\phi}_{m_0}}} \frac{Q_u(v)}{(v, v)_{L^2(\Omega)}} \geq 0. \tag{6.84}$$

We now consider the map $h : S^* \to \mathbb{R}^{m_0 - 1}$ defined by

$$h(e) = [(\psi_e, \tilde{\phi}_2)_{L^2(\Omega)}, \ldots, (\psi_e, \tilde{\phi}_{m_0})_{L^2(\Omega)}]$$

Since h is an odd and continuous map defined on a $(N - k - 1)$-dimensional sphere and $m_0 \leq N - k$, the map h must have a zero by the Borsuk–Ulam theorem. Hence there is $e \in S$ such that $(\psi_e, \tilde{\phi}_k)_{L^2(\Omega)} = 0$ for $k = 2, \ldots, m_0$ and, by construction, $(\psi_e, \phi_1)_{L^2(\Omega)} = 0$.

Since $\psi_e \in V_0$ and $Q_u(\psi_e) \leq 0$, the function ψ_e is a mimimizer for the quotient in (6.84). Consequently, it must be an eigenfunction of the operator L_0 corresponding to the eigenvalue zero. Hence $\psi_e \in C^2(\overline{\Omega})$, and ψ_e solves $-\Delta\psi_e - V_u(x)\psi_e = 0$ in Ω. Moreover, $\psi_e = 0$ on $H(e)$ by the definition of ψ_e, and $\partial_e\psi_e = 0$ on $H(e)$, since ψ_e is symmetric with respect to the reflection at $H(e)$. From this, it is easy to deduce that the function $\hat{\psi}_e$ defined by

$$\hat{\psi}_e(x) = \begin{cases} \psi_e(x), & x \in \Omega(e), \\ 0, & x \in \Omega(-e), \end{cases}$$

is also a (weak) solution of $-\Delta\hat{\psi}_e - V_u(x)\hat{\psi}_e = 0$. This however contradicts the unique continuation theorem for this equation (see, e. g., [200, p. 519]).

Now we assume that $m_0 = 1$.

Then, for any $e \in S^*$, since $\psi_e \in V_0$, $(\psi_e, \phi_1)_{L^2(\Omega)} = 0$ and $Q_u(\psi_e, \psi_e) \leq 0$, the function ψ_e must be an eigenfunction of the operator L_0 corresponding to the eigenvalue zero. This leads to a contradiction as in the previous case.

Since in both cases we reached a contradiction, there must be a direction $e \in S^*$ such that w^e does not change sign on $\Omega(e)$. Hence (6.78) holds either for e or for $-e$. Now, to end the proof, we again distinguish two cases. Suppose first that $w^e \equiv 0$ on Ω. Since $S^* \cap S_* = \varnothing$, we then have $\lambda_1(e, V_u) \geq 0$. If instead $w^e \not\equiv 0$, $w^e \not\equiv 0$, then $w^e > 0$ in $\Omega(e)$ by (6.78) and the strong maximum principle, since w^e solves (6.71).

Thus, by Theorem 6.12, u is foliated Schwarz symmetric.

Finally if $S_* = \varnothing$ we repeat all the steps starting from (6.78), taking $S^* = S$ and replacing $H_0^1(\Omega) \cap V_0$ by $H_0^1(\Omega)$. Again we reach a direction e' (possibly equal to e) such that $w_{e'} \equiv 0$, and thus $\lambda_1(e', V_u) \geq 0$ because $S_* = \varnothing$. Hence u must be foliated Schwarz symmetric. $\square$

Proof of Theorem 6.23. Let u be a solution with Morse index $m(u) \leq N$. In the proof of Theorem 6.22, we have obtained a symmetry hyperplane $H(e)$ for u such that $\lambda_1(e, V_u) \geq 0$. On the other hand, if the nodal set does not intersect the boundary, the proof of Theorem 3.19 shows that $\lambda_1(e, V_u) < 0$. Hence the assertion holds. $\square$

6.4 Symmetry of solutions of mixed boundary value problems

We study in this section elliptic problems with nonlinear mixed boundary conditions of the type

$$\begin{cases} -\Delta u = f(|x'|, x_N, u) & \text{in } \Omega \\ u = 0 & \text{on } \Gamma_0 \\ \dfrac{\partial u}{\partial v} = g(|x'|, u) & \text{on } \Gamma \end{cases} \tag{6.85}$$

where Ω is a cylindrically symmetric domain as in Definition 1.15 and its flat boundary Γ is defined in 1.33 (we refer to Chapter 1 for the corresponding notation, definitions and examples). We suppose that $f : [0,+\infty)\times(0,+\infty)\times\mathbb{R} \to \mathbb{R}$ and $g : [0,+\infty)\times\mathbb{R} \to \mathbb{R}$ are C^1 functions and a solution of (6.85) will be understood in a weak sense.

We identify the $(N-2)$-dimensional sphere in $\mathbb{R}^{N-1}$ in the following way:

$$S^{N-2} = \{v = (v_1,\ldots,v_N) \in S^{N-1} : v \cdot e_N = 0\} \tag{6.86}$$

where $e_N = (0,\ldots,1)$.

We recall that in Chapter 1 we discussed the maximum principles and the construction of the eigenvalues for mixed boundary value problems in analogy with the case of Dirichlet problems. This allows to follow quite closely what done in the previous sections. The main differences are that the set of directions that we consider are only those in S^{N-2}, and the linearized problem is a mixed boundary value problem.

In particular, the proof of Theorem 6.1 can be easily repeated to prove the cylindrical symmetry of *positive* solutions to (6.85) in cylindrical domains that are convex, namely half-balls or cylinders.

Theorem 6.26. *Let Ω be a half-ball or a cylinder and let $u \in H^1_0(\Omega \cup \Gamma) \cap C^0(\overline{\Omega})$ be a* positive *weak solution of* (6.85), *where $f = f(|x'|,x_N,s) : [0,+\infty) \times \mathbb{R} \times \mathbb{R} \to \mathbb{R}$, $g = g(|x'|,s) : [0,+\infty) \times \mathbb{R} \to \mathbb{R}$ are continuous functions that are C^1 with respect to s and nonincreasing w. r. t. the first variable.*

Then u has cylindrical symmetry, i. e., if $x = (x',x_N)$, $x' \in \mathbb{R}^{N-1}$, then $u(x',x_N) = v(|x'|,x_N)$ for some function $v = v(r,x_N)$. Moreover, $v_r(r,x_N) < 0$.

Proof. We have to prove the symmetry with respect to every hyperplane orthogonal to any direction in $\mathbf{S}^{N-2}$ (see (6.86)). The proof is exactly the same as the proof of Theorem 6.1 using Theorem 1.26 instead of Theorem 1.21, together with the strong comparison principle (Theorem 1.31), as in Theorem 6.1. $\qquad\square$

If $e \in S^{N-2}$, we define as before

$$H(e) = \{x \in \mathbb{R}^N : x \cdot e = 0\} \quad \text{and} \quad \Omega(e) = \{x \in \Omega : x \cdot e > 0\}$$

with the corresponding boundaries

$$\Gamma_0(e) = (\Gamma_0 \cap \overline{\Omega(e)}) \cup (H(e) \cap \Omega(e))$$
$$\Gamma(e) = \Gamma \cap (\overline{\Omega(e)} \setminus H(e)) \tag{6.87}$$

The rotating planes technique (see Theorem 6.3) works exactly as in the Dirichlet case, and we get the following result.

Theorem 6.27 (Rotating planes for cylindrically symmetric domains). *Let Ω be a bounded cylindrically symmetric domain in $\mathbb{R}^N$, $N \geq 3$ and $u \in H^1_0(\Omega \cup \Gamma) \cap C^0(\overline{\Omega})$ a weak solution of* (6.85). *Suppose that $e = e_{\vartheta_0} = (\cos(\vartheta_0),\sin(\vartheta_0),0,\ldots,0) \in S^{N-2}$ such* *that $u < u^{\sigma(e_{\vartheta_0})}$ in $\Omega(e_{\vartheta_0})$. Then there exists a direction $e_{\vartheta_1} = (\cos(\vartheta_1), \sin(\vartheta_1), 0, \dots, 0)$ with $\vartheta_1 > \vartheta_0$ such that*

$$u \equiv u^{\sigma(e_{\vartheta_1})} \text{ in } \Omega(e_{\vartheta_1}) \quad \text{and} \quad u < u^{\sigma(e_\vartheta)} \text{ in } \Omega(e_\vartheta) \text{ for any } \vartheta \in (\vartheta_0, \vartheta_1)$$

We now discuss a variant of the foliated Schwarz symmetry for solutions of (6.85). We will call it sectional foliated Schwarz symmetry. Since it is meaningful for $N \geq 3$, we will not consider the case $N = 2$.

Definition 6.28. Let Ω be a bounded domain with cylindrical symmetry in $\mathbb{R}^N$, $N \geq 3$, and let $u : \Omega \to \mathbb{R}$ be a continuous function. We say that u is *sectionally foliated Schwarz symmetric* if there exists a vector $p' = (p_1, \dots, p_{N-1}, 0) \in \mathbb{R}^N$, $|p'| = 1$, such that $u(x) = u(x', x_N)$ depends only on x_N, $r = |x'|$ and (if $x' \neq 0$) on $\vartheta = \arccos(\frac{x'}{|x'|} \cdot p')$, and u is nonincreasing in ϑ.

Referring to Definition 1.15, we have that the sectional foliated Schwarz symmetry of a function u just means that the h-sections of u, i. e., the functions $x' \mapsto u(x', h)$ are either

-   radial for any $h \in (0, b)$

    or

-   nonradial but foliated Schwarz symmetric for any $h \in (0, b)$ in the corresponding domain $\Omega^h = \Omega \cap \{x_N = h\}$, with the same axis of symmetry.

Let us consider the potentials

$$V_u(x) = f'(|x|, x_N, u(x)) \in C^0(\overline{\Omega}), \quad W_u(x') = g'(|x'|, u(x)) \in C^0(\overline{\Gamma}) \tag{6.88}$$

(where, as before, $f'(|x|, x_N, s)$ is the derivative of the nonlinearity $f = f(|x'|, x_N, s)$ with respect to the third variable, and analogously $g'(|x'|, s)$, is the derivative of the function $g(|x'|, s)$ appearing in the nonlinear boundary conditions, with respect to the second variable). Then we consider the corresponding quadratic form

$$Q_u(v; \Omega) = \int_\Omega (|\nabla v|^2 - V_u |v|^2)\, dx - \int_\Gamma W_u |v|^2\, dx', \quad v \in H_0^1(\Omega \cup \Gamma) \tag{6.89}$$

Let us denote by λ_k, Φ_k the eigenvalues and eigenfunctions of the problem

$$\begin{cases} -\Delta\Phi - V_u\Phi = \lambda\Phi & \text{in } \Omega \\ \Phi = 0 & \text{on } \Gamma_0 \\ \dfrac{\partial\Phi}{\partial v} - W_u\Phi = \lambda\Phi & \text{on } \Gamma \end{cases} \tag{6.90}$$

If $e \in S^{N-2}$ is a direction orthogonal to e_N, we consider also the corresponding quadratic form

$$Q_u(v; \Omega(e)) = \int_{\Omega(e)} (|\nabla v|^2 - V_u |v|^2)\, dx - \int_\Gamma W_u |v|^2\, dx', \quad v \in H_0^1(\Omega(e) \cup \Gamma(e)) \tag{6.91}$$

while we denote by $\lambda_k^e = \lambda_k(e; V_u; W_u)$ and $\varphi_k^e = \varphi_k(e; V_u; W_u)$ the eigenvalues and eigenfunctions of the problem

$$\begin{cases} -\Delta\varphi - V_u\varphi = \lambda\varphi & \text{in } \Omega(e) \\ \varphi = 0 & \text{on } (\Gamma_0(e)) \\ \dfrac{\partial\varphi}{\partial v} - W_u\varphi = \lambda\varphi & \text{on } \Gamma(e) \end{cases} \tag{6.92}$$

where $\Gamma_0(e)$ and $\Gamma(e)$ are defined in (6.87).

We also consider, for any direction $e \in S^{N-2}$, the potentials

$$V_e(x) = \int_0^1 f'(|x'|, x_N, tu(x) + (1-t)u^{\sigma(e)}(x))\, dt$$

$$= \begin{cases} \dfrac{f(|x'|, x_N, u(x)) - f(|x'|, x_N, u^{\sigma(e)}(x))}{u(x) - u^{\sigma(e)}(x)} & \text{if } u(x) \neq u^{\sigma(e)}(x) \\ f'(|x'|, x_N, u(x)) & \text{if } u(x) = u^{\sigma(e)}(x) \end{cases}, \quad x \in \Omega(e) \tag{6.93}$$

$$W_e(x) = \int_0^1 g'(|x'|, tu(x) + (1-t)u^{\sigma(e)}(x))\, dt$$

$$= \begin{cases} \dfrac{g(|x'|, u(x')) - g(|x'|, u^{\sigma(e)}(x'))}{u(x) - u^{\sigma(e)}(x')} & \text{if } u(x') \neq u^{\sigma(e)}(x') \\ g'(|x|, u(x)) & \text{if } u(x') = u^{\sigma(e)}(x') \end{cases}, \quad x' \in \Gamma(e) \tag{6.94}$$

Note that the difference $w^e(x) = u(x) - u^{\sigma(e)}(x)$ satisfies in $\Omega(e)$ the problem

$$\begin{cases} -\Delta w - V_e w = 0 & \text{in } \Omega(e) \\ w = 0 & \text{on } \Gamma_0(e) \\ \dfrac{\partial w}{\partial v} - W_e w = 0 & \text{on } \Gamma(e) \end{cases} \tag{6.95}$$

If e, η are orthogonal directions and we define the angular derivative $u_\vartheta = u_{\vartheta(e,\eta)} = \frac{\partial u}{\partial\vartheta}$ with respect to cylindrical coordinates in the plane (e, η), as in the previous sections, it is easy to see that u_ϑ weakly satisfies the mixed boundary value problem

$$\begin{cases} -\Delta u_\vartheta - V_u u_\vartheta = 0 & \text{in } \Omega \\ u_\vartheta = 0 & \text{on } \partial\Gamma_0 \\ \dfrac{\partial u_\vartheta}{\partial v} - W_u u_\vartheta = 0 & \text{on } \Gamma \end{cases} \tag{6.96}$$

Moreover, if $e \in S^{N-2}$ is a direction of symmetry for the solution, i. e., $u \equiv u^e$ in Ω, then u_ϑ weakly satisfies the analogous problem in the half-domain $\Omega(e)$, namely

$$\begin{cases} -\Delta u_\vartheta - V_u u_\vartheta = 0 & \text{in } \Omega(e) \\ u_\vartheta = 0 & \text{on } \Gamma_0(e) \\ \dfrac{\partial u_\vartheta}{\partial v} - W_u u_\vartheta = 0 & \text{on } \Gamma(e) \end{cases} \tag{6.97}$$

Proceeding exactly as in the previous sections, we get the following.

Proposition 6.29. *Let Ω be a cylindrically symmetric domain and $u \in H_0^1(\Omega \cup \Gamma) \cap C^0(\overline{\Omega})$ a weak solution of (6.85). Assume that for every unit vector $e \in S^{N-2}$ we have*

$$\text{either } u(x) \geq u(\sigma_e(x)) \; \forall x \in B(e) \quad \text{or} \quad u(x) \leq u(\sigma_e(x)) \; \forall x \in B(e) \tag{6.98}$$

Then u is sectionally foliated Schwarz symmetric.

Proof. Let $p' \in S^{N-2}$ be such that for some $r, t > 0$, $S_r^t = \{x = (x', x_N) \in \mathbb{R}^N : |x'| = r, x_N = t\} \subset \Omega$ and $u(rp', t) = \max_{S_r} u$.

We define $T_{p'}^+ = \{e \in S^{N-2} : e \cdot p' > 0\}$ and $T_p^- = \{e \in S^{N-1} : e \cdot p' < 0\}$.

As in the case of rotationally symmetric domains to prove that u is foliated Schwarz symmetric with respect to p', it is enough to show that

$$u(x) \geq u(\sigma_e(x)) \quad \text{for all } x \in \Omega(e) \tag{6.99}$$

whenever $e \in T_p^+$, and this is proved exactly as in Proposition 6.7. $\square$

Arguing as in the previous sections, we get the following.

Theorem 6.30 (Sufficient conditions for sectional FSS). *Let Ω be a bounded rotationally symmetric domain and let $u \in C^2(\overline{\Omega})$ be a solution of (6.85), where f and g are C^1 functions. Then u is sectionally foliated Schwarz symmetric provided one of the following conditions holds:*
(i)   there exists a direction $e \in S^{N-2}$ such that $\lambda_1(e; V_u; W_u) \geq 0$ and $u \equiv u^{\sigma(e)}$ in Ω
(ii) there exists a direction $e \in S^{N-2}$ such that $u < u^{\sigma(e)}$ or $u > u^{\sigma(e)}$ in $\Omega(e)$.

Using the previous theorem, we can extend the results of Section 6.3 to the case of elliptic problems in cylindrically symmetric domains. We limit ourselves to the case of convex nonlinearities, i. e., we prove the following result, analogous to Theorem 6.20.

Theorem 6.31. *Suppose that Ω is a cylindrically symmetric domain in $\mathbb{R}^N$, $N \geq 3$ and $f(|x|, x_N, s), g(x', s)$ are C^1 functions which are convex in the last variable.*

Then any solution $u \in C^2(\overline{\Omega})$ of (6.85) with Morse index $m(u) \leq N - 1$ is sectionally foliated Schwarz symmetric.

As in the case of a rotationally symmetric domain, we first prove the following preliminary result.

Lemma 6.32. *Let Ω be a cylindrically symmetric domain in $\mathbb{R}^N$, $N \geq 3$, and let u be a solution of problem (6.85) with Morse index $m(u) \leq N - 1$. Then there exists a direction $e \in S^{N-2}$ such that the first eigenvalue $\lambda_1^e \geq 0$.*

Proof. The proof is immediate if $m(u) \leq 1$ since in this case, for any direction $e \in S^{N-2}$ at least one among λ_1^e and λ_1^{-e} must be nonnegative, otherwise taking the corresponding first eigenfunctions we would obtain a 2-dimensional subspace of $H_0^1(\Omega)$ where the quadratic form Q_u^e is negative definite. So let us assume that $2 \le j = m(u) \le N - 1$.

Recall that we denote by Φ_k the eigenfunctions of the problem

$$\begin{cases} -\Delta\varphi - f'(|x'|, x_N, u)\varphi = \lambda\varphi & \text{in } \Omega \\ \varphi = 0 & \text{on } \Gamma_0 \\ \dfrac{\partial\varphi}{\partial v} - g'(|x'|, u)\varphi = \lambda\varphi & \text{on } \Gamma \end{cases} \tag{6.100}$$

and in particular we consider the first *positive* eigenfunction Φ_1. Let us denote, as in Chapter 1, $\mathbf{V} = L^2(\Omega) \times L^2(\Gamma)$ and for any direction $e \in S^{N-2}$ let us consider the function

$$\psi^e(x) = \begin{cases} \left(\dfrac{(\varphi_1^{-e}, \Phi_1)_{\mathbf{V}}}{(\varphi_1^e, \Phi_1)_{\mathbf{V}}}\right)^{\frac{1}{2}} \varphi_1^e(x) & \text{if } x \in \Omega(e) \\ -\left(\dfrac{(\varphi_1^e, \Phi_1)_{\mathbf{V}}}{(\varphi_1^{-e}, \Phi_1)_{\mathbf{V}}}\right)^{\frac{1}{2}} \varphi_1^{-e}(x) & \text{if } x \in \Omega(-e) \end{cases}$$

(in the scalar product in the space $\mathbf{V}$ we consider the trivial extension to Ω of the eigenfunctions $\varphi^e := \varphi_1^e$).

The mapping $e \mapsto \psi_e$ is a continuous odd function from S^{N-2} to $H_0^1(\Omega \cup \Gamma)$ and, by construction,

$$(\psi^e, \Phi_1)_{\mathbf{V}} = 0$$

The function $h : S^{N-2} \to \mathbb{R}^{j-1}$ defined by

$$h(e) = ((\psi^e, \Phi_2)_V, \ldots, (\psi^e, \Phi_j)V)$$

is an odd continuous mapping, and since $j - 1 < N - 1$, by the Borsuk–Ulam theorem it must have a zero. This means that there exists a direction $e \in S^{N-2}$ such that ψ^e is orthogonal to all the eigenfunctions $\Phi_1, \ldots, \Phi_j$. Since $m(u) = j$ this implies that $Q_u(\psi^e; \Omega) = B_u(\psi^e, \psi^e) \ge 0$, which in turn implies that either $Q_u(\varphi^e; \Omega(e)) \ge 0$ or $Q_u(\varphi^{-e}; \Omega(-e)) \ge 0$, i. e., either λ_1^e or λ_1^{-e} is nonnegative, so the assertion is proved. $\square$

Proof of Theorem 6.31. Once, by Lemma 6.32, we have a direction $e \in S^{N-2}$ such that $\lambda_1^e \ge 0$ the proof is similar to the proofs of Theorem 6.14 and Theorem 6.20. Indeed let $e \in S^{N-2}$ be such that $\lambda_1^e \ge 0$. If $v(x) = u^{\sigma(e)}(x)$, by the convexity of the maps $u \mapsto f(|x'|, x_N, u)$, $u \mapsto g(|x'|, u)$ we have that $f(|x|, u(x)) - f(|x|, v(x)) \le f'(|x|, u(x))(u - v)$ and $g(|x'|, u(x')) - g(|x'|, v(x')) \le g'(|x|, u(x'))(u - v)$ for any x such that $u(x) \ge v(x)$, so that

$$V_e((u - v)^+)^2 \le V_u((u - v)^+)^2, \quad W_e((u - v)^+)^2 \le W_u((u - v)^+)^2 \tag{6.101}$$

(with strict inequality if f or g is strictly convex in the last variable and $u(x) \ne v(x)$).

Testing (6.95) with the test function $w^+ = (w^e)^+ = (u - u^{\sigma(e)})^+$ we obtain that

$$0 = \int_{\Omega(e)} (|\nabla w^+|^2 - V_e|w^+|^2) - \int_\Gamma W_e|w^+|^2 \ge \int_{\Omega(e)} (|\nabla w^+|^2 - V_u|w^+|^2) - \int_\Gamma W_u|w^+|^2 \tag{6.102}$$

(with strict inequality if f or g is strictly convex in the last variable and $w^+ \not\equiv 0$). Since $\lambda_1^e \geq 0$, we deduce that either $w^+ \equiv 0$ or w is the first positive eigenfunction with zero eigenvalue of the associated mixed boundary value problem (6.92) (and this cannot happen if f is strictly convex in the last variable).

This implies in turn that either $w^e > 0$ in $\Omega(e)$ or $w^e \leq 0$ in $\Omega(e)$, so that, by the strong maximum principle, either $w^e > 0$ in $\Omega(e)$ or $w^e \equiv 0$ in $\Omega(e)$, or $w^e < 0$ in $\Omega(e)$.

By the sufficient condition given by Theorem 6.30, u is sectionally foliated Schwarz symmetric. $\qquad\square$

The same proof works if $\Gamma = \emptyset$, and gives an analogous result for the Dirichlet problem in cylindrically symmetric domains, i. e., the following result holds.

Theorem 6.33. *Let Ω be a cylindrically symmetric bounded domain in $\mathbb{R}^N$, $N \geq 3$ and let $u \in H_0^1(\Omega) \cap C^1(\overline{\Omega})$ be a weak solution of the problem*

$$\begin{cases} -\Delta u = f(|x|', x_N, u) & in\ \Omega \\ u = 0 & on\ \partial\Omega \end{cases} \qquad (6.103)$$

where $f \in C^1(\overline{\Omega} \times \mathbb{R})$. Assume that f is convex in the third variable and that u has Morse index $m(u) \leq N - 1$. Then u is sectionally foliated Schwarz symmetric.

7 Morse index and symmetry for elliptic systems in bounded domains

In this chapter, we show symmetry results for solutions of elliptic systems obtained by using Morse index bounds. As for the case of scalar equations, described in Chapter 6, we get the foliated Schwarz symmetry of solutions in bounded rotationally symmetric domains, by exploiting convexity properties of the nonlinear terms.

The results we present are mostly taken from [77] and [72]. However, we simplify some proofs of both papers. Let us remark that a crucial hypothesis for the symmetry results presented in this chapter is the cooperativity of the system (see Section 7.1). When this hypothesis fails, e. g., for competing systems, the same symmetry results are not expected. We refer to [206] and [215] for different symmetry properties holding in the noncooperative case.

The domains considered in the next sections are either balls or annuli. Symmetry results for more general symmetric domains have been obtained in [79] assuming different bounds on the Morse index.

7.1 Morse index of solutions of elliptic systems

Let us consider semilinear elliptic systems of the type

$$\begin{cases} -\Delta U = F(x, U) & \text{in } \Omega \\ U = 0 & \text{on } \partial\Omega \end{cases} \tag{7.1}$$

i. e.,

$$\begin{cases} -\Delta u_1 = f_1(x, u_1, \ldots, u_m) & \text{in } \Omega \\ \quad \cdots & \quad \cdots \\ -\Delta u_m = f_m(x, u_1, \ldots, u_m) & \text{in } \Omega \\ u_1 = \cdots = u_m = 0 & \text{on } \partial\Omega \end{cases}$$

where $U = (u_1, \ldots, u_m) : \Omega \to \mathbb{R}^m$, Ω is a bounded domain in $\mathbb{R}^N$ and $F(x, s_1, \ldots, s_m) = (f_1(x, S), \ldots, f_m(x, S)) \in C^1(\overline{\Omega} \times \mathbb{R}^m; \mathbb{R}^m)$, $S = (s_1, \ldots, s_m)$. By solution of (7.1), we mean a weak solution as in (1.111), and inequalities will be understood in a weak sense, as in (1.112), (1.113). We refer to Section 1.5 in Chapter 1 for all the notation about systems.

Let us recall the definitions of cooperativeness and full coupling.

The system (7.1) is said to be

1. **cooperative or weakly coupled** in an open set $\Omega' \subseteq \Omega$ if

$$\frac{\partial f_i}{\partial s_j}(x, s_1, \ldots, s_m) \geq 0 \quad \text{for every } (x, s_1, \ldots, s_m) \in \Omega' \times \mathbb{R}^m \tag{7.2}$$

and every $i, j = 1, \ldots, m$ with $i \neq j$.

https://doi.org/10.1515/9783110538243-007

2. **fully coupled** in an open set $\Omega' \subseteq \Omega$ along $U \in \mathbf{H}_0^1(\Omega) \cap C^0(\Omega; \mathbb{R}^m)$ if it is cooperative in Ω' and in addition $\forall I, J \subset \{1, \ldots, m\}$ such that $I \neq \emptyset, J \neq \emptyset, I \cap J = \emptyset, I \cup J = \{1, \ldots, m\}$ there exist $i_0 \in I, j_0 \in J$ such that

$$\mathrm{meas}\left(\left\{x \in \Omega' : \frac{\partial f_{i_0}}{\partial s_{j_0}}(x, U(x)) > 0\right\}\right) > 0 \tag{7.3}$$

For what concerns semilinear systems of elliptic equations, the definition of the Morse index of a solution is analogous to the one given for equations, using the corresponding quadratic form.

Definition 7.1.

(i) Let $U \in \mathbf{H}_0^1(\Omega) \cap \mathbf{L}^\infty(\Omega)$ be a weak solution of (7.1). We say that U is linearized stable (or has zero Morse index) if the quadratic form

$$Q_U(\Psi; \Omega) = \int_\Omega \left[|\nabla \Psi|^2 - J_F(x, U(x))(\Psi, \Psi) \right] dx$$

$$= \int_\Omega \left[\sum_{i=1}^m |\nabla \psi_i|^2 - \sum_{i,j=1}^m \frac{\partial f_i}{\partial s_j}(x, U(x)) \psi_i \psi_j \right] dx \geq 0 \tag{7.4}$$

for any $\Psi = (\psi_1, \ldots, \psi_m) \in C_c^1(\Omega; \mathbb{R}^m)$ where $J_F(x, U(x))$ is the Jacobian matrix of $F(x, S)$ with respect to the variables $S = (s_1, \ldots, s_m)$ computed at $S = U(x)$.

(ii) U has (linearized) Morse index equal to the integer $m = m(U) \geq 1$ if m is the maximal dimension of a subspace of $C_c^1(\Omega; \mathbb{R}^m)$ where the quadratic form is negative definite.

(iii) U has infinite (linearized) Morse index if for any integer k there exists a k-dimensional subspace of $C_c^1(\Omega; \mathbb{R}^m)$ where the quadratic form is negative definite.

The crucial, simple remark that will allow us to extend some of the symmetry results known for equations to the case of systems, is that, as observed in Section 1.5, the quadratic form associated to the linearized operator at a solution U, i. e. to the linear operator

$$L_U(V) = -\Delta V - J_F(x, U)V \tag{7.5}$$

which in general is not selfadjoint, coincides with the quadratic form corresponding to the selfadjoint operator

$$L_U^s(V) = -\Delta V - \frac{1}{2}(J_F(x, U) + J_F^t(x, U))V \tag{7.6}$$

where J_F^t is the transpose of the matrix J_F.

Therefore the *symmetric eigenvalues* of L as defined in Chapter 1, i. e. the eigenvalues of L_U^s, can be exploited to study the symmetry of the solution U, using the information on its Morse index.

However, though essential in our proofs, the only consideration of these eigenvalues is not sufficient to reach the assertions of our theorems, but we also need to use the principal eigenvalue of the linearized operator (see Definition 1.63) as stressed in Remark 7.9. This can be understood by the fact that the positivity of this eigenvalue is a necessary and sufficient condition for the (weak) maximum principle to hold (see Theorem 1.64) and the maximum principle is another key-ingredient in our proofs.

By the previous remark if $\lambda_k^s = \lambda_k(-\Delta + C; \Omega)$ and W^k, $k \in \mathbb{N}^+$, denote the *symmetric* eigenvalues and eigenfunctions of L_U, i. e., W^k satisfy

$$\begin{cases} -\Delta W^k + C W^k = \lambda_k^s W^k & \text{in } \Omega \\ W^k = 0 & \text{on } \partial\Omega, \end{cases} \tag{7.7}$$

where

$$C = c_{ij}(x), \quad c_{ij}(x) = -\frac{1}{2}\left[\frac{\partial f_i}{\partial s_j}(x, U(x)) + \frac{\partial f_j}{\partial s_i}(x, U(x)) \right] \tag{7.8}$$

as in the scalar case we can prove the following.

Theorem 7.2. *Let Ω be a bounded domain in $\mathbb{R}^N$. Then the Morse index of a solution U to (7.1) equals the number of negative symmetric eigenvalues of the linearized operator L_U.*

Proof. It is the same as the proof of Theorem 3.2, using Theorem 1.56(iii) and Theorem 1.57(i). $\qquad\square$

7.2 Symmetry results

7.2.1 Moving and rotating planes

From now on, we will consider the case of system (7.1) in bounded rotationally symmetric domains, with radial dependence on x:

$$\begin{cases} -\Delta U = F(|x|, U) & \text{in } \Omega \\ U = 0 & \text{on } \partial\Omega \end{cases} \tag{7.9}$$

where $F(r, S) = (f_1(r, S), \ldots, f_m(r, S))$ is a function belonging to $C^1([0, +\infty) \times \mathbb{R}^m; \mathbb{R}^m)$ and Ω is a bounded rotationally symmetric domain in $\mathbb{R}^N$, $m, N \geq 2$.

If Ω is a ball and U is a *positive* solution of (7.9), the radial symmetry of U holds under some mild regularity and coupling hypotheses.

Theorem 7.3. *Let $\Omega = B_R$ be a ball centered at the origin and suppose that $U \in H_0^1(\Omega) \cap C^0(\overline{\Omega}; \mathbb{R}^m)$ is a positive weak solution of (7.9), where $F = F(r, S) = (f_1(r, S), \ldots, f_m(r, S)) \in$

$C^1([0, +\infty) \times \mathbb{R}^m; \mathbb{R}^m)$ *is nonincreasing in* $r \in [0, +\infty)$ *for any* $S \in \mathbb{R}^m$. *Assume further that the system* (7.9) *is cooperative.*

Then U *is radial, i. e.,* $U(x) = V(|x|)$, *where* $V = V(r) : [0, R] \to \mathbb{R}^m$, *and* $V_r(r) < 0$.

Proof. The proof is exactly the same as that of Theorem 6.1, using the corresponding comparison principles for systems, namely Theorems 1.66 and 1.67.

Let us only remark that the strong comparison principle holds for fully coupled systems, while we assume here only the cooperativeness of the system. The reason is that we actually use the statement 1. of Theorem 1.67 (which only requires the weak coupling of the system) which asserts that for every component u_i of the vector solution U either $(u_i) < (u_i)_\lambda$ in Ω_λ or $(u_i) \equiv (u_i)_\lambda$ in Ω_λ. Then the first alternative necessarily holds for *every* component because if $\lambda < 0$ then, for *every* component u_i, there are points on the boundary of the ball where the strict inequality $u_i < (u_i)_\lambda$ holds, because they are reflected inside the ball, where all components u_i are positive by hypothesis.[1] $\qquad\square$

Let us remark that the previous extension to systems of the analogous Gidas–Ni–Nirenberg's theorem 6.1 has been proved by many authors (see [53, 94, 95, 153, 219] and the references therein) with different methods.

As observed in Section 6.1, the proof used in Theorem 6.1 is based on the comparison principle in small domains. This makes the proof flexible and easily adaptable to different problems, as in Theorem 7.3. Moreover, as in the case of equations, this technique can be used in the analogous theorem that uses the rotating plane method (see Theorem 7.4 below).

Let us recall some notation from Chapter 6. For a unit vector $e \in S^{N-1}$, we consider the hyperplane $H(e) = \{x \in \mathbb{R}^N : x \cdot e = 0\}$ orthogonal to the direction e and the open half-domain $\Omega(e) = \{x \in \Omega : x \cdot e > 0\}$. We then set $\sigma_e(x) = x - 2(x \cdot e)e$, $x \in \Omega$, i. e., $\sigma_e : \Omega \to \Omega$ is the *reflection with respect to the hyperplane* $H(e)$. Finally, if $U : \Omega \to \mathbb{R}^m$ is a continuous function we define the *reflected function* $U^{\sigma(e)} : \Omega \to \mathbb{R}^m$ defined by $U^{\sigma(e)}(x) = U(\sigma_e(x))$.

The following result can be proved exactly as Theorem 6.3 (using the comparison principles for systems given by Theorem 1.66 and Theorem 1.67 instead of the corresponding comparison principles for scalar equations).

Theorem 7.4 (Rotating planes method for systems). *Let* Ω *be a bounded rotationally symmetric domain,* $F \in C^1([0, \infty) \times \mathbb{R}^m; \mathbb{R}^m)$ *and* $U \in H_0^1(\Omega) \cap C^0(\overline{\Omega}; \mathbb{R}^m)$ *a weak solution of* (7.9). *Assume that the system* (7.9) *is fully coupled along* U *in* Ω *and there exists a direction* $\mathbf{e}_{\vartheta_0} = (\cos(\vartheta_0), \sin(\vartheta_0), 0, \ldots, 0)$ *such that*

$$U < U^{\sigma(\mathbf{e}_{\vartheta_0})} \quad \text{in } \Omega(\mathbf{e}_{\vartheta_0})$$

---

[1] Note, however, that if the solution is only assumed to be nonnegative then we can deduce that it is actually positive only for fully coupled systems.

Then there exists a direction $\mathbf{e}_{\vartheta_1} = (\cos(\vartheta_1), \sin(\vartheta_1), 0, \ldots, 0)$, *with* $\vartheta_1 > \vartheta_0$, *such that*

$$U \equiv U^{\sigma(e_{\vartheta_1})} \quad in \; \Omega(e_{\vartheta_1})$$

and

$$U < U^{\sigma(e_\vartheta)} \quad in \; \Omega(e_\vartheta) \; \forall \vartheta \in (\vartheta_0, \vartheta_1)$$

7.2.2 Foliated Schwarz symmetry

Let us now give the definition of foliated Schwarz symmetry for vector valued functions.

Definition 7.5. Let Ω be a rotationally symmetric domain in $\mathbb{R}^N$, $N \geq 2$. We say that a continuous vector valued function $U = (u_1, \ldots, u_m) : \Omega \to \mathbb{R}^m$ is foliated Schwarz symmetric if each component u_i is foliated Schwarz symmetric with respect to the same vector $p \in \mathbb{R}^N$. In other words, there exists a vector $p \in \mathbb{R}^N$, $|p| = 1$, such that $U(x)$ depends only on $r = |x|$ and (if $x \neq 0$) on $\theta = \arccos(\frac{x}{|x|} \cdot p)$ and U is (componentwise) nonincreasing in θ.

Remark 7.6. Let us observe that if U is a solution of (7.9) and the system satisfies some coupling conditions, as required in Theorem 7.12, then the foliated Schwarz symmetry of U implies that either U is radial or it is strictly decreasing in the angular variable θ as in the scalar case (see Remark 6.13).

The sufficient condition for the foliated Schwarz symmetry given by Proposition 6.7 can be readily extended.

Proposition 7.7. *Let* Ω *be a rotationally symmetric domain and* $U \in H_0^1(\Omega) \cap C^0(\overline{\Omega})$ *a weak solution of (7.9) where* $F = F(r, S) \in C^1([0, \infty) \times \mathbb{R}^m; \mathbb{R}^m)$. *Assume that the system is fully coupled along* U *in* Ω *and that for every unit vector* $e \in S^{N-1}$:

$$either \quad U(x) \geq U(\sigma_e(x)) \; \forall x \in \Omega(e) \quad or \quad U(x) \leq U(\sigma_e(x)) \; \forall x \in \Omega(e) \tag{7.10}$$

Then U *is foliated Schwarz symmetric.*

Proof. By Proposition 6.7, each component u_i of the solution U is foliated Schwarz symmetric with respect to a vector $p_i \in \mathbb{R}^N$, $i = 1, \ldots, m$, so to prove that the solution U is foliated Schwarz symmetric we only need to prove that the vectors p_i are all the same, for $i = 1, \ldots, m$.

To this aim, consider the vector p_1 and take any hyperplane $H(e) = \{x \in \mathbb{R}^N : x \cdot e = 0\}$ passing through the axis with direction p_1. Note that also the function $U^{\sigma(e)}$ satisfies the system (7.9), hence, since $u_1 \equiv u_1^{\sigma(e)}$ in $\Omega(e)$, by the strong comparison principle (Theorem 1.67) we have that $u_j \equiv u_j^{\sigma(e)}$ in $\Omega(e)$ for all $j = 1, \ldots, m$, i.e., all vectors p_j coincide with p_1. $\qquad \square$

As in the scalar case, we consider a pair of orthogonal directions η_1, η_2 and the polar coordinates (ρ, ϑ) in the plane spanned by them. Then we define for $U \in C^2(\overline{\Omega}; \mathbb{R}^m)$ the angular derivative

$$U_\vartheta = U_{\vartheta(\eta_1,\eta_2)} \tag{7.11}$$

which solves the linearized system

$$\begin{cases} -\Delta U_\vartheta - J_F(|x|, U)U_\vartheta = 0 & \text{in } \Omega \\ U_\vartheta = 0 & \text{on } \partial\Omega \end{cases} \tag{7.12}$$

and, if $e \in \text{span} \, (\eta_1, \eta_2)$ and $U \equiv U^{\sigma(e)}$ in $\Omega(e)$, also the system

$$\begin{cases} -\Delta U_\vartheta - J_F(|x|, U)U_\vartheta = 0 & \text{in } \Omega(e) \\ U_\vartheta = 0 & \text{on } \partial\Omega(e) \end{cases} \tag{7.13}$$

Then Proposition 6.8 and Proposition 6.10 extend immediately to the vectorial case and we will not repeat them.

Using the properties of the principal eigenvalue and of the corresponding eigenfunction, we deduce, as in the scalar case, the following sufficient conditions for the foliated Schwarz symmetry.

Theorem 7.8 (Sufficient conditions for FSS-Systems). *Let Ω be a bounded rotationally symmetric domain and $U \in C^2(\overline{\Omega}; \mathbb{R}^m)$ a solution of (7.9), where $F \in C^1([0, R] \times \mathbb{R}^m; \mathbb{R}^m)$. Then U is foliated Schwarz symmetric provided one of the following conditions holds:*
(i) *there exists a direction $e \in S^{N-1}$ such that $U \equiv U^{\sigma(e)}$ in $\Omega(e)$ and the principal eigenvalue $\tilde{\lambda}_1(\Omega(e))$ of the linearized operator $L_U = -\Delta - J_F(x, U)$ in $\Omega(e)$ is nonnegative;*
(ii) *there exists a direction $e \in S^{N-1}$ such that either $U < U^{\sigma(e)}$ or $U > U^{\sigma(e)}$ in $\Omega(e)$.*

Proof. If η is a direction orthogonal to e, $U_\vartheta = U_{\vartheta(\eta_1,\eta_2)}$ satisfies (7.13) as remarked. Thus if $\tilde{\lambda}_1(\Omega(e)) > 0$, since the maximum principle holds (see Theorem 1.64), we get that $U_\vartheta \equiv 0$ in $\Omega(e)$. If instead $\tilde{\lambda}_1(\Omega(e)) = 0$ and $U_\vartheta \neq 0$, by the simplicity of the principal eigenvalue we get that U_ϑ is a principal eigenfunction and hence it is positive (or negative) in $\Omega(e)$. In any case, U_ϑ does not change sign in $\Omega(e)$, and, arguing exactly as in the scalar case, we get that for any direction e' either $U \geq U^{\sigma(e')}$ or $U \leq U^{\sigma(e')}$. Thus, by Proposition 7.7, U is foliated Schwarz symmetric and (i) is proved.

To prove (ii), we proceed as in the scalar case (see Theorem 6.12 and Remark 6.13) using Theorem 7.4. $\qquad\square$

Remark 7.9. Let us observe that in Theorem 7.8 it is the nonnegativity of the *principal* eigenvalue the crucial hypothesis, while the information we get in the next section using the techniques introduced in Chapter 6 will concern the *symmetric* eigenvalues of the linearized system. Therefore, in the proofs that follow there will be an interplay and a comparison between the principal eigenvalue and the first symmetric eigenvalue in the cap $\Omega(e)$.

7.2.3 Nonlinearities having convex components

If U is a solution of (7.9) and the system is fully coupled along U in Ω, then the difference $W = W^e = U - U^{\sigma(e)} = (w_1, \ldots, w_m)$, as in the scalar case, satisfies a linear system in Ω, which is fully coupled in Ω and $\Omega(e)$.

Lemma 7.10.

(i) *Assume that $U \in C^1(\overline{\Omega}; \mathbb{R}^m)$ is a solution of (7.9) and that the system is fully coupled along U in Ω. Let us define for any direction $e \in S^{N-1}$ the matrix $B^e(x) = (b_{ij}^e(x))_{i,j=1}^m$, where*

$$b_{ij}^e(x) = -\int_0^1 \frac{\partial f_i}{\partial s_j}(|x|, tU(x) + (1-t)U^{\sigma(e)}(x))\, dt \tag{7.14}$$

Then for any $e \in S^{N-1}$ the function $W^e = U - U^{\sigma(e)}$ satisfies in $\Omega(e)$ the linear system

$$\begin{cases} -\Delta W^e + B^e(x)W^e = 0 & \text{in } \Omega(e) \\ W^e = 0 & \text{on } \partial\Omega(e) \end{cases} \tag{7.15}$$

which is fully coupled in $\Omega(e)$.

(ii) *If $\Psi = (\psi_1, \ldots, \psi_m) \in H_0^1(\Omega(e))$, let $Q^e(\Psi; \Omega(e))$ denote the quadratic form associated to the system (7.15) in $\Omega(e)$, i.e.,*

$$Q^e(\Psi; \Omega(e)) = \int_{\Omega(e)} (|\nabla\Psi|^2 + B^e(\Psi, \Psi))\, dx$$

$$= \int_{\Omega(e)} \left(\sum_{i=1}^m |\nabla\psi_i|^2 + \sum_{i,j=1}^m b_{ij}^e \psi_i \psi_j \right) dx \tag{7.16}$$

Then

$$Q^e(W^e; \Omega(e)) = \int_{\Omega(e)} [|\nabla(W^e)|^2 + B^e(W^e, W^e)]\, dx = 0 \tag{7.17}$$

while for the positive and negative parts of W^e the following holds:

$$Q^e((W^e)^\pm; \Omega(e)) = \int_{\Omega(e)} [|\nabla(W^e)^\pm|^2 + B^e((W^e)^\pm, (W^e)^\pm)]\, dx \le 0 \tag{7.18}$$

Proof. From the equation $-\Delta U = F(|x|, U(x))$ we deduce that the reflected function $U^{\sigma(e)}$ satisfies the equation $-\Delta U^{\sigma(e)} = F(|x|, U^{\sigma(e)}(x))$, and hence the difference $W^e = U - U^{\sigma(e)} = (w_1, \ldots, w_m)$ satisfies the equation

$$-\Delta W^e = F(|x|, U) - F(|x|, U^{\sigma(e)})$$

Let us set $V = U^{\sigma(e)}$. For any $i = 1, \ldots, m$, we have that

$$f_i(|x|, U(x)) - f_i(|x|, V(x)) = \sum_{j=1}^{m} \int_0^1 \frac{\partial f_i}{\partial s_j}(|x|, tU(x) + (1-t)V(x))(u_j(x) - v_j(x))\, dt$$

As a consequence, W^e satisfies (7.15). Moreover, if $i \neq j$ then $b_{ij}^e(x) \leq 0$ by (7.2), so that the linear system (7.15) is weakly coupled.

If $U \in C^1(\overline{\Omega}; \mathbb{R}^m)$ is a solution of (7.9) and the system is fully coupled along U, then the linear system associated to the matrix B^e is fully coupled in Ω. Indeed, if $i_0 \neq j_0$ and $\frac{\partial f_{i_0}}{\partial s_{j_0}}(x, U(x)) > 0$ then, since $\frac{\partial f_i(y)}{\partial s_j} \geq 0$ for every $y \in \Omega$, we get that $b_{ij}(x) = -\int_0^1 \frac{\partial f_i}{\partial s_j}[|x|, tU(x) + (1-t)V(x)]\, dt < 0$.

Since B^e is symmetric with respect to the reflection σ_e, (7.15) is fully coupled in $\Omega(e)$ as well and (i) is proved.

To get (7.17), it is enough to multiply the ith equation of the system for w_i and integrate. Instead, multiplying the ith equation of (7.15) for w_i^+, we get

$$0 = \int_{\Omega(e)} \left(|\nabla w_i^+|^2 + \sum_{j=1}^{m} b_{ij}^e w_j w_i^+ \right) dx \geq \int_{\Omega(e)} \left(|\nabla w_i^+|^2 + \sum_{j=1}^{m} b_{ij}^e w_j^+ w_i^+ \right) dx$$

since $w_i w_i^+ = |w_i^+|^2$, while $w_j w_i^+ \leq w_j^+ w_i^+$ and $b_{ij} \leq 0$ if $i \neq j$.

Summing on i, we get

$$0 \geq \int_{\Omega(e)} \sum_{i=1}^{m} |\nabla w_i^+|^2 + \sum_{i,j=1}^{m} b_{ij}^e w_j^+ w_i^+\, dx$$

i. e., (7.18) in the case of the positive part.

For the negative part, we proceed analogously multiplying the ith equation of (7.15) for w_i^- and integrating. We get

$$0 = -\int_{\Omega(e)} |\nabla w_i^-|^2 + \sum_{j=1}^{m} b_{ij}^e w_j w_i^-\, dx \leq -\int_{\Omega(e)} |\nabla w_i^-|^2 + \sum_{j=1}^{m} b_{ij}^e(-w_j^-)w_i^-\, dx$$

$$= -\int_{\Omega(e)} |\nabla w_i^-|^2 - \sum_{j=1}^{m} b_{ij}^e(w_j^-)w_i^-\, dx$$

since $w_i w_i^- = -|w_i^-|^2$, while $w_j w_i^- \geq -(w_j^-)w_i^-$ and $b_{ij} \leq 0$ if $i \neq j$.

Summing on i, we obtain

$$0 \geq \int_{\Omega(e)} \sum_{i=1}^{m} |\nabla w_i^-|^2 + \sum_{i,j=1}^{m} b_{ij}^e w_j^- w_i^-\, dx$$

i. e., (7.18) in the case of the negative part. $\qquad\square$

Remark 7.11.

1. It is standard to consider the linear system (7.15) to deduce comparison principles from maximum principles. Indeed we already considered a similar system in Chapter 1 in the proofs of the comparison principles for systems in Theorem 1.66 and Theorem 1.67, where elliptic inequalities were considered.

2. Note that the inequalities in (7.18) could be strict. Indeed the products $w_i^+ w_j^-$ could be not identically zero if $i \neq j$ and, therefore, $Q(W^e)$ does not coincide in general with $Q((W^e)^+) + Q((W^e)^-)$, as it happens in the scalar case.

Our first result is the counterpart of Theorem 6.20 for systems.

Theorem 7.12. *Let Ω be a ball or an annulus in $\mathbb{R}^N$, $N \geq 2$, and let $U \in C^2(\overline{\Omega}; \mathbb{R}^m)$ be a solution of (7.9) with Morse index $m(U) \leq N$. Moreover, assume that:*
(i) the system is fully coupled along U in Ω
(ii) for any $i = 1, \ldots, m$ the scalar function $f_i(|x|, S)$ is convex in the variable $S = (s_1, \ldots, s_m) \in \mathbb{R}^m$.

Then U is foliated Schwarz symmetric and if U is not radial then it is strictly decreasing in the angular variable.

We observe that, with respect to the corresponding result in [77], in Theorem 7.12 only the convexity of the functions f_i, $i = 1, \ldots, m$, is required, as in [79], where more general domains are considered.

Remark 7.13. Requiring that the system is fully coupled along a solution U in Ω is an assumption easily satisfied by most of the systems encountered in applications (see Section 7.3). It is essentially needed to ensure the validity of the strong maximum and comparison principles.

Obviously, the previous theorem holds in particular for stable solutions. However, in this case, as for scalar equations, it is not difficult to see that the solution is radial, even without the assumption (ii). Therefore, we have the following

Remark 7.14. If the system (7.9) is fully coupled along a stable solution U in Ω, then U is radial.

We will deduce from the proof of Theorem 7.12 that for nonradial Morse index one solutions the following condition holds.

Theorem 7.15. *Under the assumptions of Theorem 7.12 if a solution U has Morse index one and is not radial then necessarily*

$$\sum_{j=1}^m \frac{\partial f_i}{\partial s_j}(r, U(r, \vartheta)) \frac{\partial u_j}{\partial \vartheta}(r, \vartheta) = \sum_{j=1}^m \frac{\partial f_j}{\partial s_i}(r, U(r, \vartheta)) \frac{\partial u_j}{\partial \vartheta}(r, \vartheta) \qquad (7.19)$$

for any $i = 1, \ldots, m$, with (r, ϑ) as in Definition 7.5.

In particular, if $m = 2$ then (7.19) implies that

$$\frac{\partial f_1}{\partial s_2}(|x|, U(x)) = \frac{\partial f_2}{\partial s_1}(|x|, U(x)), \quad \forall x \in \Omega \tag{7.20}$$

Remark 7.16. The result of Theorem 7.15 is somewhat surprising because it asserts that under the hypotheses of Theorem 7.12, for any nonradial Morse index one solution another coupling condition, namely (7.19) (and in particular (7.20) if $m = 2$), must hold along the solution.

We will prove Theorem 7.12 by several auxiliary results.

Lemma 7.17. *Assume that U is a solution of (7.9) and that the hypotheses (i)–(ii) of Theorem 7.12 hold. Then for any direction $e \in S^{N-1}$*

$$Q_U((W^e)^+; \Omega(e)) \le 0$$

where Q_U is the quadratic form defined in (7.4) and W^e is as in Lemma 7.10.

Proof. For any $i = 1, \dots, m$ we have

$$-\Delta w_i = f_i(|x|, U) - f_i(|x|, U^{\sigma(e)}) \quad \text{in } \Omega(e)$$

Testing the equation with w_i^+, we obtain

$$\int_{\Omega(e)} |\nabla(w_i)^+|^2 \, dx = \int_{\Omega(e)} (f_i(|x|, U) - f_i(|x|, U^{\sigma(e)})) w_i^+ \, dx \tag{7.21}$$

Observe that $f_i(|x|, S)$ is convex in S, so that

$$(f_i(|x|, U(x)) - f_i(|x|, U^{\sigma(e)}(x))) w_i^+ \le (\nabla f_i(|x|, U(x)) \cdot (U(x) - U^{\sigma(e)}(x))) w_i^+$$

$$= (\nabla f_i(|x|, U(x)) \cdot W^e) w_i^+ = \sum_{j=1}^{m} \frac{\partial f_i}{\partial u_j}(|x|, U(x)) w_j w_i^+$$

where ∇ stands for the gradient of f_i with respect to the variables $S = (s_1, \dots, s_m)$. Moreover,

$$\frac{\partial f_i}{\partial s_i} w_i w_i^+ = \frac{\partial f_i}{\partial s_i} |w_i^+|^2, \quad \text{while} \quad \frac{\partial f_i}{\partial s_j} w_j w_i^+ \le \frac{\partial f_i}{\partial s_j} w_j^+ w_i^+ \quad \text{if } i \ne j$$

because $\frac{\partial f_i}{\partial s_j} \ge 0$ by the weak coupling assumption.

By (7.21), taking into account the previous inequalities, we get

$$\int_{\Omega(e)} |\nabla(w_i)^+|^2 \, dx \le \int_{\Omega(e)} \sum_{j=1}^{m} \frac{\partial f_i}{\partial s_j}(|x|, U(x)) w_j^+ w_i^+ \, dx$$

Thus, summing on $i = 1, \dots, m$, we obtain

$$\int_{\Omega(e)} \left(\sum_{i=1}^{m} |\nabla(w_i)^+|^2 - \sum_{i,j=1}^{m} \frac{\partial f_i}{\partial s_j}(|x|, U(x)) w_i^+ w_j^+ \right) dx \leq 0 \tag{7.22}$$

i. e., $Q_U((W^e)^+; \Omega(e)) \leq 0$. $\qquad\square$

Lemma 7.18. *Suppose that U is a solution of (7.9) with Morse index $m(U) \leq N$ and assume that the hypothesis (i) of Theorem 7.12 holds. Then there exists a direction $e \in S^{N-1}$ such that*

$$Q_U(\Psi; \Omega(e)) \geq 0$$

for any $\Psi \in C_c^1(\Omega(e); \mathbb{R}^m)$.

Equivalently, the first symmetric eigenvalue $\lambda_1^s(L_U, \Omega(e))$ of the linearized operator $L_U = -\Delta - J_F(x, U)$ in $\Omega(e)$ is nonnegative (and hence also the principal eigenvalue $\tilde{\lambda}_1(L_U, \Omega(e))$ is nonnegative, by Theorem 1.64).

Proof. It is analogous to the proof of Proposition 6.21. $\qquad\square$

Proof of Theorem 7.12. By Lemma 7.18, there exists a direction e such that the first symmetric eigenvalue $\lambda_1^s(L_U, \Omega(e))$ of the linearized operator is nonnegative, so that the principal eigenvalue $\tilde{\lambda}_1(\Omega(e))$ is nonnegative as well. Moreover, by Lemma 7.17 we have that $Q_U((W^e)^+) \leq 0$, so that either $(W^e)^+ \equiv 0$, or $\lambda_1^s(L_U, \Omega(e)) = 0$ and $(W^e)^+$ is the positive first symmetric eigenfunction in $\Omega(e)$. In any case, either $U \leq U^{\sigma(e)}$ or $U \geq U^{\sigma(e)}$ in $\Omega(e)$ holds.

Thus, by the strong maximum principle, either $U \equiv U^{\sigma(e)}$ in $\Omega(e)$, and the principal eigenvalue $\tilde{\lambda}_1(\Omega(e))$ is nonnegative, or $U < U^{\sigma(e)}$ in $\Omega(e)$ or $U > U^{\sigma(e)}$ in $\Omega(e)$. Hence, by Theorem 7.8, U is foliated Schwarz symmetric. $\qquad\square$

Remark 7.19.
1. In the previous proof when $(U - U^{\sigma(e)})^+ \equiv 0$, we also have by construction that $\lambda_1^s(L_u, \Omega(e)) \geq 0$ and, therefore, the principal eigenvalue satisfies $\tilde{\lambda}_1(\Omega(e)) = \tilde{\lambda}_1(\Omega(-e)) \geq 0$.
   In the case when $U < U^{\sigma(e)}$ in $\Omega(e)$ or $U > U^{\sigma(e)}$ in $\Omega(e)$, by rotating the planes we find a different direction e' such that $U \equiv U^{\sigma(e')}$ in $\Omega(e')$ and it could happen that $\lambda_1^s(\Omega(e')) < 0$. However, let us observe explicitly that the sign of the principal eigenvalue is preserved in the rotation, i. e., $\tilde{\lambda}_1(\Omega(e')) = \tilde{\lambda}_1(\Omega(-e')) \geq 0$, and actually $\tilde{\lambda}_1(\Omega(e')) = \tilde{\lambda}_1(\Omega(-e')) = 0$.
   Indeed since $U < U^{\sigma(g)}$ for any direction g between e and e', we have that 0 is the principal eigenvalue of the system satisfied by $U - U^{\sigma(g)}$, namely (7.15), with coefficients

$$b_{ij}^g(x) = -\int_0^1 \frac{\partial f_i}{\partial s_j}[|x|, tU(x) + (1 - t)U^{\sigma(g)}(x)] \, dt$$

As $g \to e'$, where e' is the symmetry position, the coefficients b_{ij} approach the coefficients of the linearized system, namely $c_{ij} = -\frac{\partial f_i}{\partial s_j}$, so by continuity $\tilde{\lambda}_1(\Omega(e')) = \tilde{\lambda}_1(\Omega(-e')) = 0$.

2. We can deduce, exactly as in Remark 6.13, point 2, that the solution U is either radial or strictly decreasing with respect to θ, where $\theta = \arccos(\frac{x}{|x|} \cdot p)$ is the angular variable that appears in Definition 7.5.

Proof of Theorem 7.15. By the proof of Theorem 7.12 and Remark 7.19, we can find a direction e such that U is symmetric with respect to the hyperplane $H(e)$ and the principal eigenvalue $\tilde{\lambda}_1(\Omega(e)) = \tilde{\lambda}_1(\Omega(-e)) \geq 0$.

As in the proof of Proposition 6.21, it is easy to see that if U is a Morse index one solution then for any direction e either $\lambda_1^s(L_U, \Omega(e))$ or $\lambda_1^s(L_U, \Omega(-e))$ must be nonnegative. On the other hand, by symmetry, $\lambda_1^s(\Omega(e)) = \lambda_1^s(\Omega(-e))$, so that $\tilde{\lambda}_1(\Omega(e)) = \tilde{\lambda}_1(\Omega(-e)) \geq \lambda_1^s(\Omega(e)) = \lambda_1^s(\Omega(-e)) \geq 0$.

As observed in the previous remark, for any couple of directions η_1, η_2, where η_2 is the direction of the axis of symmetry and η_1 is any direction orthogonal to it (η_1 could be the direction e), the derivative with respect to the angular variable in the plane (η_1, η_2) coincides in $\Omega(\eta_1)$, with the opposite of the angular derivative U_θ, where θ is as in Definition 7.5. Then, if $\tilde{\lambda}_1(\Omega(e)) > 0$, the angular derivative U_θ must vanish (since it satisfies (7.13) and the maximum principle holds in $\Omega(e)$). Hence U is radial.

So if $U_\theta \not\equiv 0$ necessarily $\tilde{\lambda}_1(\Omega(e)) = \lambda_1^s(\Omega(e)) = 0$ and by (iv) of Proposition 1.64 the derivative U_θ is a negative first eigenfunction of the symmetrized system in $\Omega(e)$, as well as a solution of (7.13). Thus we get that

$$J_F(|x|, U)U_\theta = \frac{1}{2}(J_F(|x|, U) + J_F^t(|x|, U))U_\theta$$

i. e., (7.19) and if $m = 2$ we get (7.20), since U_θ is strictly negative. $\qquad\square$

7.2.4 Nonlinearities with convex derivatives

The second symmetry result we get is the counterpart of Theorem 6.22 for systems.

Theorem 7.20. *Let Ω be a ball or an annulus in $\mathbb{R}^N$, $N \geq 2$, and let $U \in C^2(\overline{\Omega}; \mathbb{R}^m)$ be a solution of (7.9) with Morse index $m(U) \leq N - 1$.*

Moreover, assume that:

(i) *the system is fully coupled along U in Ω*

(ii) *for any $i, j = 1, \ldots m$ the function $\frac{\partial f_i}{\partial s_j}(|x|, S)$ is convex in $S = (s_1, \ldots, s_m)$:*

$$\frac{\partial f_i}{\partial s_j}(|x|, tS' + (1-t)S'') \leq t\frac{\partial f_i}{\partial s_j}(|x|, S') + (1-t)\frac{\partial f_i}{\partial s_j}(|x|, S'')$$

for any $t \in [0, 1]$, $S', S'' \in \mathbb{R}^m$ and $x \in \Omega$.

Then U is foliated Schwarz symmetric and if U is not radial then it is strictly decreasing in the angular variable.

Remark 7.21. As for the scalar case the assumption (ii) of Theorem 7.20 allows to get the symmetry of solutions in cases not covered by Theorem 7.12. However, the assumption on the Morse index is more restrictive since we require $m(U) \leq N - 1$, while in the analogous result for the scalar case (Theorem 6.22) the Morse index was assumed to be less than or equal to N. For systems, technical difficulties arise in the proof when $m(U) = N$.

Theorem 7.22. *Under the assumptions of Theorem 7.20 assume that U is a nonradial solution of (7.9) and either*
a) *U has Morse index one*

   or

b) *there exist $i_0, j_0 \in \{1, \ldots, m\}$ such that the function $\frac{\partial f_{i_0}}{\partial s_{j_0}}(|x|, S)$ satisfies the following strict convexity assumption:*

$$\frac{\partial f_{i_0}}{\partial s_{j_0}}(|x|, tS' + (1-t)S'') < t\frac{\partial f_{i_0}}{\partial s_{j_0}}(|x|, S') + (1-t)\frac{\partial f_{i_0}}{\partial s_{j_0}}(|x|, S'') \tag{7.23}$$

for any $t \in (0, 1)$, whenever $x \in \Omega$ and $S', S'' \in \mathbb{R}^m$ satisfy $s_k' \neq s_k''$ for any $k \in \{1, \ldots, m\}$.

Then

$$\sum_{j=1}^{m} \frac{\partial f_i}{\partial s_j}(r, U(r,\theta))\frac{\partial u_j}{\partial \theta}(r,\theta) = \sum_{j=1}^{m} \frac{\partial f_j}{\partial s_i}(r, U(r,\theta))\frac{\partial u_j}{\partial \theta}(r,\theta) \tag{7.24}$$

for any $i = 1, \ldots, m$, with (r, θ) as in Definition 7.5.
In particular, if $m = 2$ then (7.24) implies that

$$\frac{\partial f_1}{\partial s_2}(|x|, U(x)) = \frac{\partial f_2}{\partial s_1}(|x|, U(x)), \quad \forall x \in \Omega. \tag{7.25}$$

Remark 7.23. Note that (7.24) and (7.25) were also deduced under the assumptions of Theorem 7.12, but only for Morse index one solutions.

Remark 7.24 (Radial symmetry of stable solutions). The symmetry result of Theorem 7.20 holds in particular for stable solutions of (7.9). However, in this case it is easy to get that the solution is radial without any assumption on the nonlinearity (see also Remark 7.14).

The proof of Theorem 7.20 follows the scheme of the proof of Theorem 7.12, and it is based upon the following results.

Lemma 7.25. *Assume that U is a solution of (7.9) and the hypotheses of Theorem 7.20 hold. Let $B^e(x) = (b^e_{ij}(x))^m_{i,j=1}$ be the matrix associated to the fully coupled system (7.15) defined by (7.14), i. e.,*

$$b^e_{ij}(x) = -\int_0^1 \frac{\partial f_i}{\partial s_j}[|x|, tU(x) + (1-t)U^{\sigma(e)}(x)]\, dt$$

and let us define the matrix $B^{e,s}(x) = (b^{e,s}_{ij}(x))^m_{i,j=1}$, where

$$b^{e,s}_{ij}(x) = -\frac{1}{2}\left(\frac{\partial f_i}{\partial s_j}(|x|, U(x)) + \frac{\partial f_i}{\partial s_j}(|x|, U^{\sigma(e)}(x))\right) \tag{7.26}$$

Then the linear system with matrix $B^{e,s}$ is fully coupled in Ω and $\Omega(e)$ for any $e \in S^{N-1}$. Moreover, for any $i, j = 1, \ldots, m$ and $x \in \Omega$ it holds

$$b^e_{ij}(x) \geq b^{e,s}_{ij}(x) \tag{7.27}$$

with strict inequality for any i_0, j_0 such that $\frac{\partial f_{i_0}}{\partial s_{j_0}}$ satisfies the strict convexity assumption (7.23) if $u_k(x) \neq u^{\sigma(e)}_k(x)$ for any $k \in \{1, \ldots, m\}$.

Finally, for the quadratic forms Q^e and $Q^{e,s}$ associated to the matrices B^e and $B^{e,s}$ we have that

$$0 \geq Q^e((W^e)^{\pm}; \Omega(e)) = \int_{\Omega(e)} [|\nabla(W^e)^{\pm}|^2 + B^e((W^e)^{\pm}, (W^e)^{\pm})]\, dx$$

$$\geq \int_{\Omega(e)} [|\nabla(W^e)^{\pm}|^2 + B^{e,s}((W^e)^{\pm}, (W^e)^{\pm})]\, dx = Q^{e,s}((W^e)^{\pm}; \Omega(e)) \tag{7.28}$$

for $W^e = U - U^{\sigma(e)}$, with the strict inequality

$$Q^{e,s}((W^e)^+; \Omega(e)) = Q^{e,s}((W^e); \Omega(e)) < 0$$

if F satisfies the hypothesis b) of Theorem 7.22 and $W^e > 0$ in $\Omega(e)$, while

$$Q^{e,s}((W^e)^-; \Omega(e)) = Q^{e,s}((W^e); \Omega(e)) < 0$$

if F satisfies the hypothesis b) of Theorem 7.22 and $W^e < 0$ in $\Omega(e)$.

Proof. By hypothesis (ii) of Theorem 7.20, we get

$$-b^e_{ij}(x) = \int_0^1 \frac{\partial f_i}{\partial s_j}[|x|, tU(x) + (1-t)U^{\sigma(e)}(x)]\, dt$$

$$\leq \int_0^1 \left(t\frac{\partial f_i}{\partial s_j}[|x|, U(x)] + (1-t)\frac{\partial f_i}{\partial s_j}[|x|, U^{\sigma(e)}(x)]\right) dt$$

$$= \frac{1}{2}\left(\frac{\partial f_i}{\partial s_j}(|x|, U(x)) + \frac{\partial f_i}{\partial s_j}(|x|, U^{\sigma(e)}(x)) \right) = -b_{ij}^{e,s}(x) \qquad (7.29)$$

This implies (7.27) and the inequality is strict for any i_0, j_0 such that $\frac{\partial f_{i_0}}{\partial s_{j_0}}$ satisfies the strict convexity assumption (7.23) if $u_k(x) \neq u_k^{\sigma(e)}(x)$ for any $k \in \{1, \dots, m\}$.

This in turn implies the full coupling of the system with matrix $B^{e,s}$, since by Lemma 7.10 the system with matrix B^e is fully coupled. From (7.18) and (7.27), since $w_k^{\pm} \geq 0$, we get

$$0 \geq \int_{\Omega(e)} \left(\sum_{i=1}^m |\nabla w_i^+|^2 + \sum_{i,j=1}^m b_{ij}^e w_j^+ w_i^+ \right) dx$$

$$\geq \int_{\Omega(e)} \left(\sum_{i=1}^m |\nabla w_i^+|^2 + \sum_{i,j=1}^m b_{ij}^{e,s} w_j^+ w_i^+ \right) dx$$

i. e., (7.28) in the case of the positive part of W^e, with strict inequality if F satisfies the hypothesis b) of Theorem 7.22 and $W^e > 0$. Analogously, we get the corresponding inequality for the negative part of W^e. $\qquad \square$

Lemma 7.26. *Suppose that U is a solution of (7.9) with Morse index $m(U) \leq N - 1$ and assume that the hypothesis (i) of Theorem 7.20 holds.*

Let $Q^{e,s}$ be the quadratic form associated to the operator $L^{e,s}(V) = -\Delta V + B^{e,s} V$, $B^{e,s}$ being the matrix defined in (7.26):

$$Q^{e,s}(\Psi; \Omega') = \int_{\Omega'} \left[|\nabla \Psi|^2 + B^{e,s}(\Psi, \Psi) \right] dx$$

$$= \int_{\Omega} \left[\sum_{i=1}^m |\nabla \psi_i|^2 - \sum_{i,j=1}^m \frac{1}{2}\left(\frac{\partial f_i}{\partial s_j}(|x|, U(x)) + \frac{\partial f_i}{\partial s_j}(|x|, U^{\sigma(e)}(x)) \right) \psi_i \psi_j \right] dx \quad (7.30)$$

Then there exists a direction $e \in S^{N-1}$ such that

$$Q^{e,s}(\Psi; \Omega(e)) \geq 0 \quad \forall \Psi \in C_c^1(\Omega(e); \mathbb{R}^m)$$

Equivalently, the first symmetric eigenvalue $\lambda_1^s(L^{e,s}, \Omega(e))$ of the operator $L^{e,s}(V) = -\Delta V + B^{e,s} V$ in $\Omega(e)$ is nonnegative (and hence also the principal eigenvalue $\tilde{\lambda}_1(L^{e,s}, \Omega(e))$ is nonnegative).

Proof. Let us assume that $1 \leq j = m(U) \leq N - 1$ and let $\Phi_1, \dots, \Phi_j$ be mutually orthogonal eigenfunctions corresponding to the negative symmetric eigenvalues $\lambda_1^s(L_U, \Omega), \dots, \lambda_j^s(L_U, \Omega)$ of the linearized operator $L_U(V) = -\Delta V - J_F(x, U)V$ in Ω.

For any $e \in S^{N-1}$, let $\phi^{e,s}$ be the first positive L^2-normalized eigenfunction of the symmetric system associated to the linear operator $L^{e,s}$ in $\Omega(e)$. We observe that $\phi^{e,s}$ is uniquely determined since the corresponding system is fully coupled in $\Omega(e)$. Let $\Phi^{e,s}$

be the odd extension of $\phi^{e,s}$ to Ω, and let us observe that $\Phi^{-e,s} = -\Phi^{e,s}$, because $B^{e,s}$ is symmetric with respect to the reflection σ_e.

The mapping $e \mapsto \Phi^{e,s}$ is a continuous odd function from S^{N-1} to $H_0^1(\Omega \cup \Gamma)$, therefore, the mapping $h : S^{N-1} \to \mathbb{R}^j$ defined by

$$h(e) = \left((\Phi^{e,s}, \Phi_1)_{L^2(\Omega)}, \ldots, (\Phi^{e,s}, \Phi_j)_{L^2(\Omega)} \right)$$

is an odd continuous mapping, and since $j \leq N - 1$, by the Borsuk–Ulam theorem it must have a zero. This means that there exists a direction $e \in S^{N-1}$ such that $\Phi^{e,s}$ is orthogonal to all the eigenfunctions $\Phi_1, \ldots, \Phi_j$. This implies that $Q_U(\Phi^{e,s}; \Omega) \geq 0$, because $m(U) = j$, and since $\Phi^{e,s}$ is an odd function, we obtain that $0 \leq Q_U(\Phi^{e,s}; \Omega) = Q^{e,s}(\Phi^{e,s}, \Omega) = 2Q^{e,s}(\phi^{e,s}, \Omega(e)) = 2\lambda_1^s(L^{e,s}, \Omega(e))$ $\square$

Proof of Theorem 7.20. By Lemma 7.26, there exists a direction e such that the first symmetric eigenvalue $\lambda_1^s(L^{e,s}, \Omega(e))$ of the operator $L^{e,s}(V) = -\Delta V + B^{e,s} V$ in $\Omega(e)$ is nonnegative, and hence also the principal eigenvalue $\tilde{\lambda}_1(L^{e,s}, \Omega(e))$ is nonnegative.

Since $Q^{e,s}((W^e)^\pm; \Omega(e)) \leq 0$ by Lemma 7.25, we have two alternatives. The first one is that $(W^e)^+$ and $(W^e)^-$ both vanish, in which case $W^e \equiv 0$ in $\Omega(e)$, and this implies in turn that $L^{e,s} = L_U$. Then U is symmetric and the principal eigenvalue $\tilde{\lambda}_1(L_U, \Omega(e))$ is nonnegative, so that the hypothesis (i) of Theorem 7.8 holds and we get that U is foliated Schwarz symmetric. The second alternative is that one among $(W^e)^+$ and $(W^e)^-$ does not vanish and $\lambda_1^s(L^{e,s}, \Omega(e)) = 0$. Then either $(W^e)^+$ or $(W^e)^-$ is a first symmetric eigenfunction of the operator $L^{e,s}(V)$ in $\Omega(e)$. If $(W^e)^+$ is a first symmetric eigenfunction of the operator $L^{e,s}(V) = -\Delta V + B^{e,s} V$ in $\Omega(e)$, then it is positive in $\Omega(e)$, i. e., $U > U^{\sigma_e}$ in $\Omega(e)$. In the case when $(W^e)^-$ is the first symmetric eigenfunction, we get the reversed inequality. Then, by the sufficient condition (ii) given by Theorem 7.8, u is foliated Schwarz symmetric. $\square$

Remark 7.27 (Alternative proof of Theorem 7.20). By Lemma 7.25, we have the comparison (7.27) between the entries of the matrices B^e and $B^{e,s}$. However, this does not imply that the quadratic forms are ordered (this happens, by Theorem 1.57(iv) a), if $(B^{e,s} - B^e)$ is positive semidefinite). This is the reason why we considered in Lemma 7.25 and in the previous proof of Theorem 7.20 the positive and negative parts of W^e.

Nevertheless, the comparison (7.27) between the single entries of the matrices B^e and $B^{e,s}$ implies, by Theorem 1.57(iv) b), the inequality

$$\lambda_1^s(-\Delta + B^{e,s}; \Omega(e)) \leq \lambda_1^s(-\Delta + B^e; \Omega(e)) \tag{7.31}$$

for the first symmetric eigenvalues. Then part of the proof of Theorem 7.20 can be made using this property.

Indeed we can start with a direction e such that the first symmetric eigenvalue $\lambda_1^s(L^{e,s}, \Omega(e))$ of the operator $L^{e,s}(V) = -\Delta V + B^{e,s} V$ in $\Omega(e)$ is nonnegative, and hence also the principal eigenvalue $\tilde{\lambda}_1(L^{e,s}, \Omega(e))$ is nonnegative.

Then by (7.31) we get that the first symmetric eigenvalue $\lambda_1^s(-\Delta + B^e, \Omega(e))$ of the operator $-\Delta + B^e$ in $\Omega(e)$ (B^e being defined in (7.26)), is also nonnegative.

If $\lambda_1^s(-\Delta + B^e, \Omega(e)) > 0$, then necessarily the difference $W^e = U - U^{\sigma(e)}$ must vanish since $Q^e(W^e; \Omega(e)) = 0$ by (7.17). This implies that $B^e = B^{e,s} = J_F(|x|, U)$, so that we find a direction e with $U \equiv U^{\sigma(e)}$ in $\Omega(e)$ and since, by symmetry, $L^{e,s} = L_U$, we also have $\lambda_1^s(L^{e,s}, \Omega(e)) = \lambda_1^s(L_U, \Omega(e)) > 0$.

If instead $\lambda_1^s(-\Delta + B^e, \Omega(e)) = 0$, then necessarily $\lambda_1^s(-\Delta + B^{e,s}; \Omega(e)) = \lambda_1^s(-\Delta + B^e, \Omega(e)) = 0$. Since $Q^e(W^e; \Omega(e)) = 0$ either W^e vanishes or W^e is a first symmetric eigenfunction of the system (7.15), which is fully coupled. This implies that it does not change sign in $\Omega(e)$, and assuming that, e. g., $U \geq U^{\sigma(e)}$ then $U > U^{\sigma(e)}$, by the strong comparison principle, and again by Theorem 7.8 the solution U is foliated Schwarz symmetric.

Proof of Theorem 7.22. Let us remark that in the proof of Theorem 7.20 we found a direction e such that U is symmetric with respect to $H(e)$ and not only the principal eigenvalue $\tilde{\lambda}_1(L_U, \Omega(e'))$ of the linearized operator in $\Omega(e')$ is nonnegative, but also the first symmetric eigenvalue $\lambda_1^s(L_U, \Omega(e)) = \lambda_1^s(L_U, \Omega(-e)) \geq 0$.

If hypothesis b) of Theorem 7.22 holds, then necessarily $W^e \equiv 0$, since $Q^{e,s}((W^e)^+;$ $\Omega(e)) < 0$ if $(W^e)^+$ is positive, analogously for the negative part.

The same happens if U is a Morse index one solution since in this case for any direction e either $\lambda_1^s(L_U, \Omega(e))$ or $\lambda_1^s(L_U, \Omega(-e))$ must be nonnegative, so that the proof goes on as the one of Theorem 7.15. $\qquad\square$

7.3 Examples

A first type of elliptic systems that could be considered are those of "gradient type" (see [95]), i. e., systems of type (7.9) where $f_j(|x|, S) = \frac{\partial g}{\partial s_j}(|x|, S)$ for some scalar function $g \in C^{2,\alpha}([0, +\infty) \times \mathbb{R}^m)$.

In this case, the solutions correspond to critical points of the functional

$$\Phi(u) = \frac{1}{2}\int_\Omega |\nabla U|^2\, dx - \int_\Omega g(|x|, U)\, dx$$

in $\mathbf{H}_0^1(\Omega)$ and the linearized operator (7.5) coincides with the second derivative of Φ. Thus standard variational methods apply which often give solutions of finite (linearized) Morse index, as, for example, in the case when the mountain pass theorem can be used (see Section 3.2). So, if the hypotheses of Theorem 7.12 are satisfied, the symmetry results of Section 7.2 can be applied.

Many systems of this type have been studied (see [95] and the references therein). An example is the nonlinear Schrödinger system

$$\begin{cases} -\Delta u_1 + u_1 = |u_1|^{2q-2}u_1 + b|u_2|^q|u_1|^{q-2}u_1 & \text{in } \Omega \\ -\Delta u_2 + \omega^2 u_2 = |u_2|^{2q-2}u_2 + b|u_1|^q|u_2|^{q-2}u_1 & \text{in } \Omega \end{cases} \tag{7.32}$$

where $b > 0$, Ω is either a bounded domain or the whole $\mathbb{R}^N$, $N \geq 2$, $q > 1$ if $N = 2$ and $1 < q < \frac{N}{N-2}$ if $N \geq 3$. If $\Omega \neq \mathbb{R}^N$ the Dirichlet boundary conditions are imposed

$$u_1 = u_2 = 0 \quad \text{on } \partial\Omega \tag{7.33}$$

The system (7.32) has been considered in several papers (see [13, 170] and the references therein).

It is easy to see that the system is of gradient type and the solutions of (7.32), (7.33) are critical points of the functional

$$I(U) = I(u_1, u_2) = \frac{1}{2} \int_{\Omega} |\nabla U|^2 \, dx - \frac{1}{2q} \int_{\Omega} \left(|u_1|^{2q} + |u_2|^{2q} \right) dx - \frac{b}{q} \int_{\Omega} |u_1 u_2|^q \, dx \tag{7.34}$$

in the space $\mathbf{H}_0^1(\Omega)$. Thus the linearized operator (7.5) at a solution U corresponds to the second derivative of I in $\mathbf{H}_0^1(\Omega)$, and hence the corresponding linear system is symmetric.

Moreover, as observed in [170], the system (7.32) is cooperative, and fully coupled in $\Omega(e)$, for any $e \in S^{N-1}$, along every purely vector solution $U = (u_1, u_2)$, i. e., such that u_1 and u_2 are both not identically zero. In particular, the system is fully coupled along positive solutions in any $\Omega(e)$. Then Theorem 7.12 could be applied to any purely vector solution with Morse index less than or equal to N, in an annulus or in a ball. In particular, we can consider solutions obtained by the mountain pass theorem which have a Morse index not bigger than one which are purely vector positive solutions for suitable values of b and $q \geq 2$, as could be obtained by the same proof of [170] for the case when $\Omega = \mathbb{R}^N$.

A second type of interesting systems of two equations are the so-called "Hamiltonian type" systems (see [95] and the references therein). More precisely, we consider the system

$$\begin{cases} -\Delta u_1 = f_1(|x|, u_1, u_2) & \text{in } \Omega \\ -\Delta u_2 = f_2(|x|, u_1, u_2) & \text{in } \Omega \\ u_1 = u_2 = 0 & \text{on } \partial\Omega \end{cases} \tag{7.35}$$

with

$$f_1(|x|, u_1, u_2) = \frac{\partial H}{\partial u_2}(|x|, u_1, u_2), \quad f_2(|x|, u_1, u_2) = \frac{\partial H}{\partial u_1}(|x|, u_1, u_2) \tag{7.36}$$

for some scalar function $H \in C^{2,\alpha}([0, +\infty) \times \mathbb{R}^2)$. These systems can be studied by considering the associated functional

$$J(U) = I(u_1, u_2) = \frac{1}{2} \int_{\Omega} \nabla u_1 \cdot \nabla u_2 \, dx - \int_{\Omega} H(|x|, u_1, u_2) \, dx \tag{7.37}$$

either in $\mathbf{H}_0^1(\Omega)$ or in other suitable Sobolev spaces (see [95]).

It is easy to see that the linearized operator defined in (7.5) does not correspond to the second derivative of the functional J, which is strongly indefinite. Nevertheless, solutions of (7.35) can have finite linearized Morse index as we show with a few simple examples.

Let us consider the following system:

$$\begin{cases} -\Delta u_1 = \lambda e^{u_2} & \text{in } \Omega \\ -\Delta u_2 = \mu e^{u_1} & \text{in } \Omega \\ u_1 = u_2 = 0 & \text{on } \partial\Omega \end{cases} \tag{7.38}$$

where $\lambda, \mu \in \mathbb{R}$.

If $\lambda, \mu > 0$, then the system is fully coupled along every solution in any $\Omega(e)$, $e \in S^{N-1}$, and the hypotheses of Theorem 7.12 are satisfied.

Let us consider the function $G : \mathbb{R}^2 \times (C^{2,\alpha}(\Omega))^2 \to (C^{0,\alpha}(\Omega))^2$ defined by $G((\lambda, \mu), (u_1, u_2)) = (-\Delta u_1 - \lambda e^{u_2}, -\Delta u_2 - \lambda e^{u_1})$. We have that $G((0,0), (0,0)) = (0,0)$, $\frac{\partial G}{\partial(u_1,u_2)}((0,0), (0,0))(\phi, \psi) = (-\Delta\phi, -\Delta\psi)$, and the first (symmetric) eigenvalue of this operator is strictly positive, so that the operator is invertible. Thus, by the implicit function theorem, for small λ, μ there is a unique nontrivial solution $(u_1(\lambda, \mu), u_2(\lambda, \mu))$ close to the trivial solution of the system (7.38) corresponding to $\lambda = \mu = 0$. Moreover, the solution is positive by the maximum principle and it is linearized stable, so that it is radial. Indeed the first (symmetric) eigenvalue of the linearized operator corresponding to $\lambda = \mu = 0$, $u = v = 0$ is strictly positive, and by continuity (see Theorem 1.57) it is positive for small λ, μ. The same happens substituting the exponential with other nonlinearities $f(u_2)$, $g(u_1)$ with nonnegative derivative in a neighborhood of 0 and such that $f(0), g(0) > 0$.

We now consider, as another example, the system

$$\begin{cases} -\Delta u_1 = u_2^p & \text{in } \Omega \\ -\Delta u_2 = u_1^q & \text{in } \Omega \\ u_1, u_2 > 0 & \text{in } \Omega \\ u_1 = u_2 = 0 & \text{on } \partial\Omega \end{cases} \tag{7.39}$$

where $1 < p, q < \frac{N+2}{N-2}$ if $N \geq 3$, $p, q > 1$ if $N = 2$. The idea is to proceed as in the previous example starting from the case $p = q$ and the solution $u_1 = u_2 = z$, where z is a scalar solution of the equation

$$\begin{cases} -\Delta z = z^p & \text{in } \Omega \\ z > 0 & \text{in } \Omega \\ z = 0 & \text{on } \partial\Omega \end{cases} \tag{7.40}$$

which is nondegenerate.

Let us observe that if $p = q$ and z has Morse index equal to the integer $m(z)$, then $m(z)$ is also the Morse index of the solution $U = (u_1, u_2) = (z, z)$ of the system (7.39). Indeed the linearized equation at z for the equation (7.40) and the linearized system at (z, z) for the system (7.39) are respectively

$$\begin{cases} -\Delta\phi - pz^{p-1}\phi = 0 & \text{in } \Omega \\ \phi = 0 & \text{on } \partial\Omega \end{cases} \tag{7.41}$$

and

$$\begin{cases} -\Delta\phi_1 - pz^{p-1}\phi_2 = 0 & \text{in } \Omega \\ -\Delta\phi_2 - pz^{p-1}\phi_1 = 0 & \text{in } \Omega \\ \phi_1 = \phi_2 = 0 & \text{on } \partial\Omega \end{cases} \tag{7.42}$$

and the eigenvalues of the two operators are the same. Indeed, if ϕ is an eigenfunction for (7.41) corresponding to the eigenvalue λ_k, then taking $\phi_1 = \phi_2 = \phi$ we obtain an eigenfunction (ϕ_1, ϕ_2) for (7.42) corresponding to the same eigenvalue. Vice versa, if (ϕ_1, ϕ_2) is an eigenfunction for (7.42) corresponding to the eigenvalue λ_k, then $\phi = \phi_1 + \phi_2$ is an eigenfunction for (7.41) corresponding to the same eigenvalue. So proceeding as in the previous example we can start from a nondegenerate solution of (7.40) with a fixed exponent $\bar{p} \in (1, \frac{N+2}{N-2})$ (or $\bar{p} \in (1, +\infty)$ if $N = 2$) and find a branch of solutions of (7.39) corresponding to (possibly different) exponents p, q close to $\bar{p}$. For example, if we start with a mountain pass (positive) solution z in the ball of equation (7.40) with the exponent $\bar{p}$, knowing that its Morse index is one (see Chapter 3) and it is nondegenerate (see [74]), we get a branch of Morse index one radial solutions for p, q close to $\bar{p}$. Thus, starting from an exponent $\bar{p}$ for which the mountain pass solution z of (7.40) is not radial in the case when Ω is an annulus, we can construct solutions $U = (u_1, u_2)$ of (7.39) in correspondence of exponents p, q close to $\bar{p}$, with Morse index one. Then Theorem 7.12 applies and, in particular, we get that the coupling condition (7.20) holds, which, in this case, can be written as

$$pu_2^{p-1} = qu_1^{q-1} \quad \text{in } \Omega \tag{7.43}$$

Note that more generally, by Corollary 7.15, the equality (7.43) must hold for every solution $U = (u_1, u_2)$ of (7.39) with Morse index one giving so a sharp condition to be satisfied by the components of a solution of this type.

We point out that we must be careful when z is a nonradial foliated Schwarz symmetric function, since in this case, the solution is automatically degenerate because of the rotation invariance of the equation.

To bypass this difficulty, we could work in subspaces of symmetric functions. To be more precise, let us start with a solution z of equation (7.40), with exponent $\bar{p}$, which is nonradial but foliated Schwarz symmetric with respect to a fixed vector p and with low Morse index. This is the case, for example, of a positive minimal energy solution in the annulus for some values of the exponent and of the radii of the annulus, which has Morse index one.

In these cases, 0 is an eigenvalue of the linearized operator and there are $N - 1$ corresponding orthogonal eigenfunctions which are *odd* with respect to $N - 1$ orthogonal symmetry axes passing through p. Roughly speaking, they are the ones which induce the degeneration. So we consider the subspace H^e of $H_0^1(\Omega)$ consisting of the functions that are *even* with respect to the symmetry hyperplanes. If the solution is nondegenerate in this space, we can use the implicit function theorem and the continuation method as before and find a branch of foliated Schwarz symmetric solutions to the system (7.39) for values of p, q close to the initial exponent $\bar{p}$. Then, in the case of a positive Morse index one nonradial foliated Schwarz symmetric solution z in the annulus corresponding to the exponent $\bar{p}$, we obtain from Corollary 7.15 a branch of foliated Schwarz symmetric solutions (u, v) to the system (7.39) corresponding to the values p, q close to $\bar{p}$, which satisfy the coupling relation (7.20), which, in this case, are

$$p|v|^{p-1} = q|u|^{q-1}$$

The same considerations can be made for sign changing solutions, when the hypotheses of Theorem 7.20 are satisfied. In particular, we can consider the system

$$\begin{cases} -\Delta u_1 = |u_2|^{p-1}u_2 & \text{in } \Omega \\ -\Delta u_2 = |u_1|^{q-1}u_1 & \text{in } \Omega \\ u_1 = u_2 = 0 & \text{on } \partial\Omega \end{cases} \tag{7.44}$$

where $1 < p, q < \frac{N+2}{N-2}$. Then we start from the case $p = q$ and the solution $u_1 = u_2 = z$, where z is a scalar solution of the equation

$$\begin{cases} -\Delta z = |z|^{p-1}z & \text{in } \Omega \\ z = 0 & \text{on } \partial\Omega \end{cases} \tag{7.45}$$

Let us observe that if $p = q$ and z has Morse index equal to the integer $m(z)$, then as before $m(z)$ is also the Morse index of the solution $U = (u_1 = u_2) = (z, z)$ of the system (7.44).

So if we start from a nondegenerate solution of (7.45) with a fixed exponent $\bar{p} \in (1, \frac{N+2}{N-2})$ (or $\bar{p} \in (1, +\infty)$ if $N = 2$), using the implicit function theorem, we find a branch of solutions of (7.44) corresponding to (possibly different) exponents p, q close to $\bar{p}$. For example, if we start with a least energy nodal solution z in the ball of equation (7.45) with the exponent $\bar{p}$, knowing that its Morse index is two (see Section 3.3.1) we get a branch of Morse index two solutions for p, q close to $\bar{p}$. Note that, as proved in Chapter 3 (see also [7]), the least energy nodal solution of (7.45) is not radial but foliated Schwarz symmetric. So it is obviously degenerate, but working in the space of axially symmetric functions we could remove the degeneracy and apply the continuation method described above, if there are not other degeneracies.

Then Theorem 7.20 applies if $p, q \geq 2$ and, in particular, we get that the coupling condition (7.25) holds, which in this case can be written as

$$p|u_2|^{p-1} = q|u_1|^{q-1} \quad \text{in } \Omega \tag{7.46}$$

Note that more generally, by Theorem 7.22, the equality (7.46) must hold for every solution $U = (u_1, u_2)$ of (7.44) with Morse index $m(U) \leq N - 1$, giving so a sharp condition to be satisfied by the components of a solution of this type.

Let us remark that the symmetry results of Section 7.2 also apply when the nonlinearity depends on $|x|$ (in any way). Arguing as before, it is not difficult to construct systems having solutions with low Morse index, in particular with Morse index one or two.

An example could be the "Henon system"

$$\begin{cases} -\Delta u_1 = |x|^{\alpha} |u_2|^{p-1} u_2 & \text{in } \Omega \\ -\Delta u_2 = |x|^{\beta} |u_1|^{q-1} u_1 & \text{in } \Omega \\ u_1, u_2 > 0 & \text{in } \Omega \\ u_1 = u_2 = 0 & \text{on } \partial\Omega \end{cases} \tag{7.47}$$

with $\alpha, \beta > 0$, $p, q \geq 1$.

Starting again from a solution of the scalar equation with Morse index one or two, it is possible to construct solutions of (7.47) with the same Morse index, for p, q and α, β close to each other. Note that for some values of p, q, α, β such solutions would not be radial, even if Ω is a ball, but rather foliated Schwarz symmetric as for the scalar case (see [38]). Finally, as in the scalar case, we can have examples of nonhomogeneous systems with convex nonlinearities for which there exist solutions of Morse index one which change sign and to which our results apply.

8 Some results in unbounded domains

In this chapter, we briefly indicate some extensions to unbounded domains of the symmetry theorems considered in Chapter 6 and Chapter 7. We will focus on results obtained by moving planes and Morse index bounds, emphasizing, in both cases, the importance of assuming the solution in a suitable function space.

The results about foliated Schwarz symmetry in unbounded domains are based on the same ideas developed in the case of bounded domains but the proofs are rather technical and long and we refer to [131] for them.

In the last section, we recall several Liouville-type nonexistence results for solution with finite Morse index.

8.1 Moving planes and symmetry in unbounded domains

Soon after the famous result by Gidas, Ni and Nirenberg [122] about the radial symmetry of positive solutions of semilinear elliptic equations in balls, the same authors considered in [123] the following problem in $\mathbb{R}^N$:

$$\begin{cases} -\Delta u = f(u), & u > 0 \text{ in } \mathbb{R}^N, \\ u \to 0 & \text{when} \quad |x| \to \infty \end{cases} \tag{8.1}$$

Using again the Alexandrov–Serrin moving planes method, they proved that the C^2 solutions of (8.1) are radial provided $f \in C^{1+\alpha}[0,\infty)$, $f(0) = 0$, $f'(0) < 0$. They also obtain some symmetry results in the case $f'(0) = 0$ under appropriate assumptions on the growth of f near 0 and on the decay of u at ∞.

Later Y. Li and W. Ni [156] and C. Li [155] extended the previous symmetry results to fully nonlinear strictly elliptic equations proving, in particular, the symmetry of solutions of (8.1) when one of the following hypotheses holds:

(i) $\exists s_0 > 0 : f'(s) \leq 0 \quad \forall s \in (0, s_0)$

(ii) $\exists \alpha > 0, \, m \in \mathbb{R} : f'(s) = O(s^\alpha)(s \to 0), \, u = O(\frac{1}{|x|^m})(|x| \to \infty)$ and $m\alpha > 2$.

Note that if (ii) holds then $u \in L^{\alpha \frac{N}{2}}(\mathbb{R}^N)$.

As a matter of fact, it is possible to obtain the symmetry result under the only assumption that u belongs to the space $L^{\alpha \frac{N}{2}}(\mathbb{R}^N)$ using a technique introduced in [216, 217], which is based on Sobolev inequalities together with the moving planes method (see also [81] for a similar technique using Poincaré's and Hardy's inequalities).

We will sketch below the proof given in [9] (where this technique is exploited to obtain symmetry results for solutions of elliptic equations on unbounded manifolds) to illustrate the main difficulties that arise when considering problems in unbounded domains and the importance of assuming that the solutions belong to suitable function spaces.

https://doi.org/10.1515/9783110538243-008

More precisely, we will prove the following result.

Theorem 8.1. *Let $u \in C^1(\mathbb{R}^N)$ be a (weak) solution of the problem*

$$
\begin{cases}
-\Delta u = f(u) & \text{in } \mathbb{R}^N \\
u > 0 & \text{in } \mathbb{R}^N
\end{cases}
\tag{8.2}
$$

Suppose that one of the following set of hypotheses holds:
(H1)
    (i)   *$u(x) \to 0$ as $|x| \to \infty$*
    (ii) *$\exists s_0 > 0 : f$ is nonincreasing in $(0, s_0)$.*
(H2)
    (i)   *$u(x) \to 0$ as $|x| \to \infty$*
    (ii) *$\exists s_0, \alpha > 0$: such that $\frac{f(b)-f(a)}{b-a} \leq C(a+b)^\alpha$ when $0 < a < b < s_0$ and $u \in L^{\alpha N/2}(\mathbb{R}^N)$.*
(H3) $\exists \alpha > 0$, $p \in [2, N)$ such that
    (i)   *$|\frac{f(b)-f(a)}{b-a}| \leq C(a+b)^\alpha$ for all $a, b > 0$*
    (ii) *$u \in \mathcal{D}^{1,p}(\mathbb{R}^N) \cap L^{\alpha\frac{n}{2}}(\mathbb{R}^N)$*
    where $\mathcal{D}^{1,p}(\mathbb{R}^N) := \{u \in L^{p^*}(\mathbb{R}^N) : |\nabla u| \in L^p(\mathbb{R}^N)\}$, $p^* = \frac{Np}{N-p}$

Then, u is radially symmetric around some point $x_0 \in \mathbb{R}^N$, i.e., $u(x) = u(r)$, where $r := |x - x_0|$. Moreover, $u'(r) < 0$ for all $r > 0$.

Proof. For simplicity, we assume $N \geq 3$, though it is possible to obtain analogous results when $N = 2$, using the corresponding Sobolev embeddings.

As usual in radial symmetry results in $\mathbb{R}^N$, it is enough to fix an arbitrarily chosen direction and to prove symmetry w. r. t. that direction. Then it is easy to see that all the symmetry hyperplanes meet in a single point.

For simplicity of notation, we fix the $e_1 = (1, 0, \ldots, 0)$ direction and denote a point in $\mathbb{R}^N$ as $x = (x_1, x')$ with $x' \in \mathbb{R}^{N-1}$. Given $\lambda \in \mathbb{R}$, we set

$$
\Sigma_\lambda := \{x = (x_1, x') \in \mathbb{R}^N : x_1 < \lambda\}, \quad H_\lambda := \{x \in \mathbb{R}^N : x_1 = \lambda\},
$$
$$
x_\lambda := \sigma_\lambda(x) := (2\lambda - x_1, x'), \quad u_\lambda(x) := u(x_\lambda), \quad \Sigma^\lambda = \sigma_\lambda(\Sigma_\lambda)
$$

Here, $x_\lambda := \sigma_\lambda(x) := (2\lambda - x_1, x')$ is the image of the point $x = (x_1, x')$ under the reflection through the hyperplane H_λ.

The first step of the proof will consist in showing that the set $\Lambda := \{\lambda \in \mathbb{R} : \forall \mu > \lambda, u \geq u_\mu$ in $\Sigma_\mu\}$ is nonempty and bounded from below. The second step will then be to show that if $\lambda_0 := \inf \Lambda$, then $u \equiv u_{\lambda_0}$ in Σ_{λ_0}.

Step 1: $\Lambda \neq \emptyset$ and is bounded from below.

Case of hypotheses (H2). First, we see that Λ is bounded from below, since $u \to 0$ when $|x| \to \infty$.

We write $v = u_\lambda$ and suppose $q \geq 1$ (to be chosen below). For $\varepsilon > 0$, we let $w_\varepsilon = w_{\varepsilon,q}(x) := [(v - u - \varepsilon)^+]^q$. Using w_ε as test function (it has compact support in Σ_λ since $u \to 0$ as $|x| \to \infty$ and $v \equiv u$ on H_λ), we obtain, once we subtract the equation for u from the equation for v,

$$q \int_{\Sigma_\lambda} [(v - u - \varepsilon)^+]^{q-1} \big|D[(v - u - \varepsilon)^+]\big|^2 = \int_{\Sigma_\lambda} [f(v) - f(u)][(v - u - \varepsilon)^+]^q \qquad (8.3)$$

Define $z_\varepsilon := [(v - u - \varepsilon)^+]^{\frac{q+1}{2}}$. If λ is sufficiently large, then $v < s_0$. Hence, since we integrate in a set where $v > u + \varepsilon > u > 0$ and (H2) holds, we get that

$$\frac{4q}{(q+1)^2} \int_{\Sigma_\lambda} |Dz_\varepsilon|^2 = \int_{\Sigma_\lambda} \frac{f(v) - f(u)}{v - u} (v - u)[(v - u - \varepsilon)^+]^q \leq C \int_{\Sigma_\lambda} v^\alpha (v - u)[(v - u - \varepsilon)^+]^q.$$

Since $u \in L^\infty(\mathbb{R}^N) \cap L^{\alpha N/2}(\mathbb{R}^N)$, we can fix a sufficiently large q (say q such that $\alpha + q + 1 \geq \alpha \frac{N}{2}$) so that

$$\int_{\Sigma_\lambda} v^\alpha (v - u)[(v - u - \varepsilon)^+]^q \leq \int_{\Sigma_\lambda} v^{\alpha+q+1} \leq \int_{\mathbb{R}^N} u^{\alpha+q+1} < +\infty. \qquad (8.4)$$

Passing to the limit as $\varepsilon \to 0$, and denoting $z = [(v - u)^+]^{\frac{q+1}{2}}$, we obtain (using the dominated convergence theorem)

$$\int_{\Sigma_\lambda} |Dz|^2 \leq C \int_{\Sigma_\lambda} v^\alpha z^2.$$

By Hölder and Sobolev inequalities, it follows that

$$\int_{\Sigma_\lambda} |Dz|^2 \leq C \left(\int_{\Sigma_\lambda} v^{\frac{\alpha N}{2}} \right)^{\frac{2}{N}} \left(\int_{\Sigma_\lambda} z^{\frac{2N}{N-2}} \right)^{\frac{N-2}{N}}$$

$$= C \left(\int_{\Sigma^\lambda = \sigma_\lambda(\Sigma_\lambda)} u^{\frac{\alpha N}{2}} \right)^{\frac{2}{N}} \left(\int_{\Sigma_\lambda} z^{2^*} \right)^{\frac{2}{2^*}} \leq C_1 \left(\int_{\Sigma^\lambda} u^{\frac{\alpha N}{2}} \right)^{\frac{2}{N}} \int_{\Sigma_\lambda} |Dz|^2. \qquad (8.5)$$

Since $u^{\frac{\alpha N}{2}} \in L^1(\mathbb{R}^N)$, we have that $\lim_{\lambda \to \infty} \int_{\Sigma^\lambda} u^{\frac{\alpha N}{2}} = 0$, and thus, for sufficiently large λ,

$$C_1 \left(\int_{\Sigma^\lambda} u^{\frac{\alpha N}{2}} \right)^{\frac{2}{N}} < 1.$$

This, together with (8.5), yields that $\int_{\Sigma_\lambda} |Dz|^2 = 0$, and thus $|Dz| = 0$ in Σ_λ, and hence z is constant in Σ_λ. Since $z = 0$ on H_λ, this implies that $z = 0$ in Σ_λ, which proves step 1.

Case of hypotheses (H1). In this case, the proof is much easier: we can simply take $q = 1$ and, since f is decreasing near zero, for sufficiently large λ the r. h. s. in (8.3) is nonpositive and, therefore, is zero. Passing to the limit as $\varepsilon \to 0$, again we obtain

$$\int_{\Sigma_\lambda} |D(v - u)^+|^2 = 0,$$

for λ sufficiently large. We conclude as in the previous case.

Case of hypotheses (H3). Let us set $q = \frac{N(p-1)-p}{N-p}$, so that $p = \frac{N(q+1)}{N+q-1}$ and $p^* = (q+1)\frac{N}{N-2}$.

By the summability assumptions on the solution, it is possible to take directly the function $[(v - u)^+]^q$, $v = u_\lambda$, as a test function in the equations for u and u_λ in Σ_λ. More precisely, since u, v belong to $\mathcal{D}^{1,p}(\mathbb{R}^N)$ and they coincide on the hyperplane H_λ, there exists a sequence φ_j of functions in $C_c^\infty(\Sigma_\lambda)$ such that $\varphi_j \to [(v - u)^+]$ in $L^{p^*}(\Sigma_\lambda)$ and $D\varphi_j \to D(v - u)^+$ in $L^p(\Sigma_\lambda)$. Moreover, passing to a subsequence and substituting if necessary φ_j with φ_j^+, we can assume that $0 \le \varphi_j \to [(v - u)^+]$, $D\varphi_j \to D(v - u)^+$ a. e. in Σ_λ, and that there exist functions $\psi_0 \in L^{p^*}$, $\psi_1 \in L^p$, such that $|\varphi_j| \le \psi_0$, $|D\varphi_j| \le \psi_1$ a. e. in Σ_λ.

Taking the functions φ_j^q as test functions in the equations for v and u in Σ_λ and subtracting, we get

$$q \int_{\Sigma_\lambda} \varphi_j^{q-1} D(v - u) \cdot D\varphi_j = \int_{\Sigma_\lambda} [f(v) - f(u)] \varphi_j^q$$

If we can pass to the limit for $j \to \infty$, getting

$$q \int_{\Sigma_\lambda} [(v - u)^+]^{q-1} |D[(v - u)^+]|^2 = \int_{\Sigma_\lambda} [f(v) - f(u)][(v - u)^+]^q < \infty,$$

then the proof is exactly the same as in the previous case.

So it is enough to justify the passage to the limit, which easily follows from the dominated convergence theorem. Indeed we have

$$|[f(v) - f(u)]\varphi_j^q| \le C(u + v)^\alpha |v - u||\varphi_j|^q \le C(u + v)^\alpha |v - u||\psi_0|^q$$

and $(u + v)^\alpha |v - u||\psi_0|^q$ belongs to L^1, since $(u + v)^\alpha \in L^{r_1}$, $|v - u| \in L^{r_2}$, $|\psi_0|^q \in L^{r_3}$, where $r_1 = \frac{N}{2}$, $r_2 = p^* = (q + 1)\frac{N}{N-2}$, $r_3 = \frac{p^*}{q} = \frac{q+1}{q}\frac{N}{N-2}$ with $\frac{1}{r_1} + \frac{1}{r_2} + \frac{1}{r_3} = 1$. Analogously, we have that

$$|\varphi_j^{q-1} D(v - u) \cdot D\varphi_j| \le |\psi_0|^{q-1} |D(v - u)||\psi_1|$$

and $|\psi_0|^{q-1}|D(v - u)||\psi_1|$ belongs to L^1, since $\psi_0^{q-1} \in L^{s_1}$, $|D(v - u)|, |\psi_1| \in L^{s_2}$, where $s_1 = \frac{p^*}{q-1} = \frac{q+1}{q-1}\frac{N}{N-2}$, $s_2 = p = \frac{N(q+1)}{N+q-1}$ and $\frac{1}{s_1} + 2\frac{1}{s_2} = 1$.

Step 2: $u \equiv u_{\lambda_0}$.

Let us first consider the case of hypothesis (H2) or (H3). Since λ_0 is the infimum, by continuity we have that $u \geq u_{\lambda_0}$ in Σ_{λ_0}. Thus, if we suppose $u \not\equiv u_{\lambda_0}$ in Σ_{λ_0}, the strong comparison principle yields that $u > u_{\lambda_0}$ in Σ_{λ_0} and $u_{x_1} < 0$ on H_{λ_0}.

Since $u \in L^{\frac{\alpha N}{2}}(\mathbb{R}^N)$, we can choose a compact set $K \subset \Sigma_{\lambda_0}$ and a number $\delta > 0$ such that $\forall \lambda \in (\lambda_0 - \delta, \lambda_0)$ we have $K \subset \Sigma_\lambda$ and

$$
C_1 \left(\int\limits_{R_\lambda(\Sigma_\lambda \setminus K)} u^{\frac{\alpha N}{2}} \right)^{\frac{2}{N}} < \frac{1}{2},
\tag{8.6}
$$

where C_1 is as in (8.5).

On the other hand, since $u - u_{\lambda_0}$ is positive in Σ_{λ_0}, there exists $0 < \delta_1 < \delta$, such that

$$
u > u_\lambda, \quad \text{in } K \quad \forall \lambda \in (\lambda_0 - \delta_1, \lambda_0).
\tag{8.7}
$$

Using (8.6) and proceeding as in Step 1, since the integrals are on $\Sigma_\lambda \setminus K$, we see that $(u_\lambda - u)^+ \equiv 0$ in $\Sigma_\lambda \setminus K$. By (8.7) we get $u \geq u_\lambda$ in Σ_λ for all $\lambda \in (t_0 - \delta_1, \lambda_0)$, contradicting the definition of λ_0.

In the case of hypothesis (H1), since $u > u_{\lambda_0}$ in Σ_{λ_0} and $u_{x_1} < 0$ on H_{λ_0}, proceeding as in Remark 6.2 it is easy to show that, for any $R > 0$, there exists $\delta = \delta_R > 0$ such that $u > u_\lambda$ in $\Sigma_\lambda \cap B_R(0)$, if $\lambda_0 - \delta < \lambda < \lambda_0$. If $R > 0$ is sufficiently large, then $u(x_\lambda) < s_0$ for any $x \in \Sigma_\lambda \setminus B_{R_0}$ and, proceeding as in Step 1, it is easy to show that $u > u_\lambda$ in $\Sigma_\lambda \setminus B_R(0)$, if $\lambda_0 - \delta < \lambda < \lambda_0$, contradicting the definition of λ_0. $\qquad\square$

Remark 8.2. Note that in the case of critical problems, i. e., when $\alpha = 4/(N-2) = 2^* - 2$, it follows that $\alpha N/2 = 2N/(N-2) = 2^*$, and the radial symmetry holds for solutions in L^{2^*} vanishing at infinity, or solutions belonging to the space $D^{1,p}(\mathbb{R}^N)$ for some $p \in [2, N)$ without assuming a priori that it converges to zero at infinity, if the nonlinearity f satisfies the growth condition both at zero and at infinity. Note also that there could be solutions with infinite energy, i. e., whose gradient does not belong to L^2 (and this happens, e. g., in the supercritical case). This is why we have to take $q \geq 1$ in the proof of Theorem 8.1.

Remark 8.3. Symmetry and monotonicity properties of solutions of elliptic equations in unbounded domains other than $\mathbb{R}^N$, like half-spaces or cylinders, have been largely studied, e. g., in the series of papers [28–30] by Berestycki, Caffarelli and Nirenberg.

8.2 Foliated Schwarz symmetry

Using the Morse index of a solution, as defined in Chapter 3 for bounded or unbounded domains, it is possible to extend the symmetry results of Chapter 6 and Chapter 7 to the case of rotationally symmetric unbounded domains. The symmetry we consider is the foliated Schwarz symmetry, as defined in Chapter 6 and Chapter 7. We observe that the sufficient condition given by Proposition 6.7 holds also in unbounded domain.

Let us consider the problem

$$-\Delta u = f(|x|, u) \quad \text{in } \Sigma \tag{8.8}$$

where Σ is either $\mathbb{R}^N$ or the exterior of a ball, i. e., $\Sigma = \mathbb{R}^N \setminus B$ where B is a ball centered at the origin, $N \geq 2$, and $f : \overline{\Sigma} \times \mathbb{R} \to \mathbb{R}$ is locally a C^1-function.

When $\Sigma = \mathbb{R}^N \setminus B$, we also require the boundary condition

$$u = 0 \quad \text{on } \partial\Sigma. \tag{8.9}$$

In the sequel by a *solution* of (8.8), (8.9) we mean a classical $C^2(\overline{\Sigma})$ solution.

The following results, which extend the result of Chapter 6, have been proved in [131].

Theorem 8.4. *Suppose that $f(|x|, s)$ has a convex derivative $f'(|x|, s) = \frac{\partial f}{\partial s}(|x|, s)$ for every $x \in \Sigma$. Then every solution u of (8.8) and (8.9) with $|\nabla u| \in L^2(\Sigma)$ and Morse index $j \leq N$ is foliated Schwarz symmetric.*

Theorem 8.5. *Suppose that $f(|x|, s)$ is convex in the s-variable for every $x \in \Sigma$. Then every solution u of (8.8) and (8.9) with $|\nabla u| \in L^2(\Sigma)$ and Morse index $j \leq N$ is foliated Schwarz symmetric.*

We point out that the previous convexity assumptions are not satisfied for the Allen–Cahn nonlinearity $f(u) = u - u^3$ and its generalizations, for which, in contrast to our results, there exist in $\mathbb{R}^N$ rather complicated solutions with low Morse index (see the remarks in [90] and the references therein for a discussion on this topic). On the other hand, in the previous theorems, nonlinearities depending explicitly on the radial space variable $|x|$ are allowed.

We will not give the proofs of these theorems, for which we refer to the paper [131], but we only indicate the main steps and point out that in each of them a crucial role is played by the assumption that $|\nabla u| \in L^2(\Sigma)$.

As in the bounded domain case, an important step is to prove the following counterpart of Theorem 6.12.

Proposition 8.6. *Let u be a solution of (8.8) and (8.9) such that $|\nabla u| \in L^2(\Sigma)$. Assume that there exists $e \in S$ such that u is symmetric with respect to the hyperplane $H(e)$ and*

$$\inf_{\psi \in C_c^1(\Sigma(e))} Q_u(\psi) \geq 0. \tag{8.10}$$

where

$$Q_u(\psi) = \int_{\Sigma} \left[|\nabla\psi|^2 - \frac{\partial f}{\partial s}(|x|, u(x)) |\psi(x)|^2 \right] dx \tag{8.11}$$

Then u is foliated Schwarz symmetric.

Another important ingredient in the proof of the symmetry results is the rotating plane method introduced in Chapter 6.

We need the following definition.

Definition 8.7. The solution u of (8.8) and (8.9) is said to be stable outside a compact set $\mathcal{K} \subset \Sigma$ if $Q_u(\psi) \geq 0$ for all $\psi \in C_c^1(\Sigma \setminus \mathcal{K})$ where Q_u is the quadratic form defined in (8.11).

It is easy to see that the following holds (see [89, 90, 117]).

Remark 8.8. If u has finite Morse index, then u is stable outside a compact set $\mathcal{K} \subset \Sigma$.

Denoting, as in Chapter 6, by w_e the difference between a function u and its reflection with respect to the hyperplane $H(e)$, it is possible to prove the following result by means of a rotating plane argument.

Proposition 8.9. *Let u be a solution of (8.8) and (8.9) stable outside a compact set $\mathcal{K} \subset \Sigma$ and such that $|\nabla u| \in L^2(\Sigma)$. Suppose that either f or f' is convex in the second variable, and that there exists a direction $e \in S$ such that*

$$w_e(x) < 0 \quad \forall x \in \Sigma(e) \quad or \quad w_e(x) > 0 \quad \forall x \in \Sigma(e). \tag{8.12}$$

Then there exists another direction $e' \in S$ such that $w_{e'} \equiv 0$, (i. e., u is symmetric with respect to the hyperplane $H(e')$) and

$$\inf_{\psi \in C_c^1(\Sigma(e'))} Q_u(\psi) \geq 0. \tag{8.13}$$

Another result, which is the counterpart of Proposition 6.21, and which does not depend on the convexity of the nonlinearities, is the following.

Proposition 8.10. *Suppose that u has Morse index $j \leq N$. Then there exists $e \in S$ such that*

$$\inf_{\psi \in C_c^1(\Sigma(e))} Q_u(\psi, \psi) \geq 0. \tag{8.14}$$

Let us finally remark that although the characterization of the Morse index by means of the number of negative eigenvalues (Theorem 3.2) is not available for unbounded domains, in many proofs, this characterization is exploited in large balls, assuming that the solution has a Morse index not exceeding the dimension N.

Remark 8.11. In the paper [73], the results for systems of Chapter 7 have also been extended to rotationally symmetric unbounded domains, with suitable hypotheses on the solution, the main one being that the gradient of the solution belongs to L^2.

Besides the technical difficulties that arise when considering solutions in unbounded domains, also the various type of eigenvalues that can be considered for systems play a role, as in Chapter 7. We refer to [73] for the statements and the proofs.

8.3 Nonexistence and classification results

Information on the Morse index of a solution can also be used to deduce some Liouville-type nonexistence results. Indeed as a consequence of the foliated Schwarz symmetry the nonexistence of nontrivial solutions have been proved in some cases (see [131]). We state some of them below.

Theorem 8.12. *Every stable solution u of (8.8) and (8.9) such that $|\nabla u| \in L^2(\Sigma)$ is radial. If in addition $\Sigma = \mathbb{R}^N$ and f does not depend on $|x|$, then u is constant.*

This theorem generalizes and complements results on stable solutions obtained in [55] and [90]. We stress that in Theorems 8.4–8.12 we do *not* assume boundedness of the solution u.

In the case when f does not depend on $|x|$, some other nonexistence results can be deduced from Theorem 8.4 and Theorem 8.5.

Theorem 8.13. *Assume that $\Sigma = R^N$ and $f = f(s)$, i. e., f does not depend on x and that either f is convex or f' is convex. Then there are no sign changing solutions u of (8.8) with*

$$|\nabla u| \in L^2(\mathbb{R}^N), \quad u(x) \to 0 \quad as \ |x| \to \infty$$

and Morse index $j \le N$.

Theorem 8.14. *Assume that $\Sigma = \mathbb{R}^N \setminus B$ and $f = f(s)$, i. e., f does not depend on x and that either f is convex or f' is convex. Then there are no solutions u (neither positive nor sign changing) of (8.8) and (8.9) with*

$$|\nabla u| \in L^2(\Sigma), \quad u(x) \to 0 \quad as \ |x| \to \infty$$

and Morse index $j \le N$.

These results are quite easy corollaries of Theorems 8.4 and 8.5 except in the case when nodal solutions in $\mathbb{R}^N \setminus B$ have to be excluded. This part is more difficult and requires additional ideas (see [131]).

The previous results apply to many nonlinear problems. In particular, to semilinear elliptic equations of the type

$$-\Delta u = |u|^{\frac{4}{N-2}}u \quad in \ \mathbb{R}^N, N \ge 3 \tag{8.15}$$

and

$$-\Delta u + u = |u|^{\sigma}u, \quad u \in H^1(\mathbb{R}^N) \quad or \quad u \in H_0^1(\mathbb{R}^N \setminus B), \tag{8.16}$$

where $\sigma > 0$ and $\sigma < \frac{4}{N-2}$ if $N > 3$.

In particular, the nonexistence result of Theorem 8.13 applies to the case of the critical power nonlinearity (8.15) which satisfies the hypothesis that f' is convex for $3 \leq N \leq 6$. In this case, the assumptions

$$|\nabla u| \in L^2(\Sigma), \quad u(x) \to 0 \quad \text{as } |x| \to \infty \tag{8.17}$$

are automatically satisfied since it is proved in [117] that every classical solution u with finite Morse index belongs to the space $D^{1,2}(\mathbb{R}^N) \cap L^\infty(\mathbb{R}^N)$ where $D^{1,2}(\mathbb{R}^N)$ is defined as the completion of $C_c^\infty(\mathbb{R}^N)$ in the norm $\|v\|_{D^{1,2}(\mathbb{R}^N)} = \|\nabla v\|_{L^2(\mathbb{R}^N)}$. We believe that the properties (8.17) should hold for finite Morse index solutions corresponding to a more general class of nonlinearities, although they are certainly not satisfied in the case of the Allen–Cahn nonlinearity $f(u) = u - u^3$.

We end by mentioning that many nonexistence and classification results have been proved by using the Morse index of a solution but with other methods.

In particular, we state below some very interesting results proved by Farina in [117] for the solutions of the Lane–Emden equation

$$-\Delta u = |u|^{p-1}u \quad \text{in } \Omega \tag{8.18}$$

where Ω is an unbounded domains.

They are based on a method which exploits suitable test functions (we refer to the paper and the references therein for a detailed discussion of the results).

Theorem 8.15. *Let $\Omega = \mathbb{R}^N$ and let $u \in C^2(\mathbb{R}^N)$ be a* stable *solution of (8.18) with*

$$\begin{cases} 1 < p < \infty & \text{if } N \leq 10 \\ 1 < p < p_c(N) := \frac{(N-2)^2 - 4N + 8\sqrt{N-1}}{(N-2)(N-10)} & \text{if } N \geq 11 \end{cases} \tag{8.19}$$

Then $u \equiv 0$.

On the other hand for $N \geq 11$ and every $p \geq p_c(N)$, there exists a smooth, positive, bounded and stable radial solution of (8.18).

Theorem 8.16. *Let $\Omega = \mathbb{R}^N$ and let $u \in C^2(\mathbb{R}^N)$ be a solution of (8.18) which is* stable *outside a compact set. Assume*

$$\begin{cases} 1 < p < \infty & \text{if } N = 2 \\ 1 < p < \infty, \, p \neq \frac{N+2}{N-2} & \text{if } 3 \leq N \leq 10 \\ 1 < p < p_c(N), \, p \neq \frac{N+2}{N-2} & \text{if } N \geq 11 \end{cases} \tag{8.20}$$

Then $u \equiv 0$.

On the other hand, if $N \geq 3$ and $p = \frac{N+2}{N-2}$ then

$$\int_{\mathbb{R}^N} |\nabla u|^2 = \int_{\mathbb{R}^N} |u|^{\frac{2N}{N-2}} < +\infty$$

for any solution of (8.18) *and*

$$\lim_{|x|\to\infty} |x|^{\frac{N-2}{N}} u(x) = \lim_{|x|\to\infty} |x|^{\frac{N}{2}} |\nabla u(x)| = 0$$

Other results for finite Morse index solutions of semilinear elliptic equations in bounded or unbounded domains with general nonlinearities have been obtained by Dancer and other authors in several papers. We refer to [19, 89–93, 112, 113, 141, 142] and the references therein.

Bibliography

[1]   E. Abreu, J. Marcos do O and E. Medeiros, "Properties of positive harmonic functions on the half space with a nonlinear boundary condition". In: *J. Differ. Equ.* 248 (2010), pp. 617–637.

[2]   R. A. Adams, *Sobolev Spaces*, Academic Press, 1975.

[3]   Adimurthi and M. Grossi, "Asymptotic estimates for a two-dimensionl problem with polynomial nonlinearity". In: *Proc. Am. Math. Soc.* 132 (2004), pp. 1013–1019.

[4]   Adimurthi and S. Yadava, "An elementary proof of the uniqueness of positive radial solutions of a quasilinear Dirichlet problem". In: *Arch. Ration. Mech. Anal.* 127 (1996), pp. 219–229.

[5]   A. Aftalion and F. Pacella, "Uniqueness and nondegeneracy for some nonlinear elliptic problems in a ball". In: *J. Differ. Equ.* 195 (2003), pp. 380–397.

[6]   A. Aftalion and F. Pacella, "Morse index and uniqueness for positive solutions of radial p-Laplace equations". In: *Trans. Am. Math. Soc.* 356 (2004), pp. 4255–4272.

[7]   A. Aftalion and F. Pacella, "Qualitative properties of nodal solutions of semilinear elliptic equations in radially symmetric domains". In: *C. R. Acad. Sci. Paris* 339 (2004), pp. 339–344.

[8]   A. D. Alexandrov, "A characteristic property of spheres". In: *Ann. Mat. Pura Appl.* 58 (1962), pp. 303–354.

[9]   L. Almeida, L. Damascelli and Y. Ge, "A few symmetry results for nonlinear elliptic PDE on noncompact manifolds". In: *Ann. Inst. Henri Poincaré, Anal. Non Linéaire* 19 (2002), pp. 313–342.

[10]   A. L. Amadori and F. Gladiali, "Bifurcation and symmetry breaking for the Hénon equation". In: *Adv. Differ. Equ.* 19 (2014), pp. 755–782.

[11]   A. L. Amadori and F. Gladiali, "Nonradial sign changing solutions to Lane-Emden problem in an annulus". In: *Nonlinear Anal.* 155 (2017), pp. 294–305.

[12]   A. L. Amadori and F. Gladiali, *On a singular eigenvalue problem and its applications in computing the Morse index of solutions to semilinear PDE's*, arXiv:1805.04321v1.

[13]   A. Ambrosetti and E. Colorado, "Standing waves of some coupled nonlinear Schrödinger equations". In: *J. Lond. Math. Soc.* 75 (2007), pp. 67–82.

[14]   A. Ambrosetti and A. Malchiodi, *Nonlinear Analysis and Semilinear Elliptic Problems*, Cambridge University Press, 2007.

[15]   A. Ambrosetti and P. H. Rabinowitz, "Dual variational methods in critical point theory and applications". In: *J. Funct. Anal.* 14 (1973), pp. 349–381.

[16]   F. V. Atkinson and L. A. Peletier, "Elliptic equations with nearly critical growth". In: *J. Differ. Equ.* 70 (1987), pp. 349–365.

[17]   T. Aubin, "Problèmes isopérimetriques et espaces de Sobolev". In: *J. Differ. Geom.* 11 (1976), pp. 573–598.

[18]   G. Auchmuty, "Steklov eigenproblems and the representation of solutions of elliptic boundary value problems". In: *Numer. Funct. Anal. Optim.* 25 (2004), pp. 321–348.

[19]   A. Bahri and P. L. Lions, "Solutions of superlinear elliptic equations and their Morse indices". In: *Commun. Pure Appl. Math.* 45 (1992), pp. 1205–1215.

[20]   T. Bartsch, "Critical point theory on partially ordered Hilbert spaces". In: *J. Funct. Anal.* 186 (2001), pp. 117–152.

[21]   T. Bartsch, M. Clapp, M. Grossi and F. Pacella, "Asymptotically radial solutions in expanding annular domains". In: *Math. Ann.* 352 (2012), pp. 485–515.

[22]   T. Bartsch, A. Szulkin and T. Weth, "Morse theory and nonlinear differential equations". In: *Handbook of Global Analysis*. Elsevier, 2008, pp. 41–73.

[23]   T. Bartsch and T. Weth, "A note on additional properties of sign changing solutions to superlinear elliptic equations". In: *Topol. Methods Nonlinear Anal.* 22 (2003), pp. 1–14.

https://doi.org/10.1515/9783110538243-009

[24]   T. Bartsch, T. Weth and M. Willem, "Partial symmetry of least energy nodal solutions to some variational problems". In: *J. Anal. Math.* 96 (2005), pp. 1–18.

[25]   M. Ben Ayed, K. El Mehdi, M. Ould Amehdou and F. Pacella, "Energy and Morse index of solutions of Yamabe type problems on thin annuli". In: *J. Eur. Math. Soc.* 7 (2005), pp. 283–304.

[26]   M. Ben Ayed, K. El Mehdi and F. Pacella, "Classification of low energy sign-changing solutions of an almost critical problem". In: *J. Funct. Anal.* 250 (2007), pp. 347–373.

[27]   M. Ben Ayed, H. Fourti and A. Selmi, "Harmonic Functions with nonlinear Neumann boundary condition and their Morse index". In: *Nonlinear Anal., Real World Appl.* 38 (2017), pp. 96–112.

[28]   H. Berestycki, L. A. Caffarelli and L. Nirenberg, "Inequalities for second-order elliptic equations with applications to unbounded domains. I". In: *Duke Math. J.* 81 (1996), pp. 467–494.

[29]   H. Berestycki, L. A. Caffarelli and L. Nirenberg, "Further qualitative properties for elliptic equations in unbounded domains". In: *Ann. Sc. Norm. Super. Pisa, Cl. Sci.* 25 (1997), pp. 69–94.

[30]   H. Berestycki, L. A. Caffarelli and L. Nirenberg, "Monotonicity for elliptic equations in unbounded Lipschitz domains". In: *Commun. Pure Appl. Math.* 50 (1997), pp. 1089–1111.

[31]   H. Berestycki and L. Nirenberg, "On the method of moving planes and the sliding method". In: *Bol. Soc. Bras. Mat.* 22 (1991), pp. 1–22.

[32]   H. Berestycki, L. Nirenberg and S. R. S. Varadhan, "The principal eigenvalue and maximum principle for second order elliptic operators in general domains". In: *Commun. Pure Appl. Math.* 47 (1994), pp. 47–92.

[33]   H. Berestycki and F. Pacella, "Symmetry properties for positive solutions of elliptic equations with mixed boundary conditions". In: *J. Funct. Anal.* 87 (1989), pp. 177–211.

[34]   F. A. Berezin and M. A. Shubin, *The Schrödinger Equation*, Kluwer Academic Publishers Group, 1991.

[35]   I. Birindelli, F. Leoni and F. Pacella, "Symmetry and spectral properties for viscosity solutions of fully nonlinear equations". In: *J. Math. Pures Appl.* 107 (2017), pp. 409–428.

[36]   G. A. Bliss, "An integral inequality". In: *J. Lond. Math. Soc.* 5 (1930), pp. 40–46.

[37]   D. Bonheure, V. Bouchez, C. Grumiau and J. Van Schaftingen, "Asymptotic and symmetry of least energy nodal solutions of Lane-Emden problems with slow growth". In: *Commun. Contemp. Math.* 10 (2008), pp. 609–631.

[38]   D. Bonheure, E. Moreira dos Santos, M. Ramos and H. Tavares, "Existence and simmetry of least energy nodal solutions for Hamiltonian elliptic systems". In: *J. Math. Pures Appl.* 104 (2015), pp. 1075–1107.

[39]   R. Bott, "Lectures on Morse theory, old and new". In: *Bull. Am. Math. Soc.* 7 (1982).

[40]   G. E. Bredon, *Introduction to Compact Transformation Groups*, Academic Press, 1972.

[41]   H. Brezis, *Functional Analysis, Sobolev Spaces and Partial Differential Equations*, Springer, 2011.

[42]   H. Brezis and J. M. Coron, "Convergence de solutions de H-systèmes et application aux surfaces à courbure moyenne constante". In: *C. R. Acad. Sci. Paris* 298 (1984), pp. 389–392.

[43]   H. Brezis and J. M. Coron, "Convergence of solutions of H-systems or how to blow bubbles". In: *Arch. Ration. Mech. Anal.* 89 (1985), pp. 21–56.

[44]   H. Brezis and S. Kamin, "Sublinear elliptic equations in $\mathbb{R}^n$". In: *Manuscr. Math.* 74 (1992), pp. 87–106.

[45]   H. Brezis and L. Nirenberg, "Positive solutions of nonlinear elliptic equations involving critical Sobolev exponents". In: *Commun. Pure Appl. Math.* 36 (1983), pp. 437–477.

[46]   H. Brezis and L. Oswald, "Remarks on sublinear elliptic equations". In: *Nonlinear Anal.* 10 (1986), pp. 55–64.

[47]   F. Brock, "Continuous rearrangement and symmetry of solutions of elliptic problems". In: *Proc. Indian Acad. Sci. Math. Sci.* 110 (2000), pp. 157–204.

[48]   F. Brock, "Symmetry and monotonicity of solutions to some variational problems in cylinders and annuli". In: *Electron. J. Differ. Equ.* (2003), pp. 1–20.

[49]   F. Brock, "Positivity and radial symmetry of solutions to some variational problems in $\mathbb{R}^N$". In: *J. Math. Anal. Appl.* 296 (2004), pp. 226–243.

[50]   F. Brock and A. Y. Solynin, "An approach to symmetrization via polarization". In: *Trans. Am. Math. Soc.* 352 (2000), pp. 1759–1796.

[51]   J. Brüning and E. Heintze, "Représentations des groupes d'isométries dans les sous-espaces propres du laplacien". In: *C. R. Acad. Sci. Paris* 286 (1978), pp. 921–923.

[52]   J. Brüning and E. Heintze, "Representations of compact Lie groups and elliptic operators". In: *Invent. Math.* 50 (1979), pp. 169–203.

[53]   J. Busca and B. Sirakov, "Symmetry results for semilinear elliptic systems in the whole space". In: *J. Differ. Equ.* 163 (2000), pp. 41–56.

[54]   J. Busca and B. Sirakov, "Harnack type estimates for nonlinear ellyptic systems and applications". In: *Ann. Inst. Henri Poincaré, Anal. Non Linéaire* 21 (2004), pp. 543–590.

[55]   X. Cabré and A. Capella, "On the stability of radial solutions of semilinear elliptic equations in all of $\mathbb{R}^n$". In: *C. R. Acad. Sci. Paris* 338 (2004), pp. 769–774.

[56]   D. Castorina and F. Pacella, "Symmetry of positive solutions of an almost-critical problem in an annulus". In: *Calc. Var. Partial Differ. Equ.* 23 (2005), pp. 125–138.

[57]   A. Castro, J. Cossio and J. M. Neuberger, "A sign changing solution for a superlinear Dirichlet problem". In: *Rocky Mt. J. Math.* 27 (1997), pp. 1041–1053.

[58]   F. Catrina and Z. Q. Wang, "Nonlinear elliptic equations on expanding symmetric domains". In: *J. Differ. Equ.* 156 (1999), pp. 153–181.

[59]   T. Cazenave, F. Dickstein and F. B. Weissler, "Sign-changing stationary solutions and blowup for the nonlinear heat equation in a ball". In: *Math. Ann.* 344 (2009), pp. 431–449.

[60]   K. C. Chang, *Infinite Dimensional Morse Theory and Applications to Differential Equations*, Birkhauser, 1992.

[61]   G. Chen, W. M. Ni and J. Zhou, "Algorithms and visualization for solutions of nonlinear elliptic equations". In: *Int. J. Bifurc. Chaos Appl. Sci. Eng.* 10 (2000), pp. 1565–1612.

[62]   W. Chen and C. Li, "Classification of solutions of some nonlinear elliptic equations". In: *Duke Math. J.* 63 (1991), pp. 615–622.

[63]   M. Clapp, "Entire nodal solutions to the pure critical exponent problem arising from concentration". In: *J. Differ. Equ.* 261 (2015), pp. 3042–3060.

[64]   P. Clément, D. G. de Figueiredo and E. Mitidieri, "Quasilinear elliptic equations with critical exponents". In: *Topol. Methods Nonlinear Anal.* 7 (1996), pp. 133–170.

[65]   C. Cowan and N. Ghoussoub, "Estimates on pull-in distances in microelectromechanical systems model and other nonlinear aigenvalue problems". In: *SIAM J. Math. Anal.* 42 (2010), pp. 1949–1966.

[66]   M. G. Crandall and P. H. Rabinowitz, "Bifurcation from simple eigenvalues". In: *J. Funct. Anal.* 8 (1971), pp. 321–340.

[67]   L. Damascelli, "A remark on the uniqueness of the positive solution for a semilinear elliptic equation". In: *Nonlinear Anal.* 26 (1996), pp. 211–216.

[68]   L. Damascelli, "Comparison theorems for some quasilinear degenerate elliptic operators and applications to symmetry and monotonicity results". In: *Ann. Inst. Henri Poincaré, Anal. Non Linéaire* 15 (1998), pp. 493–516.

[69]   L. Damascelli, "Some remarks on the method of moving planes". In: *Differ. Integral Equ.* 11 (1998), pp. 493–501.

[70] L. Damascelli, "On the nodal set of the second eigenfunction of the laplacian in symmetric domains in $\mathbb{R}^N$". In: *Atti Accad. Naz. Lincei, Cl. Sci. Fis. Mat. Nat., Rend. Lincei, Suppl. 9* (2000), pp. 175–181.

[71] L. Damascelli and F. Gladiali, "Some nonexistence results for positive solutions of elliptic equations in unbounded domains". In: *Rev. Mat. Iberoam.* 20 (2004), pp. 67–86.

[72] L. Damascelli, F. Gladiali and F. Pacella, "A symmetry result for semilinear cooperative elliptic systems". In: *Recent Trends in Nonlinear Partial Differential Equations. Ii. Stationary Problems*. Contemporary Mathematics 595. AMS, 2013, pp. 187–204.

[73] L. Damascelli, F. Gladiali and F. Pacella, "Symmetry results for cooperative elliptic systems in unbounded domains". In: *Indiana Univ. Math. J.* 63 (2014), pp. 615–649.

[74] L. Damascelli, M. Grossi and F. Pacella, "Qualitative properties of positive solutions of semilinear elliptic equations in symmetric domains via the maximum principle". In: *Ann. Inst. Henri Poincaré, Anal. Non Linéaire* 16 (1999), pp. 631–652.

[75] L. Damascelli and F. Pacella, "Monotonicity and symmetry of solutions of p-Laplace equations, $1 < p < 2$, via the moving plane method". In: *Ann. Sc. Norm. Super. Pisa, Cl. Sci.* 26 (1998), pp. 689–707.

[76] L. Damascelli and F. Pacella, "Monotonicity and symmetry results for p-Laplace equations and applications". In: *Adv. Differ. Equ.* 5 (2000), pp. 1179–1200.

[77] L. Damascelli and F. Pacella, "Symmetry results for cooperative elliptic systems via linearization". In: *SIAM J. Math. Anal.* 45 (2013), pp. 1003–1026.

[78] L. Damascelli and F. Pacella, "Morse index and symmetry for elliptic problems with nonlinear mixed boundary conditions". In: *Proc. R. Soc. Edinb. A* 149 (2019), pp. 305–324.

[79] L. Damascelli and F. Pacella, "Sectional symmetry of solutions of elliptic systems in cylindrical domains". arXiv:1905.02026 (2019).

[80] L. Damascelli, F. Pacella and M. Ramaswamy, "Symmetry of ground states of p-Laplace equations via the moving plane method". In: *Arch. Ration. Mech. Anal.* 148 (1999), pp. 291–308.

[81] L. Damascelli and M. Ramaswamy, "Symmetry of C^1 solutions of p-Laplace equations in R^N". In: *Adv. Nonlinear Stud.* 1 (2001), pp. 40–64.

[82] L. Damascelli and B. Sciunzi, "Regularity, monotonicity and symmetry of positive solutions of m-Laplace equations". In: *J. Differ. Equ.* 206 (2004), pp. 483–515.

[83] L. Damascelli and B. Sciunzi, "Qualitative properties of solutions of m-Laplace systems". In: *Adv. Nonlinear Stud.* 5 (2005), pp. 197–221.

[84] L. Damascelli and B. Sciunzi, "Harnack inequalities, maximum and comparison principles, and regularity of positive solutions of m-Laplace equations". In: *Calc. Var. Partial Differ. Equ.* 25 (2006), pp. 139–159.

[85] L. Damascelli and B. Sciunzi, "Monotonicity of the solutions of some quasilinear elliptic equations in the half-plane, and applications". In: *Differ. Integral Equ.* 23 (2010), pp. 419–434.

[86] E. N. Dancer, "The effect of the domain shape on the number of positive solutions of certain nonlinear equations". In: *J. Differ. Equ.* 74 (1988), pp. 120–156.

[87] E. N. Dancer, "Some notes on the method of moving planes". In: *Bull. Aust. Math. Soc.* 46 (1992), pp. 425–434.

[88] E. N. Dancer, "Real analyticity and non-degeneracy". In: *Math. Ann.* 325 (2003), pp. 369–392.

[89] E. N. Dancer, "Stable and finite Morse index solutions on $\mathbb{R}^n$ or on bounded domains with small diffusion, II". In: *Indiana Univ. Math. J.* 53 (2004), pp. 97–108.

[90] E. N. Dancer, "Stable and finite Morse index solutions on $\mathbb{R}^n$ or on bounded domains with small diffusion". In: *Trans. Am. Math. Soc.* 357 (2005), pp. 1225–1243.

[91] E. N. Dancer, "Stable and not too unstable solutions on R^n for small diffusion". In: *Nonlinear Dynamics and Evolution Equations*. Fields Inst. Commun. 48. Providence, RI: Amer. Math. Soc., 2006, pp. 67–93.

[92] E. N. Dancer, "Finite Morse index solutions of supercritical problems". In: *J. Reine Angew. Math.* 620 (2008), pp. 213–233.

[93] E. N. Dancer, Y. Du and Z. M. Guo, "Finite Morse index solutions of an elliptic equation with supercritical exponent". In: *J. Differ. Equ.* 250 (2011), pp. 3281–3310.

[94] D. G. de Figueiredo, "Monotonicity and symmetry of solutions of elliptic systems in general domains". In: *Nonlinear Differ. Equ. Appl.* 1 (1994), pp. 119–123.

[95] D. G. de Figueiredo, "Semilinear elliptic systems: existence, multiplicity, symmetry of solutions". In: *Handbook of Differential Equations: Stationary Partial Differential Equations Vol. V.* 2008, pp. 1–48.

[96] D. G. de Figueiredo, P. L. Lions and R. D. Nussbaum, "A priori estimates and existence of positive solutions of semilinear elliptic equations". In: *J. Math. Pures Appl.* 61 (1982), pp. 41–63.

[97] D. G. de Figueiredo and E. Mitidieri, "Maximum principles for linear elliptic systems". In: *Rend. Ist. Mat. Univ. Trieste* (1992), pp. 36–66.

[98] F. De Marchis, M. Grossi, I. Ianni and F. Pacella, "L^∞-norm and energy quantization for the planar Lane-Emden problem with large exponent". In: *Arch. Math.* 111 (2018), pp. 421–429.

[99] F. De Marchis, M. Grossi, I. Ianni and F. Pacella, "Morse index and uniqueness of positive solutions of the Lane-Emden problem in planar domains". In: *J. Math. Pures Appl.* (2018), in press, doi.org/10.1016/j.matpur.2019.02.011.

[100] F. De Marchis and I. Ianni, "Blow up of solutions of semilinear heat equations in non radial domains in $\mathbb{R}^2$". In: *Discrete Contin. Dyn. Syst.* 35 (2015), pp. 891–907.

[101] F. De Marchis, I. Ianni and F. Pacella, "Sign-changing solutions of Lane Emden problems with interior nodal line and semilinear heat equation". In: *J. Differ. Equ.* 254 (2013), pp. 3596–3614.

[102] F. De Marchis, I. Ianni and F. Pacella, "Morse index and sign-changing bubble towers for Lane-Emden problems". In: *Ann. Mat. Pura Appl.* (2014).

[103] F. De Marchis, I. Ianni and F. Pacella, "Asymptotic analysis and sign-changing bubble towers for Lane-Emden problems". In: *J. Eur. Math. Soc.* 17 (2015), pp. 2037–2068.

[104] F. De Marchis, I. Ianni and F. Pacella, "Asimptotic profile of positive solutions of Lane-Emden problems in dimension two". In: *J. Fixed Point Theory Appl.* 19 (2017), pp. 889–916.

[105] F. De Marchis, I. Ianni and F. Pacella, "Exact Morse index computations for nodal radial solutions of Lane-Emden problems". In: *Math. Ann.* 367 (2017), pp. 185–227.

[106] F. De Marchis, I. Ianni and F. Pacella, "A Morse index formula for radial solutions of Lane-Emden problems". In: *Adv. Math.* 322 (2017), pp. 682–737.

[107] K. Deimling, *Nonlinear Functional Analysis*, New York: Springer, 1985.

[108] M. Del Pino, M. Musso, F. Pacard and A. Pistoia, "Large energy entire solutions for the Yamabe problem". In: *J. Differ. Equ.* 251 (2011), pp. 2568–2597.

[109] F. Dickstein, F. Pacella and B. Sciunzi, "Sign-changing stationary solutios and blowup for a nonlinear heat equation in dimension two". In: *J. Evol. Equ.* 14 (2014), pp. 617–633.

[110] W. Ding, "On a conformally invariant elliptic equation in $\mathbb{R}^N$". In: *Commun. Math. Phys.* 107 (1986), pp. 331–335.

[111] H. Donnelly, "G-spaces, the asymptotic splitting of $L^2(M)$ into irreducibles". In: *Math. Ann.* 237 (1978), pp. 23–40.

[112] L. Dupaigne and A. Farina, "Liouville theorems for stable solutions of semilinear elliptic equations with convex nonlinearities". In: *Nonlinear Anal.* 70 (2009), pp. 2882–2888.

[113] L. Dupaigne and A. Farina, "Stable solutions of $-\Delta u = f(u)$ in $\mathbb{R}^N$". In: *J. Eur. Math. Soc.* 12 (2010), pp. 855–882.

[114] K. El Mehdi and M. Grossi, "Asymptotic estimates and qualitative properties of an elliptic problem in dimension two". In: *Adv. Nonlinear Stud.* 4 (2004), pp. 15–36.

[115] L. C. Evans, *Partial Differential Equations*, Graduate Studies in Mathematics 19, AMS, 1998.

[116] L. C. Evans and R. F. Gariepy, *Measure Theory and Fine Properties of Functions*, CRC Press, 1992.

[117] A. Farina, "On the classification of solutions of the Lane-Emden equation on unbounded domains of $\mathbb{R}^N$". In: *J. Math. Pures Appl.* 87 (2007), pp. 537–561.

[118] A. Farina, L. Montoro and B. Sciunzi, "Monotonicity and one-dimensional symmetry for solutions of $-\Delta_p u = f(u)$ in half-spaces". In: *Calc. Var. Partial Differ. Equ.* 43 (2012), pp. 123–145.

[119] A. Farina, L. Montoro and B. Sciunzi, "Monotonicity of solutions of quasilinear degenerate elliptic equations in half-spaces". In: *Math. Ann.* 357 (2013), pp. 855–893.

[120] J. Garcia-Melián, J. D. Rossi and J. C. Sabina de Lis, "Existence and uniqueness of positive solutions to elliptic problems with sublinear mixed boundary conditions". In: *Commun. Contemp. Math.* 11 (2009), pp. 585–613.

[121] N. Ghoussoub, "Location, multiplicity and Morse index of min-max critical points". In: *J. Reine Angew. Math.* 417 (1991), pp. 27–76.

[122] B. Gidas, W. M. Ni and L. Nirenberg, "Symmetry and related properties via the maximum principle". In: *Commun. Math. Phys.* 68 (1979), pp. 209–243.

[123] B. Gidas, W. M. Ni and L. Nirenberg, "Symmetry of positive solutions of nonlinear elliptic equations in R^n". In: *Mathematical Analysis Applications, Part A*. Adv. Math. Suppl. Stud. 7A. 1981, pp. 369–402.

[124] B. Gidas and J. Spruck, "Global and local behavior of positive solutions of nonlinear elliptic equations". In: *Commun. Pure Appl. Math.* 34 (1981), pp. 525–598.

[125] B. Gidas and J. Spruck, "A priori bounds for positive solutions of nonlinear elliptic equations". In: *Commun. Partial Differ. Equ.* 6 (1981), pp. 883–901.

[126] D. Gilbarg and N. S. Trudinger, *Elliptic Partial Differential Equations of Second Order*. 2nd edition, Springer, 1983.

[127] F. Gladiali, M. Grossi and S. L. N. Neves, "Nonradial solutions for the Hénon equation in $\mathbb{R}^N$". In: *Adv. Math.* 249 (2013), pp. 1–36.

[128] F. Gladiali, M. Grossi and S. L. N. Neves, "Symmetry breaking and Morse index of solutions of nonlinear elliptic problems in the plane". In: *Commun. Contemp. Math.* 18 (2016).

[129] F. Gladiali, M. Grossi, F. Pacella and P. N. Srikanth, "Bifurcation and symmetry breaking for a class of semilinear elliptic equations in an annulus". In: *Calc. Var. Partial Differ. Equ.* 40 (2011), pp. 295–317.

[130] F. Gladiali and I. Ianni, *Quasi-radial nodal solutions for the Lane-Emden problem in a ball*, arXiv:1709.03315.

[131] F. Gladiali, F. Pacella and T. Weth, "Symmetry and Nonexistence of low Morse index solutions in unbounded domains". In: *J. Math. Pures Appl.* 93 (2010), pp. 536–558.

[132] M. J. Greenberg, *Lectures on Algebraic Topology*, Benjamin Inc, 1967.

[133] M. Grossi, "A uniqueness result for a semilinear elliptic equation in symmetric domains". In: *Adv. Differ. Equ.* 5 (2000), pp. 193–212.

[134] M. Grossi, "Asymptotitc behavior of the Kazdan-Warner solution in the annulus". In: *J. Differ. Equ.* 223 (2006), pp. 96–111.

[135] M. Grossi, "On the shape of solutions of an asymptotically linear problem". In: *Ann. Sc. Norm. Super. Pisa* 8 (2009), pp. 429–449.

[136] M. Grossi, C. Grumiau and F. Pacella, "Lane Emden problems: asymptotic behavior of low energy nodal solutions". In: *Ann. Inst. Henri Poincaré, Anal. Non Linéaire* 30 (2013), pp. 121–140.

[137] M. Grossi, C. Grumiau and F. Pacella, "Lane Emden problems with large exponents and singular Liouville equations". In: *J. Math. Pures Appl.* 101 (2014), pp. 735–754.

[138] Z. C. Han, "Asymptotic approach to singular solutions for nonlinear elliptic equations involving critical Sobolev exponent". In: *Ann. Inst. Henri Poincaré, Anal. Non Linéaire* 8 (1991), pp. 159–174.

[139] G. H. Hardy, "Notes on some points in the integral calculus". In: *Messenger Math.* 30 (1901), pp. 185–190.

[140] G. H. Hardy, J. E. Littlewood and G. Polya, *Inequalities*, Cambridge University Press, 1934.

[141] A. Harrabi, S. Rebbi and A. Selmi, "Solutions of superlinear elliptic equations and their Morse indices, I". In: *Duke Math. J.* 94 (1998), pp. 141–157.

[142] A. Harrabi, S. Rebbi and A. Selmi, "Solutions of superlinear elliptic equations and their Morse indices, II". In: *Duke Math. J.* 94 (1998), pp. 159–179.

[143] A. Harrabi, S. Rebbi and A. Selmi, "Existence of radial solutions with prescribed number of zeros for elliptic equations and their Morse index". In: *J. Differ. Equ.* 251 (2011), pp. 2409–2430.

[144] M. Hénon, "Numerical experiments on the stability of spherical stellar systems". In: *Astron. Astrophys.* 24 (1973), pp. 229–238.

[145] R. Kajikiya, "Sobolev norm of radially symmetric oscillatory solutions for super-linear elliptic equations". In: *Hiroshima Math. J.* 20 (1990), pp. 259–276.

[146] N. Kamburov and B. Sirakov, "Uniform a priori estimates for positive solutions of the Lane-Emden equation in the plane". In: *Calc. Var. Part. Differ. Eqn.* (2019) doi.org/10.1007/s00526-018-1435-6.

[147] T. Kato, *Perturbation Theory for Linear Operators*, Classics in Mathematics, Springer Berlin, 1995.

[148] J. Kazdan and F. W. Warner, "Remars on some quasilinear elliptic equations". In: *Commun. Pure Appl. Math.* 28 (1975), pp. 567–597.

[149] S. Kesavan, *Topics in Functional Analysis and Applications*, Wiley-Eastern, 1989.

[150] S. Kesavan, *Nonlinear Functional Analysis, A First Course*, Hindustan Book Agency, 2004.

[151] S. Kesavan, *Functional Analysis*, Hindustan Book Agency, 2009.

[152] S. Kesavan and F. Pacella, "Symmetry of positive solutions of a a quasilinear elliptic equation via isoperimetric inequality". In: *Appl. Anal.* 54 (1994), pp. 27–37.

[153] S. Kesavan and F. Pacella, "Symmetry of solutions of a system of semilinear elliptic equations". In: *Adv. Math. Sci. Appl.* 9 (1999), pp. 361–369.

[154] S. Lancellotti, "Morse index estimates for continuous functionals associated with quasilinear elliptic equations". In: *Adv. Differ. Equ.* 7 (2002), pp. 99–128.

[155] C. M. Li, "Monotonicity and symmetry of solutions of fully nonlinear elliptic equations on unbounded domains". In: *Commun. Partial Differ. Equ.* 16 (1991), pp. 585–615.

[156] Y. Li and W. M. Ni, "Radial symmetry of positive solutions of nonlinear elliptic equations in $\mathbb{R}^N$". In: *Commun. Partial Differ. Equ.* 18 (1993), pp. 1043–1054.

[157] Y. Y. Li, "Existence of many positive solutions of semilinear equations on annulus". In: *J. Differ. Equ.* 83 (1990), pp. 348–367.

[158] C. S. Lin, "Uniqueness of least energy solutions to a semilinear elliptic equation in $\mathbb{R}^2$". In: *Manuscr. Math.* 84 (1994), pp. 13–19.

[159] S. S. Lin, "Existence of positive nonradial solutions for nonlinear elliptic equations in annular domains". In: *Trans. Am. Math. Soc.* 332 (1992), pp. 775–791.

[160] S. S. Lin, "Asymptotic behavior of positive solutions to semilinear elliptic equations on expanding annuli". In: *J. Differ. Equ.* 120 (1995), pp. 255–288.

[161] P. L. Lions, "Two geometrical properties of semilinear problems". In: *Appl. Anal.* 12 (1981), pp. 267–272.

[162] P. L. Lions, "The concentration-compactness principle in the Calculus of Variations. i) The locally compact case, parts 1 and 2". In: *Ann. Inst. Henri Poincaré, Anal. Non Linéaire* 1 (1984), pp. 109–145; 223–283.

[163] P. L. Lions, "The concentration-compactness principle in the Calculus of Variations. ii) The limit case, parts 1 and 2". In: *Rev. Mat. Iberoam.* 1 (1985), pp. 45–121; 145–201.

[164] P. L. Lions, F. Pacella and M. Tricarico, "Best constants in Sobolev inequalities for functions vanishing on some part of the boundary and related questions". In: *Indiana Univ. Math. J.* 37 (1988), pp. 301–324.

[165] O. Lopes, "Radial and nonradial minimizers for some radially symmetric functional". In: *Electron. J. Differ. Equ.* 1996 (1996), pp. 1–14.

[166] O. Lopes, "Radial symmetry of minimizers for some translation and rotation invariant functionals functional". In: *J. Differ. Equu.* 124 (1996), pp. 378–388.

[167] O. Lopes and M. Maris, "Symmetry of minimizers for some nonlocal variational problems". In: *J. Funct. Anal.* 254 (2008), pp. 535–592.

[168] Z. Lou, T. Weth and Z. Zhang, "Symmetry breaking via Morse index for equations and systems of Hénon-Schrödinger type". In: *Z. Angew. Math. Phys.* (2019), doi.org/10.1007/s00033-019-1080-8.

[169] F. Maggi and C. Villani, "Balls have the worst Sobolev Inequalities". In: *J. Geom. Anal.* (2005), pp. 83–121.

[170] L. A. Maia, E. Montefusco and B. Pellacci, "Positive solutions for a weakly coupled nonlinear Schrodinger system". In: *J. Differ. Equ.* 229 (2006), pp. 743–767.

[171] V. Marino, F. Pacella and B. Sciunzi, "Blow up of solutions of semilinear heat equations in general domains". In: *Commun. Contemp. Math.* 17 (2015).

[172] M. Maris, "On the symmetry of minimizers". In: *Arch. Ration. Mech. Anal.* 192 (2009), pp. 311–330.

[173] N. Mavinga, "Generalized eigenproblem and nonlinear elliptic equations with nonlinear boundary conditions". In: *Proc. R. Soc. Edinb., Sect. A* 142 (2012), pp. 137–153.

[174] V. Maz'ya, *Sobolev Spaces*, Springer, 1985.

[175] P. J. McKenna, F. Pacella, M. Plum and D. Roth, "A uniqueness result for a semilinear elliptic proble: a computer-assisted proof". In: *J. Differ. Equ.* 247 (2009), pp. 2140–2162.

[176] P. J. McKenna, F. Pacella, M. Plum and D. Roth, "A computer-assisted uniqueness proof for semilinear elliptic boundary value problem". In: *Inequal. Appl.* 161 (2012), pp. 31–52.

[177] J. Milnor, *Morse Theory*, Princeton University Press, 1963.

[178] E. Moreira dos Santos and F. Pacella, "Hénon type equations and concentration on spheres". In: *Indiana Univ. Math. J.* (2016), pp. 273–306.

[179] E. Moreira dos Santos and F. Pacella, "Morse index of radial nodal solutions of Hénon type equations in dimension two". In: *Commun. Contemp. Math.* 19 (2017).

[180] M. Morse, *The Calculus of Variations in the Large*, New York: Amer. Math. Soc, 1934.

[181] W. M. Ni, "Uniqueness of solutions of nonlinear Dirichlet problems". In: *J. Differ. Equ.* 50 (1983), pp. 289–304.

[182] W. M. Ni and R. D. Nussbaum, "Uniqueness and nonuniqueness for positive radial solutions of $\Delta u + f(u, r) = 0$". In: *Commun. Pure Appl. Math.* 38 (1985), pp. 67–108.

[183] R. D. Nussbaum, "The fixed point index for local condensing maps". In: *Ann. Mat. Pura Appl.* 89 (1971), pp. 217–258.

[184] B. Opic and A. Kufner, *Hardy-type Inequalities*, Pitman Research Notes in Mathematics 219, Longman, 1990.

[185] F. Pacard, "Radial and nonradial solutions of $-\Delta u = \lambda f(u)$ on an annulus of $\mathbb{R}^n, n \geq 3$". In: *J. Differ. Equ.* 102 (1993), pp. 103–138.

[186] F. Pacella, "Symmetry of solutions to semilinear elliptic equations with convex nonlinearities". In: *J. Funct. Anal.* 192 (2002), pp. 271–282.

[187] F. Pacella and M. Ramaswamy, "Symmetry of solutions of elliptic equations via maximum principle". In: *Handbook of Differential Equations: Stationary Partial Differential Equations Vol. VI.* 2008, pp. 269–312.

[188] F. Pacella and D. Salazar, "Asymptotic behaviour of sign changing radial solutions of Lane Emden problems in the annulus". In: *Discrete Contin. Dyn. Syst., Ser. S* 7 (2014), pp. 793–805.

[189] F. Pacella and P. N. Srikanth, "A reduction method for semilinear eliiptic equations and solutions concentrating on spheres". In: *J. Funct. Anal.* 266 (2014), pp. 6456–6472.

[190] F. Pacella and T. Weth, "Symmetry results for solutions of semilinear elliptic equations via Morse index". In: *Proc. Am. Math. Soc.* 135 (2007), pp. 1753–1762.

[191] J. I. Palmore, "Classifying relative equilibria". In: *Bull. Am. Math. Soc.* 81 (1975), pp. 489–491.

[192] S. I. Pohožaev, "Eigenfunctions of the equation $\Delta u + \lambda f(u) = 0$". In: *Sov. Math. Dokl.* 6 (1965), pp. 1408–1411.

[193] M. H. Protter and H. F. Weinberger, *Maximum Principle in Differential Equations*, Prentice-Hall, 1967.

[194] P. Pucci and J. Serrin, *The Maximum Principle*, Birkhauser, 2007.

[195] P. H. Rabinowitz, "Some global results for nonlinear eigenvalue problems". In: *J. Funct. Anal.* 7 (1971), pp. 487–513.

[196] P. H. Rabinowitz, *Minimax Methods in Critical Point Theory with Applications to Differential Equations.* CMBS Reg. Conf. Series Math 65. Providence: AMS, 1986.

[197] X. Ren and J. Wei, "Single-point condensation and least-energy solutions". In: *Proc. Am. Math. Soc.* 124 (1996), pp. 111–120.

[198] X. Ren and J. Wei, "On a two-dimensional elliptic problem with large exponent in nonlinearity". In: *Trans. Am. Math. Soc.* 343 (2001), pp. 749–763.

[199] J. Serrin, "A symmetry problem in potential theory". In: *Arch. Ration. Mech. Anal.* 43 (1971), pp. 304–318.

[200] B. Simon, "Schrödinger semigroups". In: *Bull., New Ser., Am. Math. Soc.* 7 (1982), pp. 447–526.

[201] B. Sirakov, "Some estimates and maximum principles for weakly coupled systems of elliptic PDE". In: *Nonlinear Anal.* 70 (2009), pp. 3039–3046.

[202] D. Smets, J. Su and M. Willem, "Non-radial ground states for the Henon equation". In: *Commun. Contemp. Math.* 4 (2002), pp. 467–480.

[203] D. Smets and M. Willem, "Partial symmetry and asymptotic behaviour for some elliptic variational problem". In: *Calc. Var. Partial Differ. Equ.* 18 (2003), pp. 57–75.

[204] J. Smoller and A. Wasserman, "Symmetry-breaking for solutions of semilinear elliptic equations with general boundary conditions". In: *Commun. Math. Phys.* 105 (1986), pp. 415–441.

[205] J. Smoller and A. Wasserman, "Bifurcation and symmetry breaking". In: *Invent. Math.* 100 (1990), pp. 63–95.

[206] N. Soave and H. Tavares, "New existence and symmetry results for least energy positive solutions of Scrödinger systems with mixed competition and cooperation". In: *J. Differ. Equ.* 261 (2016), pp. 505–537.

[207] E. H. Spanier, *Algebraic Topology*, Springer-Verlag, 1967.

[208] M. Spivak, *A Comprehensive Introduction to Differential Geometry*, second ed., Publish or Perish, 1979.

[209] P. N. Srikanth, "Uniqueness of solutions of nonlinear Dirichlet problems". In: *Differ. Integral Equ.* 6 (1993), pp. 663–670.

[210] W. A. Strauss, *Partial Differential Equations*, John Wiley & Sons Inc., 1992.

[211] M. Struwe, "A global compactness result for elliptic boundary value problems involving limiting nonlinearities". In: *Math. Z.* 187 (1984), pp. 511–517.

[212] M. Struwe, *Variational Methods. Applications to Nonlinear Partial Differential Equations and Hamiltonian Systems*, Springer, 1990.

[213] G. Talenti, "Best constant in Sobolev inequality". In: *Ann. Mat. Pura Appl.* 110 (1976), pp. 353–372.

[214] G. Tarantello, *Selfdual Gauge Field Vortices: An Analytical Approach*, Progress in Nonlinear Diff. Eq. Appl., Springer, 2008.

[215] H. Tavares and T. Weth, "Existence and symmetry results for competing variational systems". In: *Nonlinear Differ. Equ. Appl.* 20 (2013), pp. 715–740.

[216] S. Terracini, "Symmetry properties of positive solutions to some elliptic equations with nonlinear boundary conditions". In: *Differ. Integral Equ.* 8 (1995), pp. 1911–1922.

[217] S. Terracini, "On positive entire solutions to a class of equations with a singular coefficient and critical exponent". In: *Adv. Differ. Equ.* 1 (1996), pp. 241–264.

[218] G. Troianiello, *Elliptic Differential Equations and Obstacle Problems*, Plenum Publishing, 1987.

[219] W. C. Troy, "Symmetry properties in systems of semilinear elliptic equations". In: *J. Differ. Equ.* 42 (1981), pp. 400–413.

[220] J. Van Schaftingen, "Approximation of symmetrizations and symmetry of critical points". In: *Nonlinear Anal.* 28 (2006), pp. 61–85.

[221] J. Van Schaftingen, "Universal approximation of symmetrizations by polarizations". In: *Proc. Am. Math. Soc.* 134 (2006), pp. 177–186.

[222] J. Van Schaftingen, "Explicit approximation of the symmetric rearrangement by polarization". In: *Arch. Math.* 93 (2009), pp. 181–190.

[223] J. Van Schaftingen and M. Willem, "Symmetry of solutions of semilinear elliptic problems". In: *J. Eur. Math. Soc.* 10 (2008), pp. 439–456.

[224] T. Weth, "Symmetry of solutions to variational problems for nonlinear elliptic equations via reflection methods". In: *Jahresber. Dtsch. Math.-Ver.* 112 (2010), pp. 119–158.

[225] M. Willem, *Minimax Theorems*. Progress in Nonlinear Diff. Eq. 24. Birkhauser, 1996.

[226] L. Zhang, "Uniqueness of positive solutions of $\Delta u + u^p + u = 0$ in a finite ball". In: *Commun. Partial Differ. Equ.* 17 (1992), pp. 1141–1164.

[227] W. P. Ziemer, *Weakly Differentiable Functions*, Springer, 1989.

[228] H. Zou, "On the effect of the domain geometry on the uniqueness of positive solutions of $\Delta u + u^p = 0$". In: *Ann. Sc. Norm. Super. Pisa* 3 (1994), pp. 343–356.

Index

De Gruyter Series in Nonlinear Analysis and Applications

Volume 29
Rafael Ortega
Periodic Differential Equations in the Plane. A Topological Perspective, 2019
ISBN 978-3-11-055040-5, e-ISBN (PDF) 978-3-11-055116-7,
e-ISBN (EPUB) 978-3-11-055042-9

Volume 28
Dung Le
Strongly Coupled Parabolic and Elliptic Systems. Existence and Regularity of Strong
and Weak Solutions, 2018
ISBN 978-3-11-060715-4, e-ISBN (PDF) 978-3-11-060876-2,
e-ISBN (EPUB) 978-3-11-060717-8

Volume 27
Maxim Olegovich Korpusov, Alexey Vital'evich Ovchinnikov, Alexey Georgievich
Sveshnikov, Egor Vladislavovich Yushkov
Blow-Up in Nonlinear Equations of Mathematical Physics. Theory and Methods, 2018
ISBN 978-3-11-060108-4, e-ISBN (PDF) 978-3-11-060207-4,
e-ISBN (EPUB) 978-3-11-059900-8

Volume 26
Saïd Abbas, Mouffak Benchohra, John R. Graef, Johnny Henderson
Implicit Fractional Differential and Integral Equations. Existence and Stability, 2018
ISBN 978-3-11-055313-0, e-ISBN (PDF) 978-3-11-055381-9,
e-ISBN (EPUB) 978-3-11-055318-5

Volume 25
Luboš Pick, Alois Kufner, Oldřich John, Svatopluk Fucík
Function Spaces. Volume 2, 2018
ISBN 978-3-11-027373-1, e-ISBN (PDF) 978-3-11-032747-2,
e-ISBN (EPUB) 978-3-11-038221-1

Volume 24
Alexander A. Kovalevsky, Igor I. Skrypnik, Andrey E. Shishkov
Singular Solutions of Nonlinear Elliptic and Parabolic Equations, 2016
ISBN 978-3-11-031548-6, e-ISBN (PDF) 978-3-11-033224-7,
e-ISBN (EPUB) 978-3-11-039008-7

www.degruyter.com